NEUROMETHODS

Series Editor
Wolfgang Walz
University of Saskatchewan,
Saskatoon, SK, Canada

For further volumes:
http://www.springer.com/series/7657

Neuromethods publishes cutting-edge methods and protocols in all areas of neuroscience as well as translational neurological and mental research. Each volume in the series offers tested laboratory protocols, step-by-step methods for reproducible lab experiments and addresses methodological controversies and pitfalls in order to aid neuroscientists in experimentation. Neuromethods focuses on traditional and emerging topics with wide-ranging implications to brain function, such as electrophysiology, neuroimaging, behavioral analysis, genomics, neurodegeneration, translational research and clinical trials. Neuromethods provides investigators and trainees with highly useful compendiums of key strategies and approaches for successful research in animal and human brain function including translational "bench to bedside" approaches to mental and neurological diseases.

Volume Microscopy

Multiscale Imaging with Photons, Electrons, and Ions

Edited by

Irene Wacker

Cryo EM, Centre for Advanced Materials (CAM), Universität Heidelberg, Heidelberg, Germany

Eric Hummel

Sector Life Science, Carl Zeiss Microscopy GmbH, Oberkochen, Germany; Carl Zeiss, Oberkochen, Germany

Steffen Burgold

BioQuant, Universität Heidelberg, Heidelberg, Germany; Cryo EM, BioQuant, Universitätsklinikum Heidelberg, Heidelberg, Germany

Rasmus Schröder

Cryo EM, BioQuant, Universitätsklinikum Heidelberg, Heidelberg, Germany; Cryo EM, Centre for Advanced Materials (CAM), Universität Heidelberg, Heidelberg, Germany

※ Humana Press

Editors
Irene Wacker
Cryo EM, Centre for Advanced
Materials (CAM)
Universität Heidelberg
Heidelberg, Germany

Steffen Burgold
BioQuant
Universität Heidelberg
Heidelberg, Germany

Cryo EM, BioQuant
Universitätsklinikum Heidelberg
Heidelberg, Germany

Eric Hummel
Sector Life Science
Carl Zeiss Microscopy GmbH
Oberkochen, Germany

Carl Zeiss
Oberkochen, Germany

Rasmus Schröder
Cryo EM, BioQuant
Universitätsklinikum Heidelberg
Heidelberg, Germany

Cryo EM, Centre for Advanced Materials (CAM)
Universität Heidelberg
Heidelberg, Germany

ISSN 0893-2336   ISSN 1940-6045   (electronic)
Neuromethods
ISBN 978-1-0716-0693-3   ISBN 978-1-0716-0691-9   (eBook)
https://doi.org/10.1007/978-1-0716-0691-9

Cover Caption: 3D rendering of 5 objects from a SBFSEM dataset in a virtual reality (VR) environment to illustrate the spatial setting of a neuron (cyan), a blood vessel (red), and three astrocytes in the context of the astrocyte neuron lactate shuttle. Rendering by Daniya Boges, image processing and VR realization Corrado Cali, see also chapter 14.

This Humana imprint is published by the registered company Springer Science+Business Media, LLC, part of Springer Nature.
The registered company address is: 1 New York Plaza, New York, NY 10004, U.S.A.

Preface to the Series

Experimental life sciences have two basic foundations: concepts and tools. The *Neuro-methods* series focuses on the tools and techniques unique to the investigation of the nervous system and excitable cells. It will not, however, shortchange the concept side of things as care has been taken to integrate these tools within the context of the concepts and questions under investigation. In this way, the series is unique in that it not only collects protocols but also includes theoretical background information and critiques which led to the methods and their development. Thus it gives the reader a better understanding of the origin of the techniques and their potential future development. The *Neuromethods* publishing program strikes a balance between recent and exciting developments like those concerning new animal models of disease, imaging, in vivo methods, and more established techniques, including, for example, immunocytochemistry and electrophysiological technologies. New trainees in neurosciences still need a sound footing in these older methods in order to apply a critical approach to their results.

Under the guidance of its founders, Alan Boulton and Glen Baker, the *Neuromethods* series has been a success since its first volume published through Humana Press in 1985. The series continues to flourish through many changes over the years. It is now published under the umbrella of Springer Protocols. While methods involving brain research have changed a lot since the series started, the publishing environment and technology have changed even more radically. Neuromethods has the distinct layout and style of the Springer Protocols program, designed specifically for readability and ease of reference in a laboratory setting.

The careful application of methods is potentially the most important step in the process of scientific inquiry. In the past, new methodologies led the way in developing new disciplines in the biological and medical sciences. For example, Physiology emerged out of Anatomy in the nineteenth century by harnessing new methods based on the newly discovered phenomenon of electricity. Nowadays, the relationships between disciplines and methods are more complex. Methods are now widely shared between disciplines and research areas. New developments in electronic publishing make it possible for scientists that encounter new methods to quickly find sources of information electronically. The design of individual volumes and chapters in this series takes this new access technology into account. Springer Protocols makes it possible to download single protocols separately. In addition, Springer makes its print-on-demand technology available globally. A print copy can therefore be acquired quickly and for a competitive price anywhere in the world.

Saskatoon, SK, Canada   *Wolfgang Walz*

Preface

Multimodal Large Volume Microscopy: Tools for Visualizing Morphology and Function of Complex Systems

Large Volume Electron Microscopy: Tools and Questions

When thinking about large volume electron microscopy, metrology may not be the first thing that comes to mind—even though it has been around for a long time and may gain interest in different fields in biomedical sciences (cf. editorial [1]). As pointed out there, "the metrological mindset ... is fundamental to the ability to draw believable conclusions in any domain of science, including biology." And applying this to electron microscopy leaves us with the question, how valid are our conceptions about ultrastructure, morphology, and structure. Cryo-Electron Microscopy has answered this question for single particle analysis, where many hundred thousand of individual complexes are imaged and reconstructed in 3D. However, it will keep us in the dark as soon as unique 3D reconstructions from cells or tissue are considered. Similarly, conventional 3D Transmission Electron Microscopy (TEM) from chemically fixed and resin-embedded tissue has hardly given us large numbers, neither in cubic micrometers nor in copies of a particular reconstructed sample. The questions is: Do we need such large numbers?

The problems to obtain statistically meaningful data have rather changed a few years ago with three events: The wider availability of Field Emission Scanning Electron Microscopes (FESEM), the rediscovery of blockface imaging in the SEM, and the rising interest in the determination of the wiring of neurons in brain, aka connectomics. The first two were mere technical advances, which at first sight are not revolutionary. Nevertheless, the authors do remember an instance, when a reviewer of a paper thought that they had mixed up the images from TEM and SEM when showing their direct comparison. The FESEM image was crisper than the TEM image. Technically, this can be explained by the smaller volume where the signal comes from, but it is enough at this stage to point out that SEM images of ultrathin sections or of a blockface can look superior to their TEM counterparts, depending on sample preparation and imaging conditions. This opened a new field of microscopy just at the same time when the idea about visualizing neuronal wiring turned viral.

In the meantime, large volume electron microscopy is at the transition to a well-established and widely applicable technology, and Fig. 1 (taken from [2]) shows the largest dense connectomic reconstruction to date. Compared, e.g., to the *C. elegans* "Mind of a worm," reconstructed from TEM images, this reconstructed piece of mouse brain is almost one order of magnitude bigger than one complete *C. elegans*. And, as is discussed in [2], the large number of neurons with all their dendritic and axonal connections gives completely new insight into the function of such a neuronal network. Here large numbers and metrology acquire a new meaning; it is not about statistics but about function, which could not be conceptualized otherwise.

With the book at hand we as editors have tried to put together information about the complete workflow of samples prepared for large volume electron microscopy (cf. Fig. 2) and imaged in different microscopes with different techniques (cf. Fig. 3). To create a concise account of possibilities in particular for quite large volumes, we deliberately left out cryo-electron microscopy—in all its different flavors. It should also be noted that so far

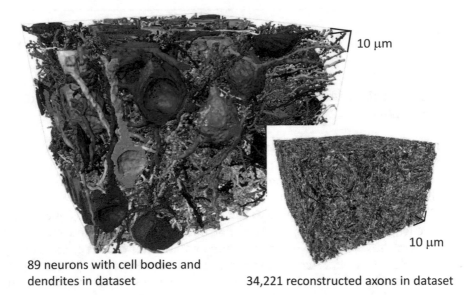

89 neurons with cell bodies and
dendrites in dataset   34,221 reconstructed axons in dataset

Fig. 1 The largest dense connectomic reconstruction (mouse primary somatosensory cortex) at present [2]. Reconstructed is a volume of about $96 \times 96 \times 64$ µm^3 using the Serial Blockface Scanning Electron Microscope (SBFSEM) technique. Figure adapted from [2] copyright American Association for the Advancement of Science, AAAS

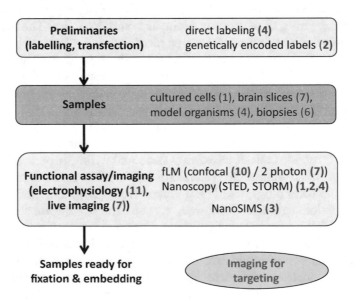

Fig. 2 A typical sample preparation workflow for large volume electron microscopy. This workflow produces samples to be used for volume imaging, but at the same time it includes crucial steps for the targeting of the sample volume selected for imaging and 3D reconstruction. Different steps in this workflow are described in detail by the chapters of this book indicated by blue numbers

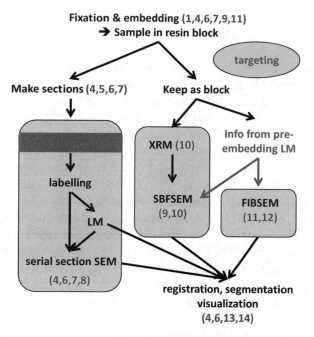

Fig. 3 After preparation of the sample, a variety of imaging techniques can be applied. Depending on the imaging tools used, different ways of targeting the volume of interest can be utilized. The details of these imaging processes are described by the chapters of this book indicated by blue numbers

nobody has succeeded in vitrifying really large pieces of tissue—and physics may prevent this forever. Also, this book should not be considered a complete compendium for any of the image processing involved in large volume reconstruction. When contemplating adding more information on this topic, two simple facts seemed to emerge: First, there is no unique processing pathway or software. It will always depend on data and the question which needs to be answered. This makes any selection of approaches and software packages arbitrary, and many scientists in the field would be legitimately reluctant to "oversell" their software. An example for a typical approach to the processing is again [2], where the authors developed many highly specialized tools for alignment and segmentation themselves—tailored exactly for their segmentation needs. Second, there is a distinct personal component to visualization in 3D: some people like surface rendering, others need to walk in 3D in a virtual reality (VR) environment. At the end of the book, two chapters on processing and visualization are included: in particular the VR approach is highly interesting, even though it is not yet widely used.

Another facet, which needs to be discussed about large volume electron microscopy, is its applicability to complex systems. In all the discussion so far, brain and connectomics seemed to dominate the field. It is true; all recent developments—both in sample preparation and instrumentation—have been pushed by neurobiology and the quest to reconstruct larger and larger pieces of brain at better and better morphological preservation and in shorter and shorter time. Automation and large-scale imaging, either using many machines in parallel or multiple electron beams in one machine (mSEM, Chapter 8), have their share in the success of large volume microscopy. And correlative imaging, i.e., functional imaging in a fluorescent light microscope before transferring the sample to the electron microscope, has been instrumental for targeting interesting structures. All these aspects are addressed in exemplary Chapters 4, 7, 10, and 11 of this book. But it should be noted that the field is moving on; new application areas such as typical cell biological questions and also 3D imaging tasks in the field of pathology are emerging. In the future, we will also see more and more studies where biomedical samples "meet" materials science—e.g., when the interaction of implants with their bio-environment needs to be visualized (Chapter 11).

We conclude this introduction also with a warning: Besides the technologies covered and their applications, there are quite a few more technological developments, which are just emerging and will set their mark only over time. One example for this is new ideas to overcome the anisotropic sampling of the 3D volume when using sectioning methods, such as serial sectioning in Array Tomography (AT) or Serial Blockface Scanning Electron Microscope imaging (SBFSEM). When using Focused Ion Beam milling for 3D imaging in FIBSEM nanotomography, this problem does not occur as milling can be adjusted such that voxels are truly cubic. In contrast to that, ultrathin sections for AT—or the layer scraped away from the blockface in SBFSEM—will always be thicker than the obtainable lateral resolution in the consecutive SEM imaging. He et al. [3] present a valuable simulation how also for normal SEM imaging "z-resolution" can be increased, and an "energy scan" implementation of this technique is also commercially available.

Another example, nanoSIMS (Secondary Ion Mass Spectroscopy imaging at the nanoscale), has been included as a teaser (Chapter 3) to illustrate the potential of this technique. It allows to analyze chemical structure and molecular turnover of the sample and by this complements conventional imaging of morphology or of fluorescently tagged molecules.

Sample Preparation and Targeting

As illustrated in Fig. 2, several chapters introduce necessary sample preparation steps and the use of labeling (Chapters 1, 2, and 4), light microscopy (Chapters 1–4, 6, 7, and 10), and X-ray microscopy (Chapter 10) for targeting.

The samples described range from cultured cells (Chapter 1) to organotypic brain slices (Chapter 7), entire brains (Chapter 2), and entire model organisms (zebrafish, Chapter 4), with human muscle biopsies (Chapter 6) and retina implants (Chapter 11) as an excursion into pathology and preclinical testing. Preliminary work such as labeling of target structures can be either rather simple—as direct labeling of accessible neurons (Chapter 4)—or more complicated by introducing genetic tags (e.g., intracranial injection to achieve viral-mediated transfection of mice, Chapter 2).

Depending on the functional assay or imaging modality used for targeting (cf. Fig. 2), several steps may be necessary before the sample can be fixed and embedded.

Examples for the different fixation and embedding protocols available are either cryo-fixation of cells (Chapter 1) or fish embryos (Chapter 4) followed by freeze substitution to preserve fluorescence or conventional chemical fixation with moderate (Chapters 6 and 11) or rather high (Chapter 9) amounts of heavy metal introduced into the block. Due to physical constraints, the sample size that can be vitrified is limited to cells or small model organisms. Smaller pieces of tissue (about 300 µm thickness as maximum)—fitting into the sample carriers—can be vitrified successfully using a high-pressure freezing device, but certainly a mouse brain is too large to fit into one carrier in its entirety.

Once the samples are embedded in resin, there are different ways to process them further (cf. Fig. 3). With the exception of X-ray microscopy (XRM), where the entire sample can be imaged without further preparation (Chapter 10), the acquisition of a 3D volume requires slicing the sample in one way or another. There are basically two distinct routes at this point: The sample block is either decomposed independently of any imaging device into an array of ultrathin sections which are placed on some type of substrate, an approach also known as array tomography (AT, Chapters 4–8), or the block is kept intact and "dismantled" inside the appropriate imaging device in conjunction with the imaging process. In that case individual sections are not imaged, but the surface of the sample block—the blockface. By alternating cycles of imaging and removal of material from the blockface, a stack of images is created forming a 3D representation of the original sample. This can happen using a diamond knife inside the SEM—in that case the approach is called serial blockface SEM (SBFSEM, sometimes also SBEM), described in Chapters 9 and 10. The alternative approach, which is using a focused ion beam to ablate material from the blockface, is called FIBSEM nanotomography (Chapters 11 and 12).

To make the samples ready for imaging in the SEM, the blocks have first to be trimmed to expose the target region. For the blockface-based techniques, the trimmed blocks are then just mounted on an appropriate sample carrier (Chapters 9–11) whereas AT arrays, i.e., serial sections, need to be produced first. This can be done in several ways: The basic instrumentation for all of them is an ultramicrotome which can be modified by different means. Manual collection of ribbons of sections, e.g., onto pieces of silicon wafer, requires a very skilled and patient operator (Chapters 4 and 6). A simple device helping with ribbon collection is demonstrated in Chapter 5, and a rather elaborate device for automated collection of thousands of sections, the ATUMtome, is introduced in Chapter 7.

Imaging Tools and Visualization

A number of imaging modalities are described/mentioned in the chapters of the book, ranging from LM in its different flavors (Chapters 1, 2, 4, 6, 7, 10, and 11) via XRM (Chapter 10) to the different varieties of 3D data generation using SEM (Chapters 4, 6–12). LM is mainly used as part of a correlated workflow (CLEM, Correlated Light and Electron Microscopy) for targeting. That means, interesting regions which have previously been characterized using functional assays (e.g., electrophysiology in Chapter 11) and/or live cell imaging of target structures, typically tagged with some kind of label, must be identified within the sample block. Some preparation methods are designed in a way to preserve fluorescent labels (Chapters 1 and 4). All others, particularly those aimed at blockface-based imaging—that introduce high amounts of heavy metal—destroy fluorescence. Here, a good characterization of the sample by LM, possibly at several levels of resolution, before embedding is necessary and will help to find back the target region by registration of the different modalities. A good example is Chapter 10 where volumes from LM, XRM, and SBFSEM are registered using a software package specifically designed for that purpose.

AT on the other hand, where the sectioning process takes place outside the imaging machine, provides a second chance for targeting: Labeling of sections (immuno or otherwise) is an option—when a hydrophilic resin has been used for embedding. In some instances, observation of unlabeled section ribbons in a conventional bright-field LM is already sufficient to identify target structures (e.g., Chapter 6).

After data acquisition, the individual images—which may well be several thousand—need to be registered. This task is relatively straightforward for the blockface-based approaches and requires a bit more effort for AT data, but there is a wide spectrum of excellent open-source software available (for references see Chapters 4, 6, and 7). For the next step, the annotation of certain features (e.g., organelles inside a given cell), automated routines would be desirable. Because even algorithms involving machine learning often fail, the last resort still remains manual segmentation, slice by slice. As mentioned above, many groups came up with specialized solutions for their own tasks. Common problems with large volume data are their handling and interpretation, which are discussed in Chapters 13 and 14. For the final step, visualization of the annotated volume, virtual reality (VR) tools are beginning to emerge (Chapter 14).

Heidelberg, Germany   *Irene Wacker*
                                                                    Rasmus Schröder

References

1. No authors (editorial) (2018) Better research through metrology. Nat Methods 15:395. https://doi.org/10.1038/s41592-018-0035-x

2. Motta A et al (2019) Dense connectomic reconstruction in layer 4 of the somatosensory cortex. Science 366:eaay3134. https://doi.org/10.1126/science.aay3134

3. He Q, et al (2018) Biological serial block face scanning electron microscopy at improved z-resolution based on Monte Carlo model. Sci Rep 8:12985. https://doi.org/10.1038/s41598-018-31231-w

Contents

Contributors

MARCO AGUS • *Visual Computing Center, King Abdullah University of Science and Technology, Thuwal, Saudi Arabia*

ANTJE BIESEMEIER • *NMI Natural and Medical Sciences Institute (NMI), University of Tübingen, Reutlingen, Germany; Luxembourg Institute of Science and Technology (LIST), Belvaux, Luxembourg*

DANIYA J. BOGES • *BESE Division, King Abdullah University of Science and Technology, Thuwal, Saudi Arabia*

CLAUS J. BURKHARDT • *NMI Natural and Medical Sciences Institute (NMI), University of Tübingen, Reutlingen, Germany*

ERIC A. BUSHONG • *National Center for Microscopy and Imaging Research, Center for Research in Biological Systems, University of California—San Diego, La Jolla, CA, USA*

CORRADO CALÌ • *BESE Division, King Abdullah University of Science and Technology, Thuwal, Saudi Arabia; Department of Neuroscience "Rita Levi Montalcini", Neuroscience Institute Cavalieri Ottolenghi, Università degli studi di Torino, Torino, Italy*

CARSTEN DITTMAYER • *Department of Neuropathology, Charité-Universitätsmedizin Berlin, Berlin, Germany*

ANNA LENA EBERLE • *Carl Zeiss Microscopy GmbH, Oberkochen, Germany*

MARK H. ELLISMAN • *National Center for Microscopy and Imaging Research, Center for Research in Biological Systems, University of California—San Diego, La Jolla, CA, USA*

TOMASZ GARBOWSKI • *Carl Zeiss Microscopy GmbH, Oberkochen, Germany*

ULRICH GENGENBACH • *Institute for Automation and Applied Informatics, Karlsruhe Institute of Technology (KIT), Karlsruhe, Germany*

DEBBIE GUERRERO-GIVEN • *Imaging Center and Electron Microscopy Facility, Max Planck Florida Institute for Neuroscience, Jupiter, FL, USA*

CHRIS HAWES • *Sector Life Science, Carl Zeiss Microscopy GmbH, Oberkochen, Germany*

KENNETH J. HAYWORTH • *Howard Hughes Medical Institute, Ashburn, VA, USA*

MANFRED HECKMANN • *Department of Neurophysiology, Institute of Physiology, University of Würzburg, Würzburg, Germany*

JANOSCH P. HELLER • *UCL Queen Square Institute of Neurology, University College London, London, UK*

FREDERIK HELMPROBST • *Biocenter, Imaging Core Facility, University of Würzburg, Würzburg, Germany; Biocenter, Department of Physiological Chemistry, University of Würzburg, Würzburg, Germany*

HARALD F. HESS • *Howard Hughes Medical Institute, Ashburn, VA, USA*

ANDREAS HOFMANN • *Institute for Automation and Applied Informatics, Karlsruhe Institute of Technology (KIT), Karlsruhe, Germany*

ERIC HUMMEL • *Sector Life Science, Carl Zeiss Microscopy GmbH, Oberkochen, Germany*

NAOMI KAMASAWA • *Integrative Biology and Neuroscience Graduate Program, Florida Atlantic University, Jupiter, FL, USA; Imaging Center and Electron Microscopy Facility, Max Planck Florida Institute for Neuroscience, Jupiter, FL, USA*

JÖRGEN KORNFELD • *Department of Brain and Cognitive Sciences, McGovern Institute for Brain Research, Massachusetts Institute of Technology, Cambridge, MA, USA; Ariadne-Service GMBH, Buchrain, Switzerland*

GRANT KUSICK • *Department of Cell Biology, School of Medicine, Johns Hopkins University, Baltimore, MD, USA; Biochemistry, Cellular and Molecular Biology Program, School of Medicine, Johns Hopkins University, Baltimore, MD, USA*

SHUO LI • *Department of Cell Biology, School of Medicine, Johns Hopkins University, Baltimore, MD, USA; Department of Biochemistry and Molecular Biology, Bloomberg School of Public Health, Johns Hopkins University, Baltimore, MD, USA*

CHRISTINA LILLESAAR • *Department of Child and Adolescent Psychiatry, Psychosomatics and Psychotherapy, Center of Mental Health, University Hospital of Würzburg, Würzburg, Germany*

PIERRE JULIUS MAGISTRETTI • *BESE Division, King Abdullah University of Science and Technology, Thuwal, Saudi Arabia*

TAKAYASU MIKUNI • *Neural Signal Transduction Research Group, Max Planck Florida Institute for Neuroscience, Jupiter, FL, USA*

RALPH NEUJAHR • *Sector Life Science, Carl Zeiss Microscopy GmbH, Oberkochen, Germany*

WILFRIED NISCH • *NMI Natural and Medical Sciences Institute (NMI), University of Tübingen, Reutlingen, Germany*

TYLER OGUNMOWO • *Department of Cell Biology, School of Medicine, Johns Hopkins University, Baltimore, MD, USA; Biochemistry, Cellular and Molecular Biology Program, School of Medicine, Johns Hopkins University, Baltimore, MD, USA*

SONG PANG • *Howard Hughes Medical Institute, Ashburn, VA, USA*

MARTIN PAULI • *Department of Neurophysiology, Institute of Physiology, University of Würzburg, Würzburg, Germany*

NHU T. N. PHAN • *Institute of Neuro- and Sensory Physiology, University of Göttingen Medical Center, Göttingen, Germany; Department of Chemistry and Molecular Biology, University of Gothenburg, Gothenburg, Sweden*

SÉBASTIEN PHAN • *National Center for Microscopy and Imaging Research, Center for Research in Biological Systems, University of California—San Diego, La Jolla, CA, USA*

SUMANA RAYCHAUDHURI • *Department of Cell Biology, School of Medicine, Johns Hopkins University, Baltimore, MD, USA*

JAMES P. REYNOLDS • *UCL Queen Square Institute of Neurology, University College London, London, UK*

SILVIO O. RIZZOLI • *Institute of Neuro- and Sensory Physiology, University of Göttingen Medical Center, Göttingen, Germany*

DMITRI A. RUSAKOV • *UCL Queen Square Institute of Neurology, University College London, London, UK*

RASMUS SCHRÖDER • *Cryo EM, BioQuant, Universitätsklinikum Heidelberg, Heidelberg, Germany; Cryo EM, Centre for Advanced Materials (CAM), Universität Heidelberg, Heidelberg, Germany*

BIRGIT SCHRÖPPEL • *NMI Natural and Medical Sciences Institute (NMI), University of Tübingen, Reutlingen, Germany*

WALDEMAR SPOMER • *Institute for Automation and Applied Informatics, Karlsruhe Institute of Technology (KIT), Karlsruhe, Germany*

CHRISTIAN STIGLOHER • *Biocenter, Imaging Core Facility, University of Würzburg, Würzburg, Germany*

MARLENE STROBEL • *Biocenter, Imaging Core Facility, University of Würzburg, Würzburg, Germany; Department of Medicine II, University Hospital of Würzburg, Würzburg, Germany*

YE SUN • *Neural Signal Transduction Research Group, Max Planck Florida Institute for Neuroscience, Jupiter, FL, USA; Integrative Biology and Neuroscience Graduate Program, Florida Atlantic University, Jupiter, FL, USA; International Max Planck Research School for Brain and Behavior, Florida Atlantic University, Jupiter, FL, USA*

FABIAN SVARA • *Ariadne-Service GMBH, Buchrain, Switzerland; Caesar, Bonn, Germany*

MARLENE THALER • *Carl Zeiss Microscopy GmbH, Oberkochen, Germany*

CONNON THOMAS • *Imaging Center and Electron Microscopy Facility, Max Planck Florida Institute for Neuroscience, Jupiter, FL, USA*

ANN-KATRIN UNGER • *Sector Life Science, Carl Zeiss Microscopy GmbH, Oberkochen, Germany*

LISA VEITH • *Cryo-EM, BioQuant, University Hospital Heidelberg, Heidelberg, Germany*

IRENE WACKER • *Cryo EM, Centre for Advanced Materials (CAM), Universität Heidelberg, Heidelberg, Germany*

ADRIAN A. WANNER • *Ariadne-Service GMBH, Buchrain, Switzerland; Princeton Neuroscience Institute, Princeton University, Princeton, NJ, USA*

SHIGEKI WATANABE • *Department of Cell Biology, School of Medicine, Johns Hopkins University, Baltimore, MD, USA; Solomon H. Snyder Department of Neuroscience, School of Medicine, Johns Hopkins University, Baltimore, MD, USA*

C. SHAN XU • *Howard Hughes Medical Institute, Ashburn, VA, USA*

RYOHEI YASUDA • *Neural Signal Transduction Research Group, Max Planck Florida Institute for Neuroscience, Jupiter, FL, USA; Integrative Biology and Neuroscience Graduate Program, Florida Atlantic University, Jupiter, FL, USA; International Max Planck Research School for Brain and Behavior, Florida Atlantic University, Jupiter, FL, USA*

Chapter 1

Correlative Super Resolution and Electron Microscopy to Detect Molecules in Their Native Cellular Context

Tyler Ogunmowo, Sumana Raychaudhuri, Grant Kusick, Shuo Li, and Shigeki Watanabe

Abstract

Nano-resolution fluorescence electron microscopy (nano-fEM) provides the precise localization of bioma-cromolecules within electron micrographs. Classically, electron microscopy has provided the highest possible cellular detail, boasting nanometer-scale resolution. However, while cellular ultrastructure is clearly defined, molecular identity is obscured even when electron dense tags in the form of antibodies or locally polymerized moieties are used. Fluorescence microscopy complements electron microscopy by providing significant molecular specificity. Further, super-resolution techniques surpass the diffraction limit and localize labelled proteins at ~20 nm resolution. However, sparse light-emitting points do little to provide the subcellular context of labeled molecules. In nano-fEM, fluorescently tagged biological samples are first high-pressure frozen and processed via freeze substitution for fixation and to preserve fluorescence. Afterwards, samples are embedded in hydrophilic resin, cut into ultrathin sections, and visualized by, for example, direct stochastic optical reconstruction microscopy (dSTORM) followed by transmission electron microscopy. Fluorescence and electron micrographs are correlated by use of fiduciary markers and post-processing. This approach also provides 3D information similar to Array Tomography by serial sectioning of ultrathin sections followed by super-resolution microscopy and electron microscopy of each section in a sequential manner, enabling 3D reconstruction of the z axis.

Key words Fluorescence EM, CLEM, Super-CLEM, Nano-fEM (nano resolution fluorescence electron microscopy), Localization microscopy, Photoactivated localization microscopy (PALM), Stochastic optical reconstruction microscopy (STORM)

1 Introduction

Understanding many biological processes requires visualization of subcellular structures and the precise localization of numerous molecules. In the last few decades, several approaches have been developed to localize proteins in electron micrographs. These approaches are largely categorized into three groups: antibody-based [1–5], diaminobenzidine-based [6–10], and both diffraction-limited and sub-diffraction [11–17] fluorescence-

Irene Wacker et al. (eds.), *Volume Microscopy: Multiscale Imaging with Photons, Electrons, and Ions*, Neuromethods, vol. 155, https://doi.org/10.1007/978-1-0716-0691-9_1, © Springer Science+Business Media, LLC, part of Springer Nature 2020

based [18–30]. Each method has its own set of advantages and disadvantages.

First, antibody-based approaches can visualize the precise locations of proteins-of-interest in electron micrographs [1–5]. This approach typically uses two sets of antibodies: primary antibodies, which recognize proteins-of-interest, and secondary antibodies conjugated with colloidal gold (5–15 nm) that recognize the primary antibody [1–5]. Electron-dense gold particles thus pinpoint the locations of proteins-of-interest. However, this approach lacks sensitivity and spatial accuracy because (1) most primary antibodies are not compatible with the preparation regime required for electron microscopy, (2) retention of antigenicity of proteins precludes preservation of fine morphological details of cells, (3) many proteins are not labelled due to the comparatively large size of antibodies (~19 nm) which may restrict access to their binding site, especially on proteins embedded in plastic, (4) and non-specific binding of the secondary antibody results in high background signals [1, 2, 5]. Given these issues, alternative methods are increasingly sought in recent years, although immuno-electron microscopy still remains as the gold standard when localizing proteins at the ultrastructural level.

Second, diaminobenzidine-based localization approaches take an advantage of polymerization of diaminobenzidine under oxidative conditions, which induces formation of brown aggregates in tissues [6–10]. These aggregates preferentially bind osmium tetroxide and can be visualized as electron-dense precipitates in electron micrographs [6–10]. By tagging proteins-of-interest with peroxidases such as horseradish peroxidase [9, 10] and ascorbate peroxidase (APEX, APEX2) [6, 7], diaminobenzidine can preferentially polymerize near the proteins-of-interest in the presence of hydrogen peroxide due to the catalytic activity of the peroxidases. Alternatively, singlet-oxygen generators (SOG) such as mini-SOG [8] can also be genetically encoded and catalyze polymerization of diaminobenzidine. This approach is particularly useful for localizing transmembrane proteins or membrane-associated proteins [8]. Due to the diffusive nature of reactive species and proteins, it is not possible to localize cytoplasmic proteins. In addition, only one protein can be localized at a time. Thus, although this approach has been widely adapted, potential applications may be limited.

Third, fluorescence-based approaches correlate fluorescence images of proteins with subcellular structures visualized by electron microscopy [18–30]. This approach combines advantages of both microscopy modalities: with fluorescence microscopy, virtually any molecule can be visualized by tagging it with fluorescent labels, while cellular ultrastructure can be resolved on the order of a nanometer with electron microscopy. There are largely two strategies to perform correlative microscopy: pre-embedding [20, 23, 24, 26, 27, 29] or post-embedding [19–22, 25, 28, 30]. The pre-embedding method correlates fluorescence images acquired from live or fixed cells before embedding specimens into resin for electron microscopy. Although fluorescence imaging is simple, registering the fluorescence images onto ultrastructure is extremely difficult with this approach because of distortion caused by dehydration of tissue during plastic embedding [18]. The post-embedding method uses the same ultrathin slices of tissue collected from resin-embedded specimens for both fluorescence and electron microscopy imaging [18]. Thus, registration of fluorescence and electron micrographs is relatively simple, but samples must be processed for electron microscopy while maintaining fluorescent signals from probes. Thus, preservation of morphology and fluorescence must be balanced. Although this approach can be used to visualize multiple proteins with ~20 nm resolution, it has been met with limited success because of difficulties associated with optimizing the balance.

Despite these challenges, we have previously developed nano-resolution fluorescence electron microscopy (nano-fEM) (Fig. 1), which is a post-embedding method that pairs excellent ultrastructure with high spatial resolution of fluorescently labeled molecules [31–33]. The key consideration of this method is to balance fluorescence and morphology preservation by high-pressure freezing (Fig. 2), freeze-substitution (Fig. 3) in the presence of 5% water and potassium permanganate and embedding in hydrophilic resin (glycol methacrylates or HM20). Here, we describe a modified protocol that confers better preservation of morphology and fluorescence. In short, cells are labeled with an organic fluorophore, e.g. TMR, conjugated to a ligand which binds to a protein of interest fused to ligand-binding tag (e.g. HaloTag [36]), frozen using high-pressure freezing and processed for freeze-substitution. During freeze-substitution, cells are fixed with 0.1% $KMnO_4$ and 0.01% osmium tetroxide with 5% water in acetone—preserving up to 90% of the original fluorescence without compromising morphology. Next, cells are embedded in hydrophilic HM20 resin at $-50\ ^\circ C$ and ultrathin sections are collected. Then, fluorescence micrographs are acquired from ultrathin sections using single molecule localization microscopy (dSTORM). Afterwards, sections are stained with 1% uranyl acetate and imaged with transmission electron microscopy. The-step by-step protocol is outlined below (cf. Fig. 1 for general workflow).

2   Materials

2.1   Cell Culture

1. Culture Media.
2. Halo Ligand-TMR (Promega).

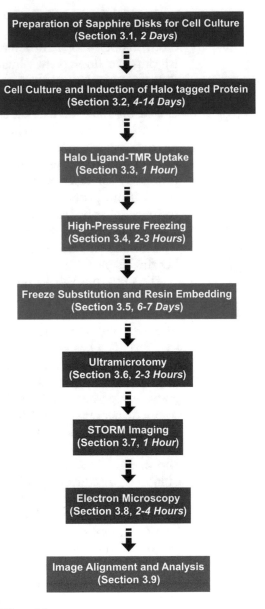

Fig. 1 Nano-fEM workflow

2.2 High-Pressure Freezing

1. High-pressure freezer (EM ICE, Leica Microsystems, 16771802).

2. Sapphire Disk 6 × 0.10 mm (TechnoTrade, 616-100).

3. Transparent half-cylinders (Leica Microsystems, 16771846).

4. CLEM Sample holder middle plate (Leica Microsystems, 16771838).

5. Spacer Ring, 100 μm (Leica Microsystems, 16770180).

6. Spacer Ring, 400 μm (Leica Microsystems, 16771840).

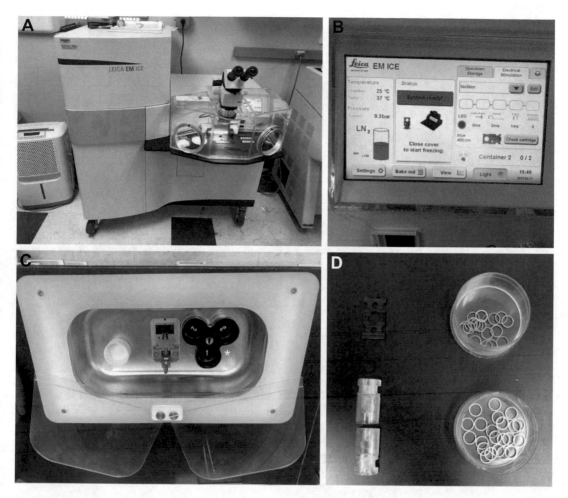

Fig. 2 High-pressure freezing—necessary tools and equipment. (**a**) Leica microsystems EM Ice high-pressure freezer. (**b**) Leica EM Ice interface. Monitor EM Ice status, set stimulation protocol and determine time delay for freezing. (**c**) Freeze-transfer station. Samples are deposited in this chamber after freezing via sample Dewar taken from EM Ice (not shown). Order of freezes determined by number in chamber device (asterisk) ejected from sample Dewar. (**d**) Freezing apparatus. *Top left.* Middle plate which holds sample during freezing. *Below.* Top and bottom component of sample-holding cylinder which carries sample through the EM Ice. *Right.* 100 and 400 µm spacers which avoid physical damage to sample

7. Freezing buffer (140 mM NaCl, 2.4 mM KCl, 10 mM HEPES, 10 mM glucose, 4 mM $CaCl_2$, 1 mM $MgCl_2$; pH adjusted to 7.5–7.6; osmolarity adjusted to ~300 mOsm).

8. Forceps for cryo techniques (Leica Microsystems, 16771872).

9. Forceps for cryo techniques (Leica Microsystems, 16701955).

10. Whatman filter paper.

11. Liquid nitrogen.

2.3 Freeze Substitution

1. Automated freeze-substitution unit (AFS2, Leica Microsystems).

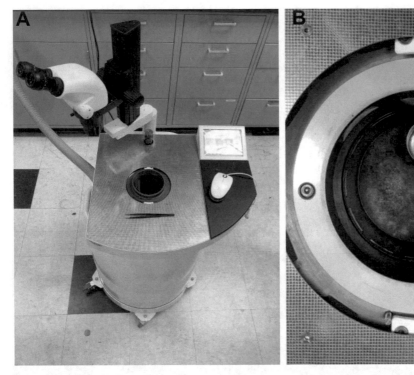

Fig. 3 Automated freeze-substitution. (**a**) Leica Microsystems Automated Freeze-Substitution Unit (AFS2). AFS2 equipped with microscope, freeze-substitution chamber, user interface and exhaust tube. (**b**) Freeze-substitution chamber. Cryo-samples incubate in this chamber until desired temperature is reached. Temperature ramp speed and plateaus are set via AFS2 interface. Cup with acetone which aides transfer of cryo-samples into vials containing freeze-substitution solution shown

2. Cryo-vials.

3. 50 ml screw cap conical tubes.

4. 15 ml screw cap conical tubes.

5. Acetone (glass distilled, Electron Microscopy Sciences).

6. Ethanol (absolute).

7. Osmium tetroxide (crystals, Electron Microscopy Sciences).

8. Potassium permanganate (Electron Microscopy Sciences).

9. Uranyl acetate (Polysciences).

10. Disposable transfer pipette (7.5 ml).

11. Disposable transfer pipette (1.5 ml).

12. Disposable Pasteur pipette (borosilicate glass).

2.4 Plastic Embedding (HM20)

1. HM20 kit (Electron Microscopy Sciences, RT14340).

2. Scintillation vials with screw caps.

3. ACLAR film (Electron Microscopy Sciences).

4. BEEM capsule (polypropylene, EBSciences, East Granby, CT).

2.5 Preparing Grids

1. PELCO® Slot Grids, 1 × 2 mm Cu (Ted Pella Inc.).
2. Pioloform, 0.7% solution in chloroform.
3. Cover glass.
4. Chloroform.

2.6 Ultramicrotomy

1. Ultramicrotome (UC6, Leica Microsystems).
2. Diamond knife (ultra jumbo, 45°, 4.0 mm; DiATOME).
3. Glass strips (Ted Pella Inc.).
4. Glass knife boats (Ted Pella Inc.).
5. Nail polish (clear).
6. Perfect loop (Electron Microscopy Sciences).
7. Hair tool for manipulation of plastic sections.
8. Razor blade (double edge; Electron Microscopy Sciences).
9. High-profile microtome blades (Leica Microsystems).
10. Canned air.
11. Super glue.

2.7 Stochastic Optical Reconstruction Microscopy (STORM)

1. Nunc Glass Base Dish (Thermo Scientific).
2. Imaging buffer (20 mM Cysteamine, 168 units glucose oxidase, 1404 units catalase, 50 mM Tris, 10 mM NaCl and 10% Glucose).
3. TetraSpeck fluorescent microspheres (0.1-μm diameter, Invitrogen).

2.8 Electron Microscopy

1. Transmission electron microscope (e.g. Philips, CM120).
2. Digital camera (e.g. AMT, XR80).

3 Methods

3.1 Sapphire Disk Preparation for Cell Culture

1. Place sapphire disks in a sputter coater and apply carbon coating (10–20 nm) [34, 35].
2. To indicate which side of sapphire disks cells are on, an asymmetric number or letter (e.g. 4) should be scratched on the disk. If the procedure requires finding a particular cell, then a finder grid should be placed on the sapphire disks prior to carbon coating. Cells should be plated on the coated side to avoid the coating being rubbed off.
3. Bake the carbon coated sapphire disks at 120 °C overnight.
4. Sterilize the sapphire disks by dipping in 70% ethanol, then dry by touching the side of the disk against a Kimwipe. Place two sapphire disks per well in a 12-well plate.

5. Apply poly-L-lysine (1 mg/ml, 1.5 ml per well) overnight.

6. Wash 3× with sterile ddH$_2$O.

3.2 Cell Plating and Tagged Protein Expression

1. Add cell culture media (1–1.5 ml) to sapphire disk-containing wells within a 12-well plate.

2. Add desired number of cells to each well to achieve preferred cell density.

3. Transfect or infect cells with vector containing a protein of interest fused to HaloTag [36] (e.g. lentiviral infection on day 7 of in vitro neuron culture). Allow time for sufficient growth of cells and expression of the protein of interest.

3.3 Halo Ligand-TMR Uptake

1. Dilute lyophilized Halo Ligand-TMR (1 mM stock in DMSO).

2. Add Halo Ligand-TMR to the cell medium for a final concentration of 1 µM and incubate for 30 min at 37 °C and 5% CO$_2$ to allow for binding of the dye to the protein of interest fused to a HaloTag [36].

3. Wash the cells with fresh medium three times.

3.4 High-Pressure Freezing (cf. Fig. 2)

1. Add liquid nitrogen to the tank of the EM AFS2 unit.

2. Enter the freeze substitution protocol (−90 °C for 5–30 h, 5 °C/h to −50 °C, pause, −50 °C for 48 h with UV). Initiate the program and pause it so that the temperature reaches −90 °C.

3. Add acetone to a small cup in the AFS specimen chamber and allow it to cool to −90 °C (30 min).

4. Prepare a solution of 1 ml milliQ water, 20 mg potassium permanganate and 20 µl of 1% osmium tetroxide, and 19 ml acetone. Aliquot 1 ml of fixative into labelled cryo-vials (using pencil; pen can be washed off by the acetone) and immediately place in liquid nitrogen. The fixatives should be kept under liquid nitrogen until use.

5. Fill the liquid nitrogen tank and sample storage Dewar of the EM ICE high-pressure freezer with liquid nitrogen.

6. Place sapphire disk containing cells in the black middle plate with cell-side facing up. Place one 100 µm ring. Dip one empty sapphire disk in the pre-warmed saline solution and place it carefully on the ring. Place one more 100 µm ring and a 400 µm ring and remove any extra liquid.

7. Place the middle plate in between two white half cylinders and close the lid to initiate freezing.

8. Once the freezing process in completed, the red cover will pop up and the sample will be dropped into the sample container.

3.5   Freeze Substitution, Infiltration, and Embedding (cf. Fig. 3)

1. After freezing, remove the sample Dewar and detach the specimen carrier under liquid nitrogen. Transfer the specimen to the AFS sample chamber. The sample must remain in liquid nitrogen during this transfer process. Additionally, forceps used to move the sample should be cooled in nitrogen before touching the sample.

2. Quickly transfer the sample to the cup with acetone. Hold onto the sample while transferring so that it settles at the bottom of the cup, rather than floating to the top. Ensure the forceps are cooled to the temperature of the AFS chamber before manipulating the sample.

3. Separate the sapphire disk from the middle plate by gently tapping. Move one of the cryovials with fixative into the AFS chamber. Quickly transfer the sapphire disk into the cryovial and seal. Click "continue" for freeze substitution to proceed.

4. Prepare 95% acetone in a scintillation vial and place inside the AFS chamber when the AFS temperature reaches −50 °C (*see* Subheading 3.3). Pre-chill for at least 30 min, with the program paused at −50 °C.

5. Using pipettes that have been pre-chilled in the AFS chamber, remove the fixative solution from the cryovials and wash 6× with pre-chilled 95% acetone.

6. Prepare 0.1% uranyl acetate in 95% acetone by mixing the tube in an end-to-end rotator in cold room for 1 h. Pre-chill to −50 °C in the AFS chamber for at least 30 min.

7. Replace 95% acetone from washing with pre-chilled 0.1% uranyl acetate and leave for 1 h.

8. Prepare 95% ethanol and pre-chill in the AFS chamber for at least 30 min.

9. Discard uranyl acetate and wash sapphire disks 6× with pre-chilled 95% ethanol.

10. Prepare HM20 resin by mixing 2.98 g Crosslinker D, 17.02 g Monomer E, and 0.1 g Initiator C.

11. Prepare 30% and 70% HM20 with 95% ethanol. Pre-chill 30%, 70% and 100% HM20 to −50 °C before use.

12. Remove all the ethanol after the last wash and leave samples in 30% HM20 for 2–3 h.

13. Replace with 70% HM20 and incubate for 2–3 h.

14. Replace with 100% HM20 and incubate overnight.

15. Transfer the sapphire disks to the caps of polypropylene BEEM capsules and add freshly prepared and pre-chilled 100% HM20.

16. Change 100% HM20 3× every 2 h.

17. After the third change, place Aclar film on top of the caps (oxygen inhibits polymerization).

18. Attach the UV lamp, and unpause the program so that UV light is turned on.

19. Allow the resin to polymerize at −50 °C in the AFS for 48 h.

20. Store the samples at −20 °C until further processing.

3.6 Ultramicrotomy

1. All the sectioning steps are performed at room temperature. Minimize exposure of the sample to bright light.

2. Scrape the thin layer of HM20 off the bottom of the sapphire disk using a razor blade. Using the pointed ends of the razor blade carve into the resin surrounding the sapphire disk; this will allow the disk to be removed from the resin.

3. Dip the bottom of the block in liquid nitrogen for 5–10 s, then insert the razor blade into the groove carved around the disk and pry the disk out of the block.

4. Find a region with cells using a microscope and mark it. Cut out the region and mount it onto a precast block with super-glue for ultramicrotomy.

5. Locate a region of interest and trim the block so that the sectioning surface is a trapezoid shape with dimensions of less than 100 µm high and 1 mm long.

6. Using a diamond knife with a cutting speed of 1.2–1.6 mm/s, start cutting sections of 50–80 nm thickness.

7. After cutting a sufficient number of sections, collect them on a grid, coated with Pioloform.

8. Once sections are dried, dip one side of the grid on a drop of TetraSpeck solution for two minutes. Flip the grids and let them sit in the solution for another two minutes.

9. Wash off the excess solution by dipping the grids in water, and dry.

10. Protect the sections from light by covering with aluminum foil. Store the sections at −20 °C until imaging.

3.7 Stochastic Optical Reconstruction Microscopy (STORM)

1. Prepare STORM buffer by mixing all components (*see* **Note 8**).

2. Add 1 ml of STORM buffer to Nunc Glass Base Dish. Place the grid facing cell-side down. The grid should touch the glass surface and should not be floating.

3. For STORM imaging, turn on the microscope and the software.

4. Turn ON 405 and 540 nm lasers to standby mode.

5. Using low power objective and bright field illumination, locate the sections and region of interest.

6. Change the objective to high power (40× water). Locate the region of interest with 540 nm laser, turned on at a low power.

7. Once the region of interest is found, increase the laser power to maximum. Collect ~20,000 frames at 20–30 ms/frame. Turn on 405 nm as necessary.

3.8 Electron Microscopy

1. Following STORM imaging, thoroughly wash grids with water and stain with 1% uranyl acetate for five minutes. Wash with grid by dipping into a 50% methanol solution (in water) 15–20 times. After, wash again by dipping grid into water 15–20 times. Let dry.

2. Image grids in a transmission electron microscope at both low and high magnification.

3.9 Image Analysis and Alignment

1. Fluorescence from TetraSpeck beads will be aligned to electron-dense particles observed in electron micrographs.

4 Notes

1. The carbon coating on sapphire disks should be dark enough to be visible at all steps, from freezing through embedding.

2. Although HaloTag is mentioned, other conjugated organic fluorophore labeling systems (e.g. SNAP-tag technology [37]) may be used instead of or in addition with HaloTag. This allows for two color super-resolution imaging.

3. Osmium tetroxide is a toxin and strong oxidizer that readily goes into the vapor phase. A container of OsO_4 should be opened only in a certified chemical hood using nitrile gloves and eye protection.

4. While setting up the sample in the high-pressure freezer, be careful not to introduce any bubbles. Any air pockets in the freezing solution can cause the sapphire discs to shatter during freezing. If any bubbles are present after the sample has been sealed, the components should be disassembled and reassembled.

5. The freezing protocol described here is for the Leica EM ICE. If using a different high-pressure freezer, follow standard procedures for that instrument.

6. We recommend freezing at least two samples for each experimental condition in case of loss of cells or poor morphology in some samples. Each chamber of the sample Dewar in the EM ICE can hold multiple samples, making it easy to keep track of duplicates.

7. After samples are frozen, ensure that sample manipulation is always done with precooled tools (in liquid nitrogen or sample cup solution) until they are embedded in resin.

8. STORM buffer must be prepared fresh for each experiment.

References

1. Morphew MK (2007) 3D immunolocalization with plastic sections. Methods Cell Biol 79:493–513

2. Rostaing P, Weimer RM, Jorgensen EM et al (2004) Preservation of immunoreactivity and fine structure of adult *C. elegans* tissues using high-pressure freezing. J Histochem Cytochem 52:1–12

3. Faulk WP, Taylor GM (1971) An immunocolloid method for the electron microscope. Immunochemistry 8:1081–1083

4. Duc-Nguyen H, Rosenblum EN (1967) Immuno-electron microscopy of the morphogenesis of mumps virus. J Virol 1:415–429

5. Roth J, Bendayan M, Carlemalm E et al (1981) Enhancement of structural preservation and immunocytochemical staining in low temperature embedded pancreatic tissue. J Histochem Cytochem 29:663–671

6. Lam SS, Martell JD, Kamer KJ et al (2015) Directed evolution of APEX2 for electron microscopy and proximity labeling. Nat Methods 12(1):51–54

7. Martell JD, Deerinck TJ, Sancak Y et al (2012) Engineered ascorbate peroxidase as a genetically encoded reporter for electron microscopy. Nat Biotechnol 30(11):1143–1148

8. Shu X, Lev-Ram V, Deerinck TJ et al (2011) A genetically encoded tag for correlated light and electron microscopy of intact cells, tissues, and organisms. PLoS Biol 9(4):e1001041

9. Hamos JE, Van Horn SC, Raczkowski D, Uhlrich DJ, Sherman SM (1985) Synaptic connectivity of a local circuit neurone in lateral geniculate nucleus of the cat. Nature 317:618–621. https://doi.org/10.1038/317618a0

10. Adams JC (1981) Heavy metal intensification of DAB-based HRP reaction product. J Histochem Cytochem 29(6):775

11. Betzig E, Patterson GH, Sougrat R et al (2006) Imaging intracellular fluorescent proteins at nanometer resolution. Science 313:1642–1645

12. Hess ST, Girirajan TPK, Mason MD (2006) Ultra-high resolution imaging by fluorescence photoactivation localization microscopy. Biophys J 91:4258–4272

13. Fölling J, Bossi M, Bock H et al (2008) Fluorescence nanoscopy by ground-state depletion and single-molecule return. Nat Methods 5:943–945

14. Rust MJ, Bates M, Zhuang X (2006) Subdiffraction- limit imaging by stochastic optical reconstruction microscopy (STORM). Nat Methods 3:793–795

15. Heilemann M, Van de Linde S, Schüttpelz M et al (2008) Subdiffraction-resolution fluorescence imaging with conventional fluorescent probes. Angew Chem Int Ed Engl 47:6172–6176

16. Hell SW, Wichmann J (1994) Breaking the diffraction resolution limit by stimulated emission: stimulated-emission-depletion fluorescence microscopy. Opt Lett 19:780–782

17. Gustafsson MG (2000) Surpassing the lateral resolution limit by a factor of two using structured illumination microscopy. J Microsc (Oxford) 198:82–87

18. Kopek BG, Paez-Segala MG, Shtengel G et al (2017) Diverse protocols for correlative superresolution fluorescence imaging and electron microscopy of chemically fixed samples. Nat Protoc 12(5):916–946

19. Koga D, Kusumi S, Shodo R et al (2015) High-resolution imaging by scanning electron microscopy of semithin sections in correlation with light microscopy. Microscopy 64:387–394

20. Kim D, Deerinck TJ, Sigal YM et al (2015) Correlative stochastic optical reconstruction microscopy and electron microscopy. PLoS One 10:e0124581–e0124520

21. Collman F, Buchanan J, Phend KD et al (2015) Mapping synapses by conjugate light-electron array tomography. J Neurosci 35:5792–5807

22. Paez-Segala MG, Sun MG, Shtengel G et al (2015) Fixation-resistant photoactivatable fluorescent proteins for CLEM. Nat Methods 12:215–218

23. Löschberger A, Franke C, Krohne G et al (2014) Correlative super-resolution fluorescence and electron microscopy of the nuclear pore complex with molecular resolution. J Cell Sci 127:4351–4355

24. Shtengel G, Wang Y, Zhang Z et al (2014) Imaging cellular ultrastructure by PALM, iPALM, and correlative iPALM-EM. Methods Cell Biol 123:273–294

25. Perkovic M, Kunz M, Endesfelder U et al (2014) Correlative light- and electron microscopy with chemical tags. J Struct Biol 186:205–213

26. Sochacki KA, Shtengel G, Van Engelenburg SB et al (2014) Correlative super-resolution fluorescence and metal-replica transmission electron microscopy. Nat Methods 11:305–308

27. Kopek BG, Shtengel G, Grimm JB et al (2013) Correlative photoactivated localization and scanning electron microscopy. PLoS One 8: e77209

28. Nanguneri S, Flottmann B, Horstmann H et al (2012) Three-dimensional, tomographic super-resolution fluorescence imaging of serially sectioned thick samples. PLoS One 7: e38098

29. Kopek BG, Shtengel G, Xu CS et al (2012) Correlative 3D superresolution fluorescence and electron microscopy reveal the relationship of mitochondrial nucleoids to membranes. Proc Natl Acad Sci U S A 109:6136–6141

30. Micheva KD, Smith SJ (2007) Array tomography: a new tool for imaging the molecular architecture and ultrastructure of neural circuits. Neuron 55:25–36

31. Watanabe S, Punge A, Hollopeter G et al (2011) Protein localization in electron micrographs using fl uorescence nanoscopy. Nat Methods 8:80–84

32. Watanabe S, Jorgensen EM (2012) Visualizing proteins in electron micrographs at nanometer resolution. Methods Cell Biol 111:283–306

33. Watanabe S, Lehmann M, Hujber E et al (2014) Nanometer-resolution fluorescence electron microscopy (nano-EM) in cultured cells. Methods Mol Biol 1117:503–526

34. Sawaguchi A, Mcdonald KL, Karvar S et al (2002) A new approach for high-pressure freezing of primary culture cells: the fi ne struc- ture and stimulation-associated transformation of cultured rabbit gastric parietal cells. J Micros (Oxford) 208:158–166

35. Hess MW, Müller M, Debbage PL et al (2000) Cryopreparation provides new insight into the effects of brefeldin A on the structure of the HepG2 Golgi apparatus. J Struct Biol 130:63–72

36. Urh M, Rosenberg M (2012) HaloTag, a platform technology for protein analysis. Curr Chem Genomics 6:72–78

37. Keppler A, Gendreizig S, Gronemeyer T et al (2003) A general method for the covalent labeling of fusion proteins with small molecules in vivo. Nat Biotechnol 21:86–89

Multicolor Superresolution Microscopy: Revealing the Nano World of Astrocytes In Situ

Janosch P. Heller, James P. Reynolds, and Dmitri A. Rusakov

Abstract

Astroglia are essential to the development, homeostasis, and metabolic support of the brain but also to the formation and regulation of synaptic circuits. Experimental evidence has been emerging that astrocytes undergo substantial structural plasticity associated with age- and use-dependent changes in neural circuitries. The underlying cellular mechanisms are poorly understood, mainly due to the extraordinary complex, essentially nanoscopic morphology of astroglia. It appears that key morphological changes occur in fine astrocytic processes that are in the vicinity of synapses. However, the characteristic size of these compartment falls below the diffraction limit of conventional optical microscopy, making the deciphering of their molecular nanostructure a challenge.

Here we detail a superresolution microscopy approach that relies on direct stochastic optical reconstruction microscopy (dSTORM) to visualize astroglial organization on the nanoscale (in fixed brain tissue). We also provide a protocol for viral infection of astroglia in vivo (aimed at monitoring the cell activity with the genetically encoded calcium indicator GCaMP), followed by tissue sectioning, immunolabeling, and the subsequent dSTORM analysis. The presented workflow can be extended to a correlational-study protocol to reconstruct the nanoscopic morphology of the imaged cells.

Key words Superresolution microscopy, dSTORM, Viral infection, Cranial window imaging, GCaMP, Brain tissue fixation, Tissue sectioning, Immunohistochemistry

1 Introduction

Astrocytes have long been acknowledged for their key role in neurotransmitter uptake and extracellular potassium buffering in the brain. Recent evidence suggests that they also contribute to the information processing in the brain while actively contributing to neural circuit formation, maintenance, and function. Astroglia express numerous plasma membrane receptors, transporters and ion channels that enable them to receive and transduce diverse physiological inputs from brain networks, in health and disease [1–9]. Even though astroglia are electrically nonexcitable, they can integrate and communicate physiological signals through intracellular calcium sparks, elevations, and regenerative waves that

Irene Wacker et al. (eds.), *Volume Microscopy: Multiscale Imaging with Photons, Electrons, and Ions*, Neuromethods, vol. 155, https://doi.org/10.1007/978-1-0716-0691-9_2, © Springer Science+Business Media, LLC, part of Springer Nature 2020

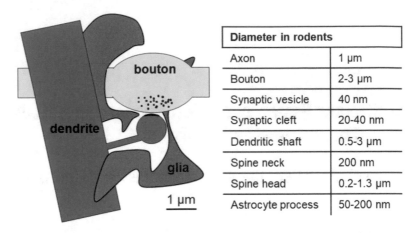

Diameter in rodents	
Axon	1 μm
Bouton	2-3 μm
Synaptic vesicle	40 nm
Synaptic cleft	20-40 nm
Dendritic shaft	0.5-3 μm
Spine neck	200 nm
Spine head	0.2-1.3 μm
Astrocyte process	50-200 nm

Fig. 1 Graphical illustration of synaptic microenvironment. Neurotransmitters are released via synaptic vesicles from an axonal bouton (gray). Neurotransmitter receptors are located in the postsynaptic density on a dendritic spine (blue). Astrocytic processes (green) are in close apposition to the synapse. The components are drawn to scale. Representative diameters of synaptic components in rodents are listed in the table. Scale bar = 1 μm

exhibit wide-ranging spatiotemporal modalities across different cellular compartments [10–16]. Excitatory synapses are often surrounded by nanoscopic astroglial processes (termed perisynaptic astrocytic processes, PAPs) (Fig. 1) which seem to enable intimate astroglia–synapse signal exchange including glutamate transport, potassium buffering, and the release of signaling "gliotransmitters" molecules [12, 17–19]. While PAPs can be found in all brain regions, their synapse coverage varies depending on the region, synaptic identity, and local circuit activity [18, 20–26].

The direct visualization of nanoscopic astrocyte structures has been technically difficult as the sponge-like astrocytic processes can be as fine as 50–100 nm in diameter (Fig. 1), a scale that lies below the diffraction limit of conventional light microscopy (200–300 nm). Historically, electron microscopy (EM) has been the only tool to successfully resolve astroglial structure on the nanoscale [20, 26–33]. However, EM has a limited ability to integrate nanoscale and microscale molecular-specific information pertinent to cellular environment: large-scale 3D EM reconstructions at high resolution require extensive human and instrumental resources and have not yet been successful in revealing protein expression patterns at single-molecule resolution beyond individual ultrathin sections (or individual frozen replicas).

The recent advent of superresolution microscopy techniques enables resolution of up to 10–70 nm while imaging structurally intact cells in vitro and in vivo [34–38]. Through the use of stimulated-emission depletion (STED), photoactivated localization microscopy (PALM) or stochastic optical reconstruction microscopy (STORM) some important aspects of astroglial nano-organization have been revealed already [15, 18, 39–43].

Here, we detail a direct STORM (dSTORM) approach that uses stochastic excitation of sparsely labeled target molecules with conventional fluorophore-labeled antibodies [44–46]. With this method we are able to reconstruct fine morphology of astroglia below the diffraction limit of conventional microscopy. We describe two protocols. The first deals with the labeling and imaging of the common astroglia marker glial fibrillary acidic protein (GFAP) [47], and the calcium-binding protein S100β [48, 49] which have been used previously [50] to reveal the nanostructure of fine astrocytic processes in fixed brain sections. We also provide a protocol which deals with the viral infection of astroglia in vivo (to monitor cell activity with cytosolic or membrane-bound calcium indicators, such as cyto-GCaMP or lymphocyte-specific protein tyrosine kinase-GCaMP (lck-GCaMP), respectively) [51–53]. These two protocols can potentially be combined to a correlational-study protocol in which the imaged tissue is sectioned, immunolabeled, and subsequently analyzed with dSTORM to reconstruct the nanoscopic morphology of the transduced and recorded cells.

2   Materials

2.1   Viral-Mediated Labeling of Cortical Astrocytes

Adult male and female C57BL/6J mice (Charles River, UK) are used for imaging of astrocytes. Animal procedures are conducted as mandated by the European Commission Directive (86/609/EEC) and the United Kingdom Home Office (Scientific Procedures) Act (1986). We make use of a commercially available adeno-associated virus (AAV) to specifically target astroglia in the somatosensory cortex for in situ multiphoton imaging and follow-up superresolution imaging. Mice are anesthetized using isoflurane (5% induction, then 1.5–2.5% v/v). For perioperative analgesia, we use topical lidocaine cream (applied to the scalp) and subcutaneous buprenorphine (60 µg/kg). Dexamethasone is also administered (2 mg/kg) to reduce inflammation and cortical stress response. Following the craniotomy and exposure of the cortex, warmed artificial cerebrospinal fluid (aCSF; in mM: 125 NaCl, 2.5 KCl, 10 HEPES, 10 glucose, 2 CaCl$_2$, 2MgSO$_4$) is applied to the skull and cortical surface throughout the procedure. AAV5 GfaABC1D-LckGCaMP6f (catalog number AV-5-PV3107) and AAV5 GfaABC1D-cytoGCaMP6f (catalog number AV-5-52925, both from Penn Vector Core, PA, USA) are pressure-injected through a pulled glass pipette, labeling astrocytes in all cortical layers at the target site. Meloxicam (subcutaneous, 1 mg/kg) is administered once daily for up to 2 days following surgery. After a 2–6 week AAV incubation period, animals are prepared for superresolution imaging as described below. In some cases, we also perform multiphoton imaging through a cranial window implantation as described elsewhere [54], before processing tissue for superresolution imaging.

2.2 Tissue Preparation

The animals are anesthetized with a terminal dose of pentobarbital (100 mg/kg) and transcardially perfused using phosphate-buffered saline (PBS, Sigma, #4417, stored at room temperature) followed by 4% (w/v) paraformaldehyde (PFA, Sigma, #P6148, stored at 4 °C) in PBS. The brains are removed and incubated in 4% PFA in PBS overnight at 4 °C. The tissue is then sectioned using a vibratome (Leica, #VT1000S) into 30 μm coronal sections and immediately used for immunohistochemistry or stored in PBS at 4 °C. For prolonged storage, the tissue is incubated in PBS supplemented with 0.01% (w/v) NaN_3 (Sigma, #S2002) and 100 mM glycine (Sigma, #G8898). The sodium azide is added to avoid contamination of the specimen and the addition of glycine is used to quench autofluorescence from any residual fixatives.

2.3 Immunohistochemistry

Brain sections (see above) are incubated in 0.1% (w/v) $NaBH_4$ (Sigma, #71320) in PBS (made fresh on the day). For the permeabilization of the sections we use saponin (Bio Basic, #SB4521). As a blocking agent to prevent nonspecific binding of the antibodies we use bovine serum albumin (BSA, Sigma #A7906 at 3% w/v). The saponin solution can be made up in advance in PBS and stored at 4 °C for several days. The primary antibodies used can be found in Table 1, and fluorescently labeled secondary antibodies used are listed in Table 2. Antibodies are diluted in PBS supplemented with

Table 1
Primary antibodies used

Antigen	Host	Clone	Supplier	Product code	RRID	IHC
GFP	Chicken	Polyclonal	Thermo	A10262	AB_2534023	1:500
GFAP	Mouse	GA5	Novus	NBP2–29415	AB_2631231	1:500
S100β	Rabbit	Polyclonal	Synaptic systems	287,003	AB_2620024	1:200

RRID research resource identifier, IHC dilution factor used in immunohistochemistry

Table 2
Secondary antibodies used

Antigen	Feature	Host	Supplier	Product code	RRID	IHC
Chicken IgY	Alexa 647-conjugated	Donkey	Millipore	AP194SA6	AB_2650475	1:1000
Mouse IgG	CF 568-conjugated	Donkey	Biotium	20,105	AB_10557030	1:200
Rabbit IgG	Alexa 647-conjugated	Goat	Thermo	A21245	AB_2535813	1:500

Ig immunoglobulin, RRID research resource identifier, IHC dilution factor used in immunohistochemistry

Table 3
Scale U2 buffer [55]

Ingredient	Concentration	Supplier and product code
Urea	4 M	Sigma, #U6504
Glycerol	30%	Fisher, #BP229–1
Triton X-100	0.1%	Sigma, #T9284

saponin and BSA just before use on the day. Postfixation of the stained tissue is performed using 4% PFA in PBS. Sections are incubated in Scale U2 buffer (Table 3) [55] and stored covered at 4 °C until being prepared for imaging. The Scale U2 buffer can be prepared in advance and kept at 4 °C.

2.4 Microscopes and Sample Preparation

Here, we employ the superresolution imaging technique direct stochastic optical reconstruction microscopy (dSTORM) [44–46]. Superresolution images are recorded with a Vutara 350 commercial microscope (Bruker Corp., Billerica, US-MA) based on the single molecule localization (SML) biplane technology [56, 57]. The targets are imaged using 647 nm (for Alexa 647) and 561 nm (for Alexa 568 and CF 568) excitation lasers, respectively, and a 405 nm activation laser in a photoswitching buffer containing 100 mM cysteamine and oxygen scavengers (glucose oxidase and catalase) (Table 4) [58]. Images are recorded using a 60×-magnification, 1.2-NA water immersion objective (Olympus) and a Flash 4.0 scientific complementary metal-oxide semiconductor (sCMOS) camera (Hamatasu) with frame rate at 50 Hz. In addition, a semiconductor charge-coupled devices (CCD) camera (Photometrics) is used for standard widefield imaging. Total number of frames acquired per channel ranged from 5000 to 20,000 frames. Data are analyzed using the Vutara SRX software (version SRX 6.00). Single molecules are identified by their brightness frame by frame after removing the background. Identified particles are then localized in three dimensions by fitting the raw data with a 3D model function, which is obtained from recorded bead data sets (see below). With our methods we routinely achieve a lateral resolution (x and y) of 50 ± 10 nm and an axial resolution (z) of 70 ± 20 nm in tissue sections (*see* **Note 1**).

3 Methods

3.1 Targeted Labeling of Cortical Astrocytes Using AAVs

It is crucial that all animal procedures be carried out in accordance with institutional and local government guidelines, with diligent consideration of the animal's welfare at each step throughout.

Table 4
Photoswitching buffer [58]

Ingredient	Concentration	Supplier and product number
Enzyme stock solution (A)		
10 μl catalase	20 μg/ml	Sigma, #C40
20 μl 1 M TCEP	4 mM	Sigma, #C4706
2.5 ml glycerol	50%	Fisher, #BP229-1
125 μl 1 M KCl	25 mM	Sigma, #P9333
100 μl 1 M Tris–HCl pH 7.5	20 mM	Sigma, #33742
5 mg glucose oxidase	1 mg/ml	Sigma, #G2133
Top up to 5 ml with distilled water and dispense into 50 μl aliquots and store frozen at −20 °C (for up to 1 year)		
Glucose stock solution (B)		
4 g glucose	100 mg/ml	Sigma, #G8270
4 ml glycerol	10%	Fisher, #BP229–1
Top up to 40 ml with distilled water and dispense into 400 μl aliquots and store at −20 °C (for up to 1 year)		
Reducing agent stock solution (C)		
113.6 mg MEA–HCl	1 M	Sigma, #M6500
Top up to 1 ml with distilled water and store at 4 °C on the day of imaging. This solution can also be prepared in advance and stored at −20 °C for up to 1 year (do not refreeze)		
Just prior to imaging mix the above solutions in the following ratio: 50 μl Solution A 400 μl Solution B 100 μl Solution C 450 μl PBS		

3.1.1 AAV Injection Protocol

1. The procedure area is prepared for sterile surgery. This includes careful disinfection of the area, the use of autoclaved instruments (with bead sterilization for batch surgeries), surgical drapes, sterile gloves, and surgical gown. Autoclaved aluminum foil is used to cover the dials of instruments that could not be autoclaved, such as the vaporizer, stereotaxic frame, and microscope.

2. The AAV solution is prepared for injection (*see* **Note 2**).

3. The animal is prepared for surgery. Mice are anesthetized using 5% (v/v) isoflurane initially, the scalp is shaved and the animal is secured in a stereotaxic frame. The anesthetic dose is reduced to 1.5–2% (v/v). Adequate anesthesia is confirmed by the loss of pedal withdrawal reflexes. The scalp is disinfected using three

topical applications of chlorhexidine, each applied with a sterile cotton swab. Other preoperative steps included the use of analgesia (topical lidocaine, applied to the scalp, and subcutaneous buprenorphine, 60 μg/kg), administration of an anti-inflammatory agent (dexamethasone, 2 mg/kg) to reduce cortical stress responses, and application of eye ointment (Lacri-Lube, Allergan, UK) to prevent dehydration. Body temperature is kept at 37 °C throughout using a rectal thermometer and feedback-controlled heating blanket.

4. A midline scalp incision is made to expose the approximate target site. The fascia is gently parted and removed using a scalpel and curved forceps. The target region, S1FL, is identified by coordinates from Bregma (0.1 mm anterior, and 2.0 mm lateral) and a craniotomy is performed. We use a 0.4 mm ball bur to drill an approximately 1 mm craniotomy. Bone dust is periodically removed with compressed air. The skull is thinned until fracturing occurred and a bone flap is carefully removed with extrafine forceps to expose the cortex. Where possible, the exposed cortex is continuously superfused with warmed aCSF from this point on.

5. The AAV-containing pipette is secured on the stereotaxic arm and lowered to the cranial surface. The z-coordinate at the surface is noted to calibrate the penetration depth and a small amount of positive pressure is applied to the pipette.

6. The pipette is slowly lowered through the dura and into the tissue, to a depth of 500 μm. Positive pressure is applied while observing the meniscus in the pipette. The pressure is carefully adjusted to generate a slow but continuous elution from the pipette approximating 50 nl per minute. The injection bolus contains between 0.1 and 1×10^{10} genomic copies, in a volume not exceeding 500 nl. Once the desired bolus volume has been reached, the pressure is released and the pipette left in place for 5 min before slowly retracting it from the tissue.

7. The scalp is sutured using 7–0 absorbable sutures and the animal removed from the stereotaxic frame. The animal is left to recover in a heated chamber and then returned to the home cage once ambulatory. Meloxicam (1 mg/kg) is administered once daily for 2 days following surgery and the animal's health status observed in line with institutional guidelines.

3.1.2 Optional: Confirmation of Viral Labeling and Correlative Multiphoton Imaging

In certain cases, we implanted a cranial window over the right S1FL region and imaged dynamic changes in astrocytic calcium in vivo (Fig. 2). Cranial window implantation is a challenging but common technique that has been described in detail elsewhere [54, 59, 60]. The use of GCaMP-encoding AAVs permits in situ verification of viral labeling and multiphoton imaging of astrocytic calcium

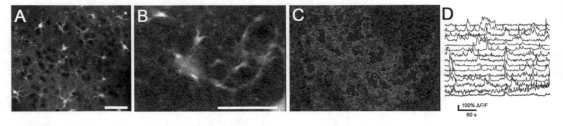

Fig. 2 Optical measurement of astrocytic calcium transients using multiphoton microscopy. (**a**) Astrocytes in layer 2/3 of the somatosensory cortex were labeled using pressure-injected sulforhodamine 101 (SR101, <500 nl bolus of 5 μM), for visualization and identification purposes. Scale bar = 50 μm. (**b**) A layer 2/3 astrocyte, expressing GCaMP6f and labeled with SR101, was imaged at 2 Hz. Illumination light (λ = 920 nm) was set to 25 mW in the back-focal plane. Scale bar = 20 μm. (**c**) Image segmentation was performed by mapping the cross-correlation value of pixel intensity (GCaMP6f emission) with respect to its immediately adjacent neighbors and selecting the largest contiguous regions exhibiting high correlation (top 5% percentile). (**d**) $\Delta F/F$ was computed for each pixel and the means values for all pixels in a given region of interest were plotted

activity. We confirmed the expression of lck-GCaMP6f and cyto-GCaMP6f in mice following AAV injection and recorded both spontaneous and sensory-evoked astrocytic calcium transients in anesthetized mice (in mg/kg: medetomidine, 0.5; midazolam 5; fentanyl, 0.05). When combined with near-infrared branding [61], single-cell labeling through electroporation or patch-clamp approaches [62], or some other means of relocating imaged cells [63], this technique can enable the performance of correlative multiphoton and superresolution imaging to directly relate biodynamic activity with superresolved distribution of immunolabeled targets.

3.1.3 Transcardial Perfusion and Tissue Sectioning

Fixation of the tissue is achieved by transcardial perfusion of the animal. The quality of the perfusion is instrumental to obtain good tissue preservation. Brains can be stored in PBS at 4 °C for up to 3 months prior to sectioning. After sectioning, slices can be stored long-term (for several months) in PBS with 0.01% (w/v) NaN_3 and 100 mM glycine.

1. A perfusion pump system is set up to elute exsanguinate and fixative from a 26-G needle, at a rate of 5 ml/min.

2. The animals are terminally anesthetized with pentobarbital (i.p., 100 mg/kg) and laid supine. We ensure a stable and deep anesthesia before continuing. The pleural cavity is exposed, the sternum is lifted away and the heart isolated from surrounding connective tissue.

3. The 26-G needle is inserted into the left ventricle and clamped with a hemostat. The right atrium is cut and the animal is exsanguinated with 10 ml PBS. The animal is then transcardially perfused with 4% PFA, noting the onset of fixation tremors.

4. After transcardial perfusion, the brain is removed from the skull and postfixed overnight.

Tissue incorporating the target injection site is dissected out and prepared for sectioning using a vibratome. We glue a solidified block of 4% agarose to the vibratome stage and adjacent to the tissue, to secure it during sectioning. The tissue itself is also glued to the stage, ensuring a minimum amount is used and taking care to prevent any adhesive contacting the area to be sectioned. The tissue is then sectioned into 30 μm sections while immersed in PBS, and stored for later use.

3.2 Revealing the Nanostructure of Astrocytes In Situ Using Superresolution Microscopy

3.2.1 Immunolabeling Protocol

1. Brain sections (see above) are briefly washed free-floating in PBS to remove any residual PFA.

2. Then, they are incubated in 0.1% $NaBH_4$ in PBS for 15 min gently rocking at room temperature to quench the autofluorescence of residual aldehydes following the fixation procedure using PFA. NH_4Cl or glycine can also be used for the quenching.

3. The sections are thereafter washed thrice for 5 min with PBS gently rocking at room temperature.

4. Permeabilization and blocking are carried out using blocking buffer (PBS supplemented with 0.1% saponin (PBS-S) and 3% BSA) for at least 2 h gently rocking at room temperature. Saponin is an amphipathic glycoside that acts as a mild detergent. In comparison to Triton—the most commonly used detergent for permeabilization—saponin reversibly permeabilizes the cells and has to be added to every solution. Instead of saponin or triton, for example digitonin or leucoperm can be used. BSA is added to the solution to block the unspecific binding of the antibodies. BSA can be replaced by serum (use serum of the animal in which the secondary antibody has been raised), milk powder, or other blocking agents.

5. The brain sections are then treated with primary antibodies (Table 1) in PBS-S supplemented with a diluted solution of blocking agent (1% BSA) overnight gently rocking at 4 °C. This incubation step can also be performed at room temperature for only a few hours.

6. Samples are washed briefly with PBS-S (~1 min) and then with PBS-S thrice for 10 min gently rocking at room temperature to remove surplus primary antibody.

7. Afterward, the sections are incubated with fluorescently labeled secondary antibodies (Table 2) diluted in PBS-S for 2 h gently rocking at room temperature. To avoid bleaching of the fluorophores, the sections have to be shielded from light (e.g., using aluminum foil to wrap the plate) from this point on. To avoid

bleaching of overexpressed fluorescent proteins such as the calcium indicators the brain sections should be kept covered in aluminum foil from the beginning.

8. The sections are then washed with PBS-S twice for 10 min and thrice with PBS for 5 min gently rocking at room temperature.

9. Lastly, the sections are postfixed using 4% PFA in PBS for 30 min gently rocking at room temperature to immobilize the antibodies in place to avoid movement during imaging.

10. This is followed by washing with PBS thrice for 10 min gently rocking at room temperature.

11. For storage, sections are incubated in Scale U2 buffer (Table 3) [55] and stored covered at 4 °C. This buffer also clears the tissue with minimal to no extension to minimize autofluorescence [55].

3.2.2 Superresolution Imaging

The superresolution imaging should be performed in switching buffer (*see* **Note 3**). Therefore, stained samples should not be mounted on microscope slides in mounting medium. The switching buffer will provide optimal "blinking" of the fluorophores. Just prior to imaging, the brain sections are mounted on glass coverslips.

1. The tissue is set on top a no. 1.5 coverslip (25 mm in diameter, SLS #MIC3350) and let slightly dry. Care has to be taken to not completely dry out the tissue (Fig. 3a).

2. Warmed 2% agarose (Lonza, #98200) is put on top of the tissue to immobilize the brain sections on the coverslip (Fig. 3b). The agarose gel is porous so that the switching buffer can penetrate the tissue. The agarose needs to dry and set before buffer is added.

3. The coverslip is then inserted into a 25 mm circular stage adaptor (Thermo, #A7816) and imaging buffer (~1 ml) (*see* **Note 3**) is pipetted on top (Fig. 3c). Then a smaller coverslip (18 mm diameter) is set on top to seal the chamber.

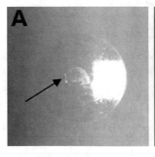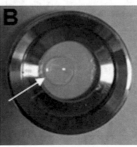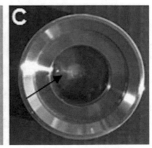

Fig. 3 Tissue preparation for superresolution imaging. (**a**) The brain section is placed on top of a coverslip. (**b**) Warmed 2% agarose is used to immobilize the tissue. (**c**) The imaging chamber is filled with buffer (*see* **Note 3**). The arrows are pointing at the tissue

3.2.3 Imaging Protocol

1. As a first step, a region of interest is selected by scanning the sample in standard epifluorescence mode (*see* **Note 4**) using the CCD camera (Figs. 4a, d and 5a, d).

2. Then, the sCMOS camera is used to confirm the region for superresolution imaging (Figs. 4b, e and 5b, e).

3. Now, the laser power is greatly increased to 0.6–6 mW/μm^2 to induce photoswitching of the fluorophores.

4. An image series of several thousand (usually 20,000) frames is recorded until most of the fluorophores have been documented and blinking is diminished. To maximize signal-to-noise ratio and to not split the blinking over multiple frames the exposure/recording time should on average match the time a single fluorophore emits photons - usually between 10 and 30 ms.

5. The biplane mode produces a 3D image with a *z* range of 2 μm (*see* **Note 4**). However, to create an image with a longer *z* range a z stack can be imaged by moving the piezo stage in 500 nm steps.

6. When performing multicolor imaging, recording should start in the red range of the spectrum and end in the blue range to avoid bleaching and activating the other fluorophores (*see* **Note 5**).

3.3 Image Reconstruction and Visualization

3.3.1 Experimental Point Spread Function Generation

For the analysis of the dSTORM raw data it is helpful to generate an experimental point spread function (PSF) (*see* **Note 6**) instead of using a theoretical PSF to (1) fit the fluorophore localizations and (2) calibrate chromatic aberrations (*see* **Note 5**). The experimental PSFs for the laser lines/fluorophores used can be generated by means of a TetraSpeck bead sample (Thermo, #T7279).

1. Aliquot 1 μl TetraSpeck microspheres into a tube and sonicate for 10 min.

2. Add 500 μl dH_2O, vortex, sonicate again for 10 min, and then vortex again.

3. Add 100 μl poly-DL-lysine solution (1 mg/ml; Sigma, #P9011) to the center of a no. 1.5 cover glass. The coverslip should have the same quality and thickness as the ones used for imaging.

4. After 10 min aspirate the lysine solution and let the coverslip air-dry.

5. Add 10–30 μl of the prepared 1:500 bead sample onto the center of the dried lysine spot and let stand for 10 min and then aspirate the remaining solution.

6. Let sample air-dry completely.

7. Aliquot 3–5 μl dH_2O or Zeiss Immersol W 2010 onto a glass slide (e.g., Henso, #7107) and invert the coverslip on top.

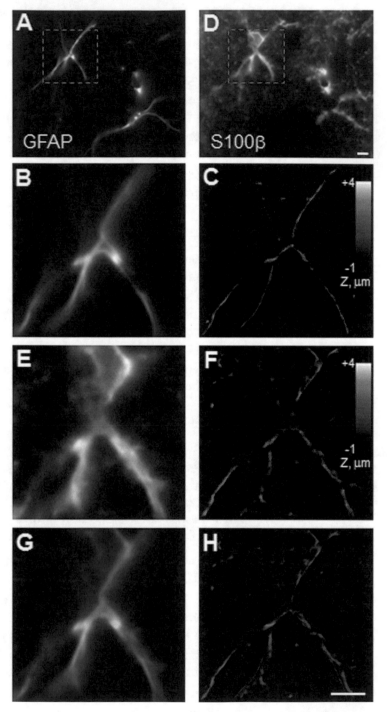

Fig. 4 Superresolution imaging of astrocytes in situ. (**a**) Wide-field image of astrocytes in mouse brain sections expressing GFAP. (**b**) Higher magnification image of area shown by rectangle in (**a**), wide-field fluorescence mode. (**c**) dSTORM image of area shown in (**b**). (**d**) Widefield image of S100β in the same field of view as in (**a**). (**e**) Higher magnification image of area shown by rectangle in (**d**), wide-field fluorescence mode. (**f**) dSTORM image of area shown in (**e**). (**g**) Merged image of (**b**) and (**e**), with GFAP shown in green and S100β in magenta. (**h**) Merged image of (**c**) and (**f**), with GFAP shown in green and S100β in magenta. Scale bars = 5 μm

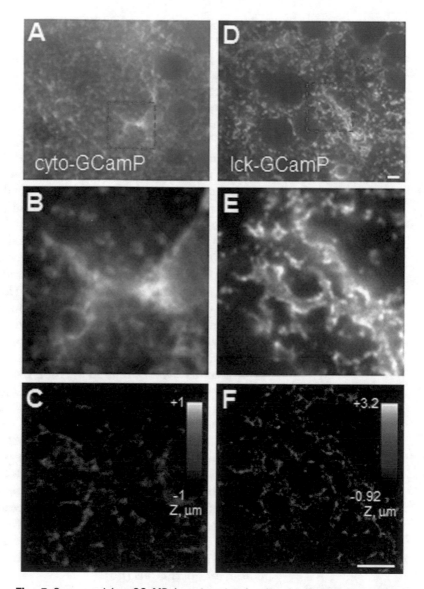

Fig. 5 Superresolving GCaMP in astrocytes in situ. (**a**) Wide-field image of astrocytes expressing cyto-GCaMP in mouse brain sections. (**b**) Higher magnification image of area shown by rectangle in (**a**), wide-field fluorescence mode. (**c**) dSTORM image of area shown in (**b**). (**d**) Widefield image of astrocytes expressing Ick-GCaMP in mouse brain sections (**e**) Higher magnification image of area shown by rectangle in (**d**), wide-field fluorescence mode. (**f**) dSTORM image of area shown in (**e**). Scale bars = 5 μm

Immersol W 2010 has the same refractive index as water but it is more viscous and less evaporative.

8. Seal the coverslip with nail polish. The bead sample can also be used with the above mentioned circular stage adaptor. In that case, the bead sample does not have to be mounted but stored in PBS at 4 °C. The imaging should also take place in PBS.

To generate the experimental PSFs, a z stack (fifty 100 nm steps) of the TetraSpeck sample is imaged and the resulting PSFs are calculated. Care has to be taken to calibrate the system with beads that fluoresce in the wavelengths that are used for imaging. When performing multicolor imaging (*see* **Note 5**) chromatic aberration can be a key problem. The acquired experimental PSFs can be used to align the individual channels.

3.3.2 Image Analysis

The acquired raw data consists of thousands of frames and up to millions of PSFs. During the analysis, the 3D position of every emitted organic dye needs to be determined as accurately as possible. This can either be achieved by using commercial software (Figs. 4c, f and 5c, f) or one of the many freely available software packages, (e.g., QuickPALM [64] or ThunderSTORM [65]). Most localization algorithms fit a two- or three-dimensional Gaussian distribution at the center of every detected fluorophore position and then sample the surrounding pixels (*see* **Note 6**). Therefore, labeling density and localization precision (number of photons/ single fluorophore, pixel size, background signal/signal-to-noise ratio, and emission wavelength) are important factors to consider when analyzing experiments [66]. Sparse switching of fluorophores is desired to avoid overlapping of PSFs. Moreover, the fitting accuracy of the Gaussian models depends greatly on the noise in the image as unspecifically bound fluorophores are identified with the same accuracy and intensity as correctly labeled ones [66]. Mechanical drift of the sample is also a problem that can occur when imaging. This is especially a problem when electron multiplying charged coupled device (EMCCD) cameras are used instead of faster sCMOS cameras. One option to eliminate the effects of drift is to add multispectral beads (the same that are used for the calibration; see above) to the sample. These beads fluoresce throughout the experiment and can hence be easily detected and traced. This approach is also helpful to counteract chromatic aberrations [66].

4 Notes

1. One has to keep in mind that the protocol described here uses complexes of primary and secondary antibodies, and the detected signals represent the position of the fluorophores rather than that of the labeled proteins. Therefore, the distance between label and protein can measure more than 15 nm, which is in the range of the achievable resolution of dSTORM and has to be considered during analysis. Some strategies exist to reduce this localization precision error. For example, primary antibodies can directly be tagged with fluorophores or smaller fragments of antibodies can be used such as nanobodies [67]. Moreover, aptamers [68], monomeric streptavidin [69],

or the pore-forming bacterial toxin streptolysin O can be used for labeling [70, 71]. Furthermore, instead of using fluorescent proteins, the protein of interest can be modified with much smaller tags, such as hexahistidine [72, 73] or click chemistry approaches [74, 75].

2. Intracranial injection of viral vectors, such as AAVs, is a widely used and versatile technique for labeling, imaging, and augmentation of brain cells. There are numerous approaches to the administration of viral vectors, but most make use of pulled glass micropipettes to inject the vector solution. Positive pressure can be applied manually or using an automated pump. This technique allows for a carefully controlled injection with minimal damage to the surrounding tissue. It is crucial that the parameters of the injection be tailored to the specific experiment, making consideration of the age of the animal, the brain region, desired cell targets, injection titers, and serotype of the virus. In our case, the micropipette is pulled using a P-97 micropipette puller (Sutter Instrument Company, CA, USA) to pull a pipette with an approximate diameter of 30 μm. The pipettes are clamped and lowered into the AAV solution, and manually backfilled by negative pressure using a 50 ml syringe. A small drop of mineral oil is applied to the tip to prevent clogging of the pipette tip during tissue penetration. Positive pressure is applied manually to the AAV solution via noncompressible PEEK tubing and a 1 ml syringe. We typically used an unprimed syringe without issue. However, the syringe can be primed with mineral oil or with H_2O. In the case of H_2O priming, the pipette is then backfilled with a small volume of mineral oil to prevent mixing of the AAV solution and the priming solution. When using an automated pump, priming is recommended. Alternatively, a wire plunger can be used to fill and inject, without the need for priming, as described elsewhere [76]. As with any reagent conferring biosafety risks, AAVs must be handled at the appropriate biosafety level (BSL). Most commercial AAVs are produced in the absence of a helper virus (and are replication defective), and those in which the transgene does not encode tumorigenic or toxin gene products can be handled at BSL-1. It is recommended that users follow the guidelines of the institution and vendor closely when working with AAVs, and consult with the chemical safety and biosafety officers of the institution prior to commencement of work.

3. The here described method uses dSTORM that makes use of immunolabeling with antibodies labeled with organic fluorophores. Even though the labeling protocol is very similar to that of standard immunochemistry, one has to take care when selecting the organic dyes as they must exhibit photoswitching [77]. This is typically induced through the switching buffer,

which contains thiols such as mercaptoethylamine (MEA) or β-mercaptoethanol (β-ME) and oxygen scavenging enzymes such as glucose oxidase and catalase [78–80]. Due to their chemical structures different fluorophores possess different redox potentials and thus might require slight changes to the imaging buffer. In particular, multicolor experiments can be challenging, as pairs of fluorophores with matching photo-switching conditions need to be identified [78, 81] (*see* **Note 6**). Several changes to the original buffer protocols have been proposed such as the addition of cyclooctatetraene (COT) [82, 83] or Oxyfluor [84].

4. Besides Bruker Vutara, other commercial systems are available from NIKON, Leica and Zeiss. Moreover, many home-built systems have been used for SMLM and lower-cost approaches have been implemented [85]. The base of most dSTORM microscopes is a conventional widefield microscope equipped with high-power lasers, a high numerical aperture objective and sensitive cameras. Even though the Bruker Vutara system used here does not provide changing the angle of illumination, most superresolution imaging systems, especially when imaging of cells is needed, use total internal reflection (TIR) illumination, or for higher contrast, especially in thicker samples, highly inclined and laminated optical sheet (HILO) illumination [86]. The biplane method of 3D imaging has been patented for commercial use by Bruker but other methods exist [86] such as creating astigmatism through inserting a cylindrical lens with a large focal length into the detection pathway of the fluorescence microscope [87], double-helically arranged PSFs [88] or interferometry [89].

5. Arguably the best dye to use for dSTORM is Alexa647 or its derivative Cy5 [78, 81]. It not only exhibits excellent photo-switching characteristics but also emits in the far-red spectrum, where tissue autofluorescence is comparatively low. For dual labeling we recommend using Alexa647 in combination with CF568. However, also Cy3B or Alexa568 can be used. When triple labeling is attempted, Atto488 should be used in combination with the aforementioned dyes. The use of Alexa750 is also possible but 750 nm lasers are not normally used in commercial microscopes. Another option for multicolor imaging is spectral demixing [90], which offers the advantage of negligible chromatic aberration [66].

6. In a diffraction-limited image, taken with a conventional fluorescence microscope, the intensity distribution follows a so-called point-spread-function (PSF). The profile of a PSF can be fitted with a two-dimensional Gaussian function with an uncertainty of only a few nanometers [91, 92]. The diffraction limit of visible light lies typically at around 200–300 nm, which corresponds to the full width at half maximum (FWHM)

of the PSF, which in turn depends on the emission wavelength and the number of detected photons. When performing standard fluorescence imaging in crowded biological tissue, the PSFs of individual molecules overlap and can hence not be distinguished as individual particles. SML microscopy (SMLM) can circumvent this diffraction limit by separating the emission of fluorophores in space and time through photo-activation or photoswitching [34, 46, 79, 93]. The final image is then a reconstruction of every detected, localized, and fitted PSF.

Acknowledgments

This research was supported by European Union's Horizon 2020 research and innovation program (Marie Skłodowska-Curie grant agreement 798644-AstroMiRimage), Wellcome Trust Principal Fellowship (101896), European Research Council Advanced Grant (323113-NETSIGNAL), FP7 ITN (606950 EXTRA-BRAIN), and European Commission NEUROTWIN Grant (857562).

References

1. Agulhon C, Petravicz J, McMullen AB, Sweger EJ, Minton SK, Taves SR, Casper KB, Fiacco TA, McCarthy KD (2008) What is the role of astrocyte calcium in neurophysiology? Neuron 59(6):932–946. https://doi.org/10.1016/j.neuron.2008.09.004

2. Dityatev A, Rusakov DA (2011) Molecular signals of plasticity at the tetrapartite synapse. Curr Opin Neurobiol 21(2):353–359. https://doi.org/10.1016/j.conb.2010.12.006

3. Halassa MM, Haydon PG (2010) Integrated brain circuits: astrocytic networks modulate neuronal activity and behavior. Annu Rev Physiol 72:335–355. https://doi.org/10.1146/annurev-physiol-021909-135843

4. Haydon PG (2001) GLIA: listening and talking to the synapse. Nat Rev Neurosci 2(3):185–193. https://doi.org/10.1038/35058528

5. Matyash V, Kettenmann H (2010) Heterogeneity in astrocyte morphology and physiology. Brain Res Rev 63(1–2):2–10. https://doi.org/10.1016/j.brainresrev.2009.12.001

6. Porter JT, McCarthy KD (1997) Astrocytic neurotransmitter receptors in situ and in vivo. Prog Neurobiol 51(4):439–455

7. Seifert G, Carmignoto G, Steinhauser C (2010) Astrocyte dysfunction in epilepsy. Brain Res Rev 63(1–2):212–221. https://doi.org/10.1016/j.brainresrev.2009.10.004

8. Verkhratsky A, Sofroniew MV, Messing A, deLanerolle NC, Rempe D, Rodriguez JJ, Nedergaard M (2012) Neurological diseases as primary gliopathies: a reassessment of neurocentrism. ASN Neuro 4(3). https://doi.org/10.1042/AN20120010

9. Volterra A, Meldolesi J (2005) Astrocytes, from brain glue to communication elements: the revolution continues. Nat Rev Neurosci 6(8):626–640. https://doi.org/10.1038/nrn1722

10. Zheng K, Bard L, Reynolds JP, King C, Jensen TP, Gourine AV, Rusakov DA (2015) Time-resolved imaging reveals heterogeneous landscapes of nanomolar Ca(2+) in neurons and astroglia. Neuron 88(2):277–288. https://doi.org/10.1016/j.neuron.2015.09.043

11. Araque A, Carmignoto G, Haydon PG, Oliet SH, Robitaille R, Volterra A (2014) Gliotransmitters travel in time and space. Neuron 81(4):728–739. https://doi.org/10.1016/j.neuron.2014.02.007

12. Bazargani N, Attwell D (2016) Astrocyte calcium signaling: the third wave. Nat Neurosci 19(2):182–189. https://doi.org/10.1038/nn.4201

13. Khakh BS, Sofroniew MV (2015) Diversity of astrocyte functions and phenotypes in neural circuits. Nat Neurosci 18(7):942–952. https://doi.org/10.1038/nn.4043

14. Rusakov DA (2015) Disentangling calcium-driven astrocyte physiology. Nat Rev Neurosci 16(4):226–233. https://doi.org/10.1038/nrn3878

15. Volterra A, Liaudet N, Savtchouk I (2014) Astrocyte Ca(2)(+) signalling: an unexpected complexity. Nat Rev Neurosci 15 (5):327–335. https://doi.org/10.1038/nrn3725

16. Zorec R, Araque A, Carmignoto G, Haydon PG, Verkhratsky A, Parpura V (2012) Astroglial excitability and gliotransmission: an appraisal of Ca2+ as a signalling route. ASN Neuro 4(2). https://doi.org/10.1042/AN20110061

17. Danbolt NC (2001) Glutamate uptake. Prog Neurobiol 65(1):1–105

18. Heller JP, Rusakov DA (2015) Morphological plasticity of astroglia: understanding synaptic microenvironment. Glia 63(12):2133–2151. https://doi.org/10.1002/glia.22821

19. Rusakov DA, Bard L, Stewart MG, Henneberger C (2014) Diversity of astroglial functions alludes to subcellular specialisation. Trends Neurosci 37(4):228–242. https://doi.org/10.1016/j.tins.2014.02.008

20. Bernardinelli Y, Randall J, Janett E, Nikonenko I, Konig S, Jones EV, Flores CE, Murai KK, Bochet CG, Holtmaat A, Muller D (2014) Activity-dependent structural plasticity of perisynaptic astrocytic domains promotes excitatory synapse stability. Curr Biol 24 (15):1679–1688. https://doi.org/10.1016/j.cub.2014.06.025

21. Haber M, Zhou L, Murai KK (2006) Cooperative astrocyte and dendritic spine dynamics at hippocampal excitatory synapses. J Neurosci 26 (35):8881–8891. https://doi.org/10.1523/JNEUROSCI.1302-06.2006

22. Hirrlinger J, Hulsmann S, Kirchhoff F (2004) Astroglial processes show spontaneous motility at active synaptic terminals in situ. Eur J Neurosci 20(8):2235–2239. https://doi.org/10.1111/j.1460-9568.2004.03689.x

23. Perez-Alvarez A, Navarrete M, Covelo A, Martin ED, Araque A (2014) Structural and functional plasticity of astrocyte processes and dendritic spine interactions. J Neurosci 34 (38):12738–12744. https://doi.org/10.1523/JNEUROSCI.2401-14.2014

24. Bernardinelli Y, Muller D, Nikonenko I (2014) Astrocyte-synapse structural plasticity. Neural Plast 2014:232105. https://doi.org/10.1155/2014/232105

25. Theodosis DT, Poulain DA, Oliet SH (2008) Activity-dependent structural and functional plasticity of astrocyte-neuron interactions. Physiol Rev 88(3):983–1008. https://doi.org/10.1152/physrev.00036.2007

26. Medvedev N, Popov V, Henneberger C, Kraev I, Rusakov DA, Stewart MG (2014) Glia selectively approach synapses on thin dendritic spines. Philos Trans R Soc Lond B Biol Sci 369(1654):20140047. https://doi.org/10.1098/rstb.2014.0047

27. Lushnikova I, Skibo G, Muller D, Nikonenko I (2009) Synaptic potentiation induces increased glial coverage of excitatory synapses in CA1 hippocampus. Hippocampus 19(8):753–762. https://doi.org/10.1002/hipo.20551

28. Patrushev I, Gavrilov N, Turlapov V, Semyanov A (2013) Subcellular location of astrocytic calcium stores favors extrasynaptic neuron-astrocyte communication. Cell Calcium 54(5):343–349. https://doi.org/10.1016/j.ceca.2013.08.003

29. Witcher MR, Kirov SA, Harris KM (2007) Plasticity of perisynaptic astroglia during synaptogenesis in the mature rat hippocampus. Glia 55(1):13–23. https://doi.org/10.1002/glia.20415

30. Medvedev NI, Popov VI, Rodriguez Arellano JJ, Dallerac G, Davies HA, Gabbott PL, Laroche S, Kraev IV, Doyere V, Stewart MG (2010) The N-methyl-D-aspartate receptor antagonist CPP alters synapse and spine structure and impairs long-term potentiation and long-term depression induced morphological plasticity in dentate gyrus of the awake rat. Neuroscience 165 (4):1170–1181. https://doi.org/10.1016/j.neuroscience.2009.11.047

31. Popov VI, Davies HA, Rogachevsky VV, Patrushev IV, Errington ML, Gabbott PL, Bliss TV, Stewart MG (2004) Remodelling of synaptic morphology but unchanged synaptic density during late phase long-term potentiation (LTP): a serial section electron micrograph study in the dentate gyrus in the anaesthetised rat. Neuroscience 128(2):251–262. https://doi.org/10.1016/j.neuroscience.2004.06.029

32. Sherpa AD, Xiao F, Joseph N, Aoki C, Hrabetova S (2016) Activation of beta-adrenergic receptors in rat visual cortex expands astrocytic processes and reduces extracellular space volume. Synapse 70(8):307–316. https://doi.org/10.1002/syn.21908

33. Witcher MR, Park YD, Lee MR, Sharma S, Harris KM, Kirov SA (2010) Three-dimensional relationships between perisynaptic astroglia and human hippocampal synapses.

Glia 58(5):572–587. https://doi.org/10.1002/glia.20946

34. Betzig E, Patterson GH, Sougrat R, Lindwasser OW, Olenych S, Bonifacino JS, Davidson MW, Lippincott-Schwartz J, Hess HF (2006) Imaging intracellular fluorescent proteins at nanometer resolution. Science 313 (5793):1642–1645. https://doi.org/10.1126/science.1127344

35. Huang B, Wang W, Bates M, Zhuang X (2008) Three-dimensional super-resolution imaging by stochastic optical reconstruction microscopy. Science 319(5864):810–813. https://doi.org/10.1126/science.1153529

36. Klar TA, Jakobs S, Dyba M, Egner A, Hell SW (2000) Fluorescence microscopy with diffraction resolution barrier broken by stimulated emission. Proc Natl Acad Sci U S A 97 (15):8206–8210

37. Patton BR, Burke D, Owald D, Gould TJ, Bewersdorf J, Booth MJ (2016) Three-dimensional STED microscopy of aberrating tissue using dual adaptive optics. Opt Express 24(8):8862–8876. https://doi.org/10.1364/OE.24.008862

38. Bates M, Huang B, Dempsey GT, Zhuang X (2007) Multicolor super-resolution imaging with photo-switchable fluorescent probes. Science 317(5845):1749–1753. https://doi.org/10.1126/science.1146598

39. Panatier A, Arizono M, Nagerl UV (2014) Dissecting tripartite synapses with STED microscopy. Philos Trans R Soc Lond B Biol Sci 369(1654):20130597. https://doi.org/10.1098/rstb.2013.0597

40. Rossi A, Moritz TJ, Ratelade J, Verkman AS (2012) Super-resolution imaging of aquaporin-4 orthogonal arrays of particles in cell membranes. J Cell Sci 125 (Pt 18):4405–4412. https://doi.org/10.1242/jcs.109603

41. Smith AJ, Verkman AS (2015) Superresolution imaging of aquaporin-4 cluster size in antibody-stained paraffin brain sections. Biophys J 109(12):2511–2522. https://doi.org/10.1016/j.bpj.2015.10.047

42. Gucek A, Jorgacevski J, Singh P, Geisler C, Lisjak M, Vardjan N, Kreft M, Egner A, Zorec R (2016) Dominant negative SNARE peptides stabilize the fusion pore in a narrow, release-unproductive state. Cell Mol Life Sci 73 (19):3719–3731. https://doi.org/10.1007/s00018-016-2213-2

43. Sakers K, Lake AM, Khazanchi R, Ouwenga R, Vasek MJ, Dani A, Dougherty JD (2017) Astrocytes locally translate transcripts in their peripheral processes. Proc Natl Acad Sci U S A. https://doi.org/10.1073/pnas.1617782114

44. van de Linde S, Loschberger A, Klein T, Heidbreder M, Wolter S, Heilemann M, Sauer M (2011) Direct stochastic optical reconstruction microscopy with standard fluorescent probes. Nat Protoc 6(7):991–1009. https://doi.org/10.1038/nprot.2011.336

45. Endesfelder U, Heilemann M (2015) Direct stochastic optical reconstruction microscopy (dSTORM). Methods Mol Biol 1251:263–276. https://doi.org/10.1007/978-1-4939-2080-8_14

46. Heilemann M, van de Linde S, Schuttpelz M, Kasper R, Seefeldt B, Mukherjee A, Tinnefeld P, Sauer M (2008) Subdiffraction-resolution fluorescence imaging with conventional fluorescent probes. Angew Chem Int Ed Engl 47(33):6172–6176. https://doi.org/10.1002/anie.200802376

47. Oberheim NA, Goldman SA, Nedergaard M (2012) Heterogeneity of astrocytic form and function. Methods Mol Biol 814:23–45. https://doi.org/10.1007/978-1-61779-452-0_3

48. Grosche A, Grosche J, Tackenberg M, Scheller D, Gerstner G, Gumprecht A, Pannicke T, Hirrlinger PG, Wilhelmsson U, Huttmann K, Hartig W, Steinhauser C, Pekny M, Reichenbach A (2013) Versatile and simple approach to determine astrocyte territories in mouse neocortex and hippocampus. PLoS One 8(7):e69143. https://doi.org/10.1371/journal.pone.0069143

49. Nishiyama H, Knopfel T, Endo S, Itohara S (2002) Glial protein S100B modulates long-term neuronal synaptic plasticity. Proc Natl Acad Sci U S A 99(6):4037–4042. https://doi.org/10.1073/pnas.052020999

50. Heller JP, Michaluk P, Sugao K, Rusakov DA (2017) Probing nano-organization of astroglia with multi-color super-resolution microscopy. J Neurosci Res. https://doi.org/10.1002/jnr.24026

51. Jiang R, Haustein MD, Sofroniew MV, Khakh BS (2014) Imaging intracellular Ca(2)(+) signals in striatal astrocytes from adult mice using genetically-encoded calcium indicators. J Vis Exp 93:e51972. https://doi.org/10.3791/51972

52. Shigetomi E, Patel S, Khakh BS (2016) Probing the complexities of astrocyte calcium signaling. Trends Cell Biol 26(4):300–312. https://doi.org/10.1016/j.tcb.2016.01.003

53. Srinivasan R, Lu TY, Chai H, Xu J, Huang BS, Golshani P, Coppola G, Khakh BS (2016) New transgenic mouse lines for selectively targeting astrocytes and studying calcium signals in astrocyte processes in situ and in vivo. Neuron 92(6):1181–1195. https://doi.org/10.1016/j.neuron.2016.11.030

54. Holtmaat A, de Paola V, Wilbrecht L, Trachtenberg JT, Svoboda K, Portera-Cailliau C (2012) Imaging neocortical neurons through a chronic cranial window. Cold Spring Harb Protoc 2012(6):694–701. https://doi.org/10.1101/pdb.prot069617

55. Hama H, Kurokawa H, Kawano H, Ando R, Shimogori T, Noda H, Fukami K, Sakaue-Sawano A, Miyawaki A (2011) Scale: a chemical approach for fluorescence imaging and reconstruction of transparent mouse brain. Nat Neurosci 14(11):1481–1488. https://doi.org/10.1038/nn.2928

56. Juette MF, Gould TJ, Lessard MD, Mlodzianoski MJ, Nagpure BS, Bennett BT, Hess ST, Bewersdorf J (2008) Three-dimensional sub-100 nm resolution fluorescence microscopy of thick samples. Nat Methods 5(6):527–529. https://doi.org/10.1038/nmeth.1211

57. Mlodzianoski MJ, Juette MF, Beane GL, Bewersdorf J (2009) Experimental characterization of 3D localization techniques for particle-tracking and super-resolution microscopy. Opt Express 17(10):8264–8277

58. Metcalf DJ, Edwards R, Kumarswami N, Knight AE (2013) Test samples for optimizing STORM super-resolution microscopy. J Vis Exp 79. https://doi.org/10.3791/50579

59. Packer AM, Russell LE, Dalgleish HW, Hausser M (2015) Simultaneous all-optical manipulation and recording of neural circuit activity with cellular resolution in vivo. Nat Methods 12(2):140–146. https://doi.org/10.1038/nmeth.3217

60. Srinivasan R, Huang BS, Venugopal S, Johnston AD, Chai H, Zeng H, Golshani P, Khakh BS (2015) Ca(2+) signaling in astrocytes from Ip3r2(−/−) mice in brain slices and during startle responses in vivo. Nat Neurosci 18(5):708–717. https://doi.org/10.1038/nn.4001

61. Bishop D, Nikic I, Brinkoetter M, Knecht S, Potz S, Kerschensteiner M, Misgeld T (2011) Near-infrared branding efficiently correlates light and electron microscopy. Nat Methods 8(7):568–570. https://doi.org/10.1038/nmeth.1622

62. Hausser M, Margrie TW (2014) Two-photon targeted patching and electroporation in vivo. Cold Spring Harb Protoc 2014(1):78–85. https://doi.org/10.1101/pdb.prot080143

63. Blazquez-Llorca L, Hummel E, Zimmerman H, Zou C, Burgold S, Rietdorf J, Herms J (2015) Correlation of two-photon in vivo imaging and FIB/SEM microscopy. J Microsc 259(2):129–136. https://doi.org/10.1111/jmi.12231

64. Henriques R, Lelek M, Fornasiero EF, Valtorta F, Zimmer C, Mhlanga MM (2010) QuickPALM: 3D real-time photoactivation nanoscopy image processing in ImageJ. Nat Methods 7(5):339–340. https://doi.org/10.1038/nmeth0510-339

65. Ovesny M, Krizek P, Borkovec J, Svindrych Z, Hagen GM (2014) ThunderSTORM: a comprehensive ImageJ plug-in for PALM and STORM data analysis and super-resolution imaging. Bioinformatics 30(16):2389–2390. https://doi.org/10.1093/bioinformatics/btu202

66. Herrmannsdorfer F, Flottmann B, Nanguneri S, Venkataramani V, Horstmann H, Kuner T, Heilemann M (2017) 3D d STORM imaging of fixed brain tissue. Methods Mol Biol 1538:169–184. https://doi.org/10.1007/978-1-4939-6688-2_13

67. Pleiner T, Bates M, Trakhanov S, Lee CT, Schliep JE, Chug H, Bohning M, Stark H, Urlaub H, Gorlich D (2015) Nanobodies: site-specific labeling for super-resolution imaging, rapid epitope-mapping and native protein complex isolation. elife 4:e11349. https://doi.org/10.7554/eLife.11349

68. de Castro MA, Rammner B, Opazo F (2016) Aptamer stainings for super-resolution microscopy. Methods Mol Biol 1380:197–210. https://doi.org/10.1007/978-1-4939-3197-2_17

69. Chamma I, Rossier O, Giannone G, Thoumine O, Sainlos M (2017) Optimized labeling of membrane proteins for applications to super-resolution imaging in confined cellular environments using monomeric streptavidin. Nat Protoc 12(4):748–763. https://doi.org/10.1038/nprot.2017.010

70. Teng KW, Ishitsuka Y, Ren P, Youn Y, Deng X, Ge P, Belmont AS, Selvin PR (2016) Labeling proteins inside living cells using external fluorophores for microscopy. elife 5. https://doi.org/10.7554/eLife.20378

71. Teng KW, Ishitsuka Y, Ren P, Youn Y, Deng X, Ge P, Lee SH, Belmont AS, Selvin PR (2017) Labeling proteins inside living cells using external fluorophores for fluorescence microscopy. elife 6. https://doi.org/10.7554/eLife.25460

72. Wieneke R, Raulf A, Kollmannsperger A, Heilemann M, Tampe R (2015) SLAP: small labeling pair for single-molecule super-resolution imaging. Angew Chem Int Ed Engl 54(35):10216–10219. https://doi.org/10.1002/anie.201503215

73. Lotze J, Reinhardt U, Seitz O, Beck-Sickinger AG (2016) Peptide-tags for site-specific protein labelling in vitro and in vivo. Mol BioSyst

12(6):1731–1745. https://doi.org/10.1039/c6mb00023a

74. Raulf A, Spahn CK, Zessin PJ, Finan K, Bernhardt S, Heckel A, Heilemann M (2014) Click chemistry facilitates direct labelling and super-resolution imaging of nucleic acids and proteinsdaggerElectronic supplementary information (ESI) available. RSC Adv 4 (57):30462–30466. https://doi.org/10.1039/c4ra01027b

75. Mateos-Gil P, Letschert S, Doose S, Sauer M (2016) Super-resolution imaging of plasma membrane proteins with click chemistry. Front Cell Dev Biol 4:98. https://doi.org/10.3389/fcell.2016.00098

76. Lowery RL, Majewska AK (2010) Intracranial injection of adeno-associated viral vectors. J Vis Exp 45. https://doi.org/10.3791/2140

77. Furstenberg A, Heilemann M (2013) Single-molecule localization microscopy-near-molecular spatial resolution in light microscopy with photoswitchable fluorophores. Phys Chem Chem Phys 15(36):14919–14930. https://doi.org/10.1039/c3cp52289j

78. Dempsey GT, Vaughan JC, Chen KH, Bates M, Zhuang X (2011) Evaluation of fluorophores for optimal performance in localization-based super-resolution imaging. Nat Methods 8(12):1027–1036. https://doi.org/10.1038/nmeth.1768

79. Rust MJ, Bates M, Zhuang X (2006) Sub-diffraction-limit imaging by stochastic optical reconstruction microscopy (STORM). Nat Methods 3(10):793–795. https://doi.org/10.1038/nmeth929

80. Chozinski TJ, Gagnon LA, Vaughan JC (2014) Twinkle, twinkle little star: photoswitchable fluorophores for super-resolution imaging. FEBS Lett 588(19):3603–3612. https://doi.org/10.1016/j.febslet.2014.06.043

81. Turkowyd B, Virant D, Endesfelder U (2016) From single molecules to life: microscopy at the nanoscale. Anal Bioanal Chem 408 (25):6885–6911. https://doi.org/10.1007/s00216-016-9781-8

82. Minoshima M, Kikuchi K (2017) Photostable and photoswitching fluorescent dyes for super-resolution imaging. J Biol Inorg Chem. https://doi.org/10.1007/s00775-016-1435-y

83. Olivier N, Keller D, Gonczy P, Manley S (2013) Resolution doubling in 3D-STORM imaging through improved buffers. PLoS One 8(7):e69004. https://doi.org/10.1371/journal.pone.0069004

84. Nahidiazar L, Agronskaia AV, Broertjes J, van den Broek B, Jalink K (2016) Optimizing imaging conditions for demanding multi-color super resolution localization microscopy. PLoS One 11(7):e0158884. https://doi.org/10.1371/journal.pone.0158884

85. Kwakwa K, Savell A, Davies T, Munro I, Parrinello S, Purbhoo MA, Dunsby C, Neil MA, French PM (2016) easySTORM: a robust, lower-cost approach to localisation and TIRF microscopy. J Biophotonics 9 (9):948–957. https://doi.org/10.1002/jbio.201500324

86. Herbert S, Soares H, Zimmer C, Henriques R (2012) Single-molecule localization super-resolution microscopy: deeper and faster. Microsc Microanal 18(6):1419–1429. https://doi.org/10.1017/S1431927612013347

87. Kao HP, Verkman AS (1994) Tracking of single fluorescent particles in three dimensions: use of cylindrical optics to encode particle position. Biophys J 67(3):1291–1300. https://doi.org/10.1016/S0006-3495(94)80601-0

88. Pavani SR, Thompson MA, Biteen JS, Lord SJ, Liu N, Twieg RJ, Piestun R, Moerner WE (2009) Three-dimensional, single-molecule fluorescence imaging beyond the diffraction limit by using a double-helix point spread function. Proc Natl Acad Sci U S A 106 (9):2995–2999. https://doi.org/10.1073/pnas.0900245106

89. Shtengel G, Galbraith JA, Galbraith CG, Lippincott-Schwartz J, Gillette JM, Manley S, Sougrat R, Waterman CM, Kanchanawong P, Davidson MW, Fetter RD, Hess HF (2009) Interferometric fluorescent super-resolution microscopy resolves 3D cellular ultrastructure. Proc Natl Acad Sci U S A 106(9):3125–3130. https://doi.org/10.1073/pnas.0813131106

90. Lampe A, Haucke V, Sigrist SJ, Heilemann M, Schmoranzer J (2012) Multi-colour direct STORM with red emitting carbocyanines. Biol Cell 104(4):229–237. https://doi.org/10.1111/boc.201100011

91. Smith CS, Joseph N, Rieger B, Lidke KA (2010) Fast, single-molecule localization that achieves theoretically minimum uncertainty. Nat Methods 7(5):373–375. https://doi.org/10.1038/nmeth.1449

92. Thompson RE, Larson DR, Webb WW (2002) Precise nanometer localization analysis for individual fluorescent probes. Biophys J 82 (5):2775–2783. https://doi.org/10.1016/S0006-3495(02)75618-X

93. Hess ST, Girirajan TP, Mason MD (2006) Ultra-high resolution imaging by fluorescence photoactivation localization microscopy. Biophys J 91(11):4258–4272. https://doi.org/10.1529/biophysj.106.091116

Chapter 3

High-Resolution Molecular Imaging and Its Applications in Brain and Synapses

Nhu T. N. Phan and Silvio O. Rizzoli

Abstract

The molecular organization of the brain and its synapses is highly regulated and closely related to their biological functions. In this chapter, we introduce several super-resolution imaging technologies for brain and synapses, including optical microscopy (STED, STORM), expansion microscopy, and secondary ion mass spectrometry (SIMS, NanoSIMS). Super-resolution microscopy allows for visualization of the localization and dynamics of fluorescently labeled molecules whereas mass spectrometry imaging provides information on chemical structure and molecular turnover of the brain and synapses. The general principle, pros and cons of each technology as well as experimental considerations, such as labeling and sample preparation methods, are presented. In addition, correlative optical and mass spectrometry imaging, which appears as a recent trend of brain and synaptic imaging, is also discussed together with selected relevant applications in this research area.

Key words Super-resolution imaging, STED, SIMS, NanoSIMS, Brain, Synapse

1 Introduction to the Brain, Neurotransmission, and Synapses

The brain is the most critical part of the body, as it controls all the processes from single cells to organs, leading to physical motion, cognition, emotion, perception, and thought. The neurons, the most important cells of the brain, are organized in networks in which they communicate to each other to transfer signals, and to direct other target cells. The basis of this process is synaptic transmission, the transfer of information between two adjacent neurons (the presynaptic and postsynaptic neurons). This takes place between specialized compartments termed synaptic boutons, on the presynaptic side, and dendritic spines, on the postsynaptic side. When neurons are depolarized, their synaptic boutons are activated, and small vesicles containing neurotransmitters, the so-called synaptic vesicles, fuse with the membrane of the boutons to release their contents in the synaptic cleft, a space of 12–20 nm between the neurons. The neurotransmitters then diffuse to the

Irene Wacker et al. (eds.), *Volume Microscopy: Multiscale Imaging with Photons, Electrons, and Ions*, Neuromethods, vol. 155, https://doi.org/10.1007/978-1-0716-0691-9_3, © Springer Science+Business Media, LLC, part of Springer Nature 2020

postsynaptic side and bind to neurotransmitter receptors, thereby propagating the signals. One neuron can communicate to thousands of other neurons with a millisecond time scale. As has often been stated in the literature, the molecular organization of the brain and of the synapse are under a strong level of control, and perturbations lead to brain disorders. Reversely, changes in neurological activity induce alterations in brain molecular organization and chemistry, which are often diagnostic for the diseases [1–5]. Therefore, it is important to study the brain molecular and organelle level organization, in order to understand its function and dysfunction.

To successfully image the brain and synapse, several requirements must be fulfilled. The imaging technique must provide highly specific molecular information, sufficient sensitivity and spatial resolution. Especially for synaptic imaging, nanoscopic resolution (less than the diffraction limit of light) is necessary, as a large proportion of the synapses in the mammalian brain have a diameter that is comparable to the diffraction limit (~200–300 nm). There have been increasing numbers of high spatial resolution imaging techniques suitable for brain and synapse imaging, such as super-resolution optical microscopy, electron microscopy, and mass spectrometry imaging [5–11]. Each utilizes different principles, and is well suited for specific applications. However, combinations of these techniques, in the form of correlative microscopy, provide much more information than any of the tools on their own, and have been therefore increasingly used.

In this chapter, we introduce several super-resolution molecular imaging techniques applicable for brain and synapses, particularly stimulated emission depletion (STED), stochastic optical reconstruction microscopy (STORM), and secondary ion mass spectrometry (SIMS), and their correlative microscopy (combined) applications.

Overall, the optical techniques, such as STORM and STED, are well designed to study the functional organization of the cell—the connection between the topological distribution of cellular elements (proteins or organelles) and their function. At the same time, they cannot be easily implemented to analyze, for example, the turnover of cellular structures, or the nature of nonfluorescently labeled components such as the neuronal lipids. To answer this type of questions, one needs a different technique, one that can image, at high spatial resolution and with high sensitivity, the turnover of the structures of interest, and/or the chemical nature of the molecules of interest. As explained in detail below, this can now be achieved by different SIMS implementations, with extraordinarily high precision.

Ultimately, a combination of optical and secondary ion super-resolution imaging will probably open the gate into a functional, structural, chemical, and turnover-based analysis of the brain, which will provide substantially more information than any of these technologies used separately. Below we discuss the different tools, with a special emphasis on SIMS, which is less familiar to the general public, and we present some of the most relevant recent applications.

2 Methods for Measuring the Structure and Activity of the Brain and Synapse

2.1 Super-resolution Microscopy (STED, STORM)

Optical fluorescence microscopy, which is capable of visualizing specific biomolecules, and of imaging living cells, has been a common technique in cell biology and neurobiology. However, it has long been limited by the diffraction of light, which has been recently overcome by a group of super-resolution microscopy techniques. Among those, STED and STORM have been the most common super-resolution techniques in the field of neuroscience and cell biology as demonstrated by a number of achievements [7, 12–15]. Continuous developments of these techniques especially for live imaging, multicolor, and 3D imaging ensures their extended horizon of applications [16–21]. Both are based on the same fundamental principle, revealing fluorophores that are in different states—critically in fluorescent and nonfluorescent states [22]—but their technical implementations differ.

2.1.1 Super-resolution Microscopy Implementations

STED belongs to the group of techniques named reversible saturable optical fluorescence transition (RESOLFT). The principle of STED was first proposed and demonstrated by Hell and Wichmann [23]. It surpasses the diffraction resolution limit by using stimulated emission to inhibit the spontaneous emission of excited fluorophores located at the outer region of an excitation center. STED utilizes two laser beams, an excitation beam that turns on the fluorophores, and a so-called STED beam that induces the energy transition of the fluorophores from the excited state to the ground state, without emission of fluorescence at the normal wavelength. The STED laser has a donut shape (meaning zero stimulated emission at the beam center), which is superimposed onto the excitation beam. The fluorescence emission therefore occurs only at the center of the two beams, which strongly reduces the size of the fluorescence emission point (Fig. 1a). Like any other scanning fluorescence microscopy, the laser beams are scanned across the sample pixel by pixel to produce images which are then assembled to an entire sample image. STED offers excellent lateral resolution ~20 nm, with the axial resolution at ~ 500 nm for simple STED setups [13, 24], but approaching the lateral resolution in setups employing the second depletion beam in the z direction [25], or the combination of STED and 4 pi microscopy [19].

STORM is the common name for a group of super-resolution microscopy tools based on single molecule localization microscopy (SMLM). STORM is based on stochastic activation of a small

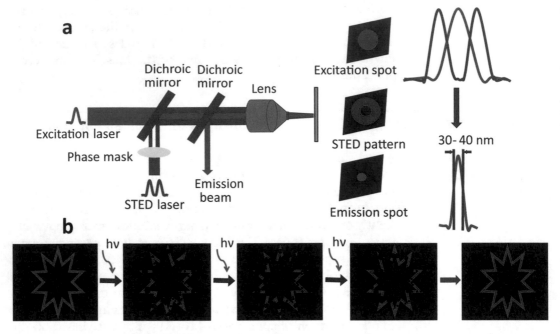

Fig. 1 (**a**) Principle of STED. The donut shaped STED beam is superimposed onto the excitation beam. Due to the zero intensity of the STED beam at the donut center, all the fluorophores locating outside the beam center are depleted resulting in a subdiffraction fluorescence emission point. (**b**) Principle of STORM. Small subsets of photoswitchable or photoactivatable dyes are turned on and off through hundreds or thousands of cycles, and the positions of individual fluorophores are registered. The final STORM image is constructed based on all the registered positions of the fluorophores

subset of densely distributed fluorophores in order to determine their individual locations [26] (Fig. 1b). The fluorophores must be photoswitchable between the fluorescence and dark states. To achieve this, one can use different strategies, ranging from pairs of activator and reporter dyes conjugated to the molecules to be imaged to photoactivable fluorescent proteins [27]. As only a small subset of fluorophores is activated at any time, the chance of two emitting fluorophores locating within the same diffraction-limited volume is negligible, which enables the localization of the different single molecules. After hundreds or thousands of activation cycles, the final super-resolution image can be generated by combining all the registered positions of the fluorophores. A lateral resolution of ~10–20 nm, and a comparable axial resolution can be achieved.

When comparing STED and STORM, it is evident that each technique possesses its own advantages and disadvantages. STED requires high intensity excitation and depletion lasers to obtain high resolutions, which implies that it may result in photobleaching, and in possible damage to living samples. However, the use of gated STED, which selects photons arriving at the detector within a set time interval, allows for reducing the laser intensity, meaning reducing possible damage for samples, while still maintaining the resolving power of imaging and high signal to background ratio [28]. At the same time, STED is suitable for in vivo imaging, owing to its high speed of acquisition. STORM utilizes a wide field imaging approach, by which the laser irradiates the entire area of analysis and multiple fluorescence emitting fluorophores in this area are imaged at once. Therefore the sample is exposed to lower laser intensity, which prolongs the lifetime of the fluorophores, and limits sample damage. However, the strategy of turning on/off a subset of fluorophores through thousands of cycles is time-consuming for the acquisition of one STORM image, when compared to STED, which makes classical STORM procedures far slower in vivo.

For a STED experiment, wavelength selection must fulfill the requirement that the depletion wavelength must be outside the excitation band, but lie within the emission spectrum of the fluorophore, and at a longer wavelength compared to the emission peak. The choice of fluorophores is thus relatively restricted. However, the development of STED diode lasers [29] and continuous wave STED lasers [30] enable more flexibility in selection of fluorophores. Another critical condition is that the two laser beams of STED must be perfectly aligned in order to obtain zero point intensity at the center of the donut. In STORM, the fluorophores must be photoswitchable or photoactivatable in a controlled manner, and should emit as many photons as possible. The fact that many fluorophores can be induced to switch [31], in what is termed also dSTORM (direct STORM), makes this technology widely applicable.

2.1.2 Expansion Microscopy as a Means to Obtain Super-resolution Imaging

An alternative direction in super-resolution fluorescence microscopy has recently come into prominence, based not on improving the optics but on modifying the specimen. This approach, termed expansion microscopy, employs polyelectrolyte gels, which expand strongly when dialyzed in water [32–35]. This technique entails that the sample of interest is first fixed, permeabilized, and immunostained, and then embedded in sodium acrylate, which is a typical compound for the production of polyelectrolyte gels. Acrylamide was used as a comonomer, and N-N'-methylenebisacrylamide as the cross-linker of the gel, in the initial implementation. The polymerization is triggered with ammonium persulfate (APS) and tetramethylethylenediamine (TEMED). The fluorophores used in immunostaining are covalently linked to the gel structure, and the tissue is afterward digested using proteases. Subsequent dialysis in water induces a 4.5-fold expansion in all directions, with no disruption of the sample aspect ratio [32]. The tissue structure disappears due to the digestion, but the fluorophores, which are covalently bound to the gel, maintain their relative positions, although they

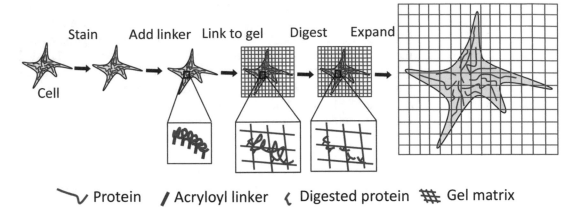

Fig. 2 Workflow of expansion microscopy. First, the cells are fixed, permeabilized and immunostained for the proteins of interest. The cells are then incubated with acryloyl linkers and are embedded in sodium acylate and acrylamide gel to which the cells covalently link through the acryloyl linkers. The gel containing cells is then digested, and expanded in water

are now positioned 4.5-fold farther away from each other than in the initial sample (Fig. 2). The preparation can then be imaged in a conventional microscope. This technique therefore produces super-resolution images with a conventional setup, by simply enlarging the samples. The initial lateral resolution obtained was ~70 nm, that is, not as high as that of optimal STED or STORM microscopes, but still sufficient to determine phenotypic changes in neuronal samples [32], and was more recently improved to ~25 nm, by iterative expansion [36].

2.1.3 Fluorescence Labeling for Super-resolution Microscopy

The success of a super-resolution microscopic experiment significantly depends on the choice of fluorescence labeling probe. There are several critical properties of labeling probes for STED and STORM. For biological systems, the dyes should be membrane permeable and nontoxic, exhibit high brightness, and be photostable. Especially for STORM, the dye must be photoswitchable or photoactivatable in a controlled manner. In addition, the dye should be highly specific for the molecules of interest. Furthermore, the excitation and emission wavelength of the dye must be suitable for the available excitation and depletion lasers. While our work here does not aim at a thorough review of the labeling literature, we would like to point out a few aspects. First, the probes used should be as small in size as possible. It is obvious that large probes would place the fluorophore away from the intended target, thereby limiting the practical resolution. Packages of primary and secondary antibodies are ~25 nm in diameter [15], and therefore it is currently desired to replace them with fluorescently conjugated small probes, such as nanobodies [37] or aptamers [38]. Second, while a large palette of fluorescent proteins are available [15, 39–

41], the ability to place a chemical fluorophore, designed for high stability and brightness, on the protein of interest is still actively sought for. One of the main avenues in this direction has been the use of click chemistry, in which an unnatural amino acid is incorporated into the protein of interest, and is then conjugated to a fluorophore that can be chosen freely, typically in a copper-catalyzed "click" reaction [42]. Another direction is the use of self-labeling enzymes, that can be conjugated to different fluorophores, and can be expressed as tags on various cellular proteins, including the HALO and SNAP tags [43, 44]. Third, the palette of small molecules that label membranes or different cellular structures such as the actin and tubulin cytoskeleton is constantly increasing [45, 46].

As these techniques are reasonably well understood in the literature, we chose to focus mainly on the mass spectrometry methods, below, due to space limitations. A number of excellent reviews can be consulted for further information on labeling for super-resolution microscopy [39, 47–49].

2.2 Secondary Ion Mass Spectrometry (SIMS)

SIMS is a surface sensitive imaging technique capable of visualizing chemical distribution of a given sample possibly at submicrometer spatial resolution. The method possesses unique characteristics of label-free detection, high chemical specificity, and applicability to almost all kinds of sample materials. The applications of SIMS have been increasing significantly in biological communities especially for the studies of brain and brain diseases.

2.2.1 Principle of SIMS

In SIMS, a beam of highly energetic ions (10–40 keV), or the so-called primary ion beam, sputters the sample surface pixel by pixel, generating the secondary ions from the sample (Fig. 3). The secondary ions are then extracted to the ion optics, separated by a mass analyzer, and eventually detected by a detector by their different mass per charge (m/z). The most common mass analyzer in SIMS is time-of-flight (ToF) due to its high speed and parallel detection; however, there have been efforts in developing SIMS instruments using other analyzers particularly magnetic sector, orbitrap, or electrostatic analyzer to obtain better mass resolution and tandem mass spectrometry capability to elucidate chemical structures of the molecules of interest [50–54]. SIMS typically is suitable for detection of biomolecules up to 1500 Da, particularly elements, metabolites, lipids, and small peptides.

There are several important considerations for a SIMS experiment including the sample materials, the target analytes, sensitivity, spatial resolution, and analysis time. First, sample materials certainly influence the analysis, as different matrices and interferences exist in different kinds of sample. Changes in either endogenous or exogenous factors can lead to artifacts in the results. Therefore, appropriate sample preparation is needed for specific samples such

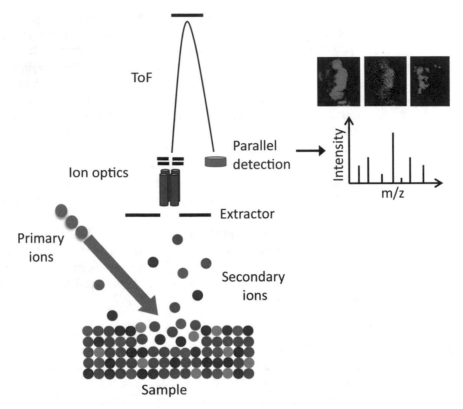

Fig. 3 Principle of ToF-SIMS imaging. The primary ion beam sputters the sample surface, pixel by pixel, producing secondary ions from the sample. These ions are extracted to the ion optics and a mass analyzer, typically time-of-flight, where they are separated by m/z, and are subsequently detected by a detector. A mass spectrum is obtained for each pixel. Eventually, the ion images of the entire surface are constructed from all the mass spectra

as for biological, inorganic, or organic samples. Second, the mass range of the analytes of interest and their abundances in the sample directly affect the sensitivity of the measurement. Higher mass or intact molecules have less sensitivity compared to small ones due to their high tendency to be fragmented during the sputtering and ionization. Various available primary ion sources are favorable for different target analytes, including the atomic primary ion gun (Cs^+, Ga^+, In^+) for detecting elements and small fragments, liquid metal ion guns (LMIG) (Bi_n^+, Au_n^+), bucky ball C_{60}^+ gun for small ions (up to ~600 Da), and gas cluster ion beams (GCIB) (Ar_n^+, $(CO_2)_n^+$, $(H_2O)_n^+$) are suitable for higher mass ions (>600 Da) such as intact lipids and small peptides [55–60]. These primary ion sources in turn determine the spatial resolution of the imaging depending on the energy and the size of their ion clusters. For examples, the obtainable spatial resolution by the 20 keV LMIG Bi_3^+ is better than 500 nm [55], by 40 keV C_{60}^+ gun is ~1 μm [53], and by 40 keV Ar_{4000}^+ GCIB is ~3 μm [61]. However, the spatial resolution is also determined by the abundance of the analytes in sample. Therefore, when selecting experimental conditions for SIMS imaging, it is important to remember that there is always a compromise among the three parameters spatial resolution, sensitivity of the target analytes, and analysis time.

To accommodate different demands regarding spatial resolution, sensitivity, and surface information of the studied samples, two main approaches in SIMS, static and dynamic SIMS, can be used. Static SIMS utilizes the primary ions at the dose density below 10^{13} ions/cm^2, which is the so-called static limit, to analyze the molecular compositions on the sample surface. Ideally only 1% of the material on the top layer of the sample is ionized, thereby ensuring that the secondary ions come from an intact sample area. The main challenge in static SIMS is sensitivity due to the restricted primary ion dose. In contrast, dynamic SIMS uses the primary ion dose above the static limit that produces much higher amount of secondary ions. Sensitivity significantly increases but the surface layer can be eroded completely. Dynamic SIMS typically employs energetic monoatomic primary ion sources therefore only atomic or small fragmented ions are detected. Furthermore, 3D imaging can be obtained using dual beam or single beam configurations [62, 63] to produce three dimensional molecular distribution of the sample at subcellular spatial resolution.

2.2.2 Sample Preparation for SIMS

One of the critical factors which determine the success of a SIMS experiment is to use an appropriate sample preparation protocol in order to highly preserve the morphology and molecular structure of the samples. Several protocols have been developed including freeze-drying, frozen hydrated, freeze fracture, and chemical fixation. For brain and neuronal analysis, fixation, freeze-drying, and frozen hydrated sample preparations can be used.

Freeze-Drying

Freeze-drying is the most commonly used method due to its simplicity. The analysis is performed at room temperature—that makes it simple and easy for sample handling. For brain tissue, the sample is plunge frozen in liquid nitrogen, liquid propane, or liquid pentane or by using a high pressure freezer. This step must be carried out as quickly as possible in order to minimize the ice crystallization which will damage the structure of the tissue. The frozen tissue is then sectioned into slices of 10–25 μm thickness using a cryomicrotome at around $-20\ ^\circ$C. The slices are quickly thaw mounted—the method used to quickly and partially thaw sample in order to adhere the sample onto a substrate for imaging—on a conductive substrate such as silicon wafers or indium tin oxide (ITO) glass, which are subsequently dried in high vacuum (10^{-2}–10^{-3} mbar) overnight. It is critical that freeze-drying takes place in high vacuum where water in the tissue slowly sublimates to minimize the delocalization of sample constituents. The dried sample is ready for

SIMS analysis or can be stored in vacuum for a few days; however, it is recommended to preform analysis as soon as possible to avoid possible biomolecular degradation.

The molecular distribution of lipids in mouse brain tissue was successfully examined using ToF-SIMS with Au_3^+ LMIG [59]. The tissue was prepared using the freeze-drying process as described above. Various lipids were found at specific localizations in the brain, for example cholesterol localized to the white matter whereas phosphatidylinositols (PIs) were primarily located in the gray matter. Phosphatidylcholines (PCs) were distributed over the entire brain section. Sulfatides were found to have complementary localization with phosphate/palmitate within the gray matter. In addition, the relation between the distribution of fatty acids and their corresponding molecular lipids was also observed, particularly for palmitate and palmitate-containing phospholipids.

To study the relation of brain lipids with the activation of antisecretory factor (AF), freeze-dried rat brain tissue was imaged with ToF-SIMS equipped with 25 keV Bi_3^+ LMIG [64]. AF is an endogenous protein regulating inflammation and fluid secretion in cells, induced by dietary supplement, especially processed cereal. The results showed a decrease in the amount of cholesterol, vitamin E in contrast to an increased content of PC, phosphatidylethanolamines (PEs) and several fatty acids in the rat brain after feeding with the supplement. This indicated structural changes of the plasma membrane that could be involved in the mechanism of AF activation in the brain.

For cultured neuronal cells, the procedure is similar to that for tissue; however, it is important to eliminate the salt contents from the cell medium before freeze-drying. High salt content could interfere with the detection of other biomolecules which are present at much lower concentrations in the sample. Several buffer solutions compatible to MS, such as ammonium formate, ammonium acetate, and HEPES can be used to quickly rinse the cells before freezing [59, 65, 66]. Vitamin E has been known to play a role in lipid oxidation and affect membrane enzyme activity by changing the properties of lipid membrane [67]. Vitamin E was imaged in single isolated neurons of *Aplysia californica* using ToF-SIMS with 22 keV Au_n^+ LMIG and the freeze-drying method [68]. The ion image of vitamin E showed that its localization was dominant at the junction of cell soma and neurite compared to other compartments whereas the signal of the choline group was unchanged across the junction. This finding supports the active role of vitamin in transport mechanisms and neuronal signaling.

Freeze-drying is well suited for brain tissue imaging; however, for single cell imaging, great care must be taken as the method could potentially cause rearrangement of biomolecules at subcellular scale.

Chemical Fixation

Chemical fixation using paraformaldehyde, glutaraldehyde, and osmium tetroxide can be used for brain and neuronal cells. Fixation preserves well the morphological structure of brain tissue and cells and fixed samples can be stored for a long time; however, there is a risk that unfixed biomolecules are lost during the washing steps of fixation. Therefore, it is critical to select good fixation reagents and protocols suitable for the molecules of interest.

The effect of fixation with glutaraldehyde for lipid imaging was examined on the multiple myeloma cell line U266 using ToF-SIMS and Au_3^+ LMIG [66]. After centrifuging and removing the supernatant, the cells were placed on ITO glass and incubated with glutaraldehyde for 15 min, rinsed with ammonium acetate and air dried. The analysis showed several cellular ion species such as phosphate and palmitic, oleic, and stearic fatty acids well localized inside the cell area. This demonstrated that membrane phospholipids were not altered by glutaraldehyde fixation at cellular level. To investigate subcellular distribution of lipids in individual neurons of *Aplysia californica* using SIMS imaging, fixation was optimized using paraformaldehyde and glycerol [69].

Frozen Hydrated Preparation

In frozen hydrated preparation, brain tissue or cell samples are kept frozen after plunge freezing and during SIMS analysis. After plunge freezing, the sample is quickly transferred into the main chamber of SIMS instrument which has been previously cooled down to around liquid nitrogen temperature (< -180 °C). To avoid the formation of ice crystals on top of the samples that will inhibit the imaging of the sample surface, brain tissue and cells are kept in argon atmosphere after plunge freezing and before analysis. This is the safest approach to preserve the molecular structures of the samples, to prevent any contamination, and increase secondary ion yield due to the enhancement effect of the water matrix [70]. However, it has a high risk of measuring artifacts caused by sample topography. In addition, sample handling and analysis at frozen temperature are complicated and difficult.

Drosophila melanogaster is a well-known model for research in drug abuse and neuroscience. The fly was used to investigate the effects of the psychostimulant drug methylphenidate on the molecular structure of the brain using ToF-SIMS equipped with 40 keV Ar_{4000}^+ GCIB [9]. After freezing, the fly brains were sectioned into 20 μm slices under argon atmosphere and analyzed at the frozen hydrated state. Different biomolecules exhibited different localizations across the fly brain especially in the central brain, optical lobes, and salivary gland area. Moreover, methylphenidate was shown to alter the amounts of phospholipids in the central brain and that could imply an important role of these lipids in neuronal functions. Experiments using frozen hydrated sample preparation were also successfully carried out on cell cultures such as *Tetrahymena* and PC12 cells [71, 72].

2.3   Nanoscale SIMS (NanoSIMS)

Recently developed NanoSIMS and isotopic labeling approaches have offered the possibility of its applications in biological imaging. The use of stable isotopes incorporated into brain and synapses and subsequent imaging by NanoSIMS at a possible spatial resolution ~ 50 nm is very useful for understanding molecular turnover of these structures at the organelle level.

2.3.1   Principle of NanoSIMS

NanoSIMS belongs to the dynamic SIMS category that utilizes monoatomic and diatomic primary ion sources particularly Cs^+ and O_2^- to erode the sample surface. These highly reactive ion sources cause intensive fragmentations producing only atomic and small fragmented ions; however, they can be focused to a very small beam size. One of the significant features of the NanoSIMS is that the primary ion beam is normal to the sample surface and coaxial with the secondary ion path that shortens the focal length of both the primary ion focal lens and the secondary ion extraction lens (Fig. 4). This results in minimal aberration and therefore a spatial resolution of ~40 nm can be achieved. In addition, the use of a magnetic sector mass analyzer with parallel detection of up to seven masses at a mass resolution $m/dm > 5000$ is possible [73]. To obtain biomolecular images using NanoSIMS, stable isotopes are

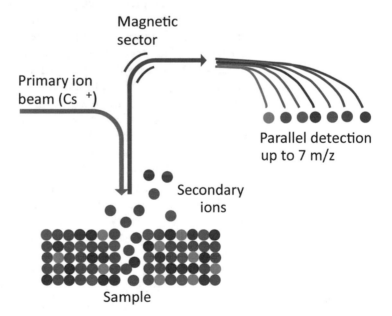

Fig. 4 Configuration of NanoSIMS. The primary ion and secondary ion paths are coaxial and normal to the sample surface in order to produce better lateral resolution compared to other SIMS instruments. The sample is labeled with stable isotopes before analysis. The primary ion beam Cs^+ (or O_2^-) erodes the sample surface, pixel by pixel. The resulting secondary ions are extracted to the ion optics and the magnetic sector analyzer. Different isotopic ions (up to 7 m/z) can be simultaneously detected on separated detectors

needed to label the molecules of interest, which then provides an isotopic ratio between incorporated and naturally occurring ones, allowing for quantification of the biomolecular turnover in the samples (or the so-called pulse and chase approach) [74–76].

The success of a NanoSIMS experiment is determined by several factors including the labeling specificity, sensitivity, and spatial resolution. The selection of the stable isotopic compound is critical in order to obtain specific labeling and sensitivity. Isotopes rarely present in brain tissue and neurons (e.g., ^{19}F [77]) produce lower background compared to isotope ^{13}C, ^{15}N and therefore have higher contrast and sensitivity. In addition, the sensitivity can be improved by increasing the number of isotopes per labeled molecule; however, this encounters technical challenges in labeling probe development. Finally, although NanoSIMS offers spatial resolution comparable to super-resolution microscopy, it still heavily depends on the abundance of the interested molecules, especially for the demand to localize specific proteins.

2.3.2 Sample Preparation for NanoSIMS

The sample preparation for a NanoSIMS experiment typically includes labeling with stable isotopes and further treatment with one of the strategies freeze-drying, chemical fixation, or cryogenic fixation. To study the protein turnover in brain tissues, animals are fed with food containing isotopically labeled amino acid such as ^{15}N leucine, ^{13}C lysine for a desired period of time before preparing for experiments [74, 75]. For neuronal cell experiments, the cells are cultured in medium containing isotopically labeled amino acids. Genetically encoded targets enabling click chemistry can be used to specifically label the protein of interest [77, 78]. This approach incorporates unnatural amino acids into proteins of interest. These amino acids are then coupled with isotopic element such as ^{15}N, ^{19}F and a fluorophore by a copper catalyzed azide-alkyne cycloaddition reaction. The probe can be used for both NanoSIMS imaging and fluorescence microscopy, for comparison purposes. The probe was successfully applied to visualize membrane proteins syntaxin 1, syntaxin 13, and SNAP-25 in mammalian cells using NanoSIMS. To image lipid turnover with NanoSIMS, isotopic lipid precursors are used, for example ^{13}C fatty acids, ^{13}C lipoprotein [79], or more specifically the isotopic precursors of sphingolipid, ^{15}N sphingosine and ^{15}N sphinganine, for labeling sphingolipid [76].

After pulsing the relevant isotopic compound, the samples undergo further preparation. Freeze-drying is an option for Nano-SIMS experiments [80, 81]. The samples are plunge frozen and freeze-dried similarly for SIMS imaging as described above. However, one of the common sample preparations for NanoSIMS is chemical fixation with glutaraldehyde followed by osmium tetroxide (OsO_4) [75, 82]. The concentration of the fixative agents and incubation time can be adjusted depending on the size and thickness of the samples, in general glutaraldehyde 2.5–4% in phosphate buffer or cacodylate buffer pH 7.4, and OsO_4 0.1–1% in the same buffer solution are used [76, 79]. Glutaraldehyde produces more efficient crosslinking with proteins than formaldehyde whereas OsO_4 creates multivalent crosslinking that helps immobilize lipids to retain the lipid organization of the cell membrane. Further dehydration with acetone or ethanol can be added.

For small molecules, easily diffusible molecules and metals, cryogenic fixation and substitution is recommended. The procedure is adopted from that used for electron microscopy [83–85]. In this approach, the sample is snap frozen using a high pressure freezing device and then gradually substituted with fixative solutions such as glutaraldehyde and then OsO_4 at −90 °C. The liquid is then replaced with acetone while slowly raising the temperature. Embedding samples with resin can start after acetone substitution at temperature between −50 and −20 °C depending on the type of resin, by mixing samples with increasing amount of resin. The sample is subsequently warmed up to room temperature for further resin embedding and hardening. Cryofixation was shown to better preserve the ultrastructure of mouse neo cortex including docked synaptic vesicles, glial volume, blood vessels compared to chemical fixation [86]. To track the delivery of the anticancer drug cisplatin to ovarian cancer cells using NanoSIMS and transmission electron microscopy (TEM), the cells were incubated with cisplatin, high pressure frozen, and freeze-substituted [87]. Pt was found to accumulate in different cellular compartments, most noticeable in nucleus, mitochondria, and autophagosomes. This method preserves ultrastructure of biological materials for NanoSIM very well, especially for easily diffusible compounds; however, the entire procedure takes days to accomplish.

2.4 Multimodal Nanoscopic and Mass Spectrometry Imaging

An exciting current trend in biomolecular imaging is the combination of different imaging techniques in order to gain the strength of all approaches, while overcoming their weaknesses. Multimodal imaging provides complementary information on the samples therefore ensuring better elucidation of molecular and cellular mechanisms. Samples can be imaged by each technique at different times, or can be imaged simultaneously. Ideally, one would ensure that the state of the sample is the same for both techniques. However, there are limitations in this respect, especially as many techniques are not compatible for combination in a single instrument. Multimodal imaging at different times is a more common solution, accepting the problem that sample properties may be changed, and the need to obtain a reliable and accurate correlation method among images obtained by different instruments. A huge effort has been currently invested into multimodal imaging, particularly to develop suitable protocols including sample preparation, labeling, as well as to design labeling probes that exhibit high specificity to the molecules of interest, high stability for imaging, and compatibility to different imaging techniques.

The most common approach is combination of fluorescence microscopy and electron microscopy (EM) by which protein localizations can be related to the morphological context of the cells or tissues obtained from EM images [6, 9]. Correlative imaging of STED, or photoactivated localization microscopy (PALM), and scanning EM (SEM) were carried out to localize mitochondrial outer membrane protein TOM20 and presynaptic dense projection protein α-liprin in *Caenorhabditis elegans* [88]. The sample preparation was optimized to ensure high fluorescence with minimal autofluorescence from the background while well preserving membrane morphology and localization of proteins. This includes high pressure freezing and freeze substitution with glutaraldehyde, followed by a mixture of OsO_4 and potassium permanganate and subsequent embedding in glycol methacrylate, before sectioning for imaging. The microscopic and SEM images were aligned using fluorescent silica beads as markers. The correlative images confirmed that TOM20 localized in the membrane of mitochondria, and α-liprin was observed in the area of presynaptic dense projections (Fig. 5a).

To study the protein turnover in different subcellular compartments of cultured hippocampal neurons, a combination of STED and NanoSIMS imaging, or the so-called correlated optical and isotopic nanoscopy, which allows for identifying cellular structures and quantifying their protein turnover within the cells was used [89]. The neurons were pulsed with ^{15}N leucine, which was then incorporated into newly synthesized proteins in the cells, and thus could be used to quantify the relative protein turnover based on the isotopic ratio $^{15}N/^{14}N$. The cells were immunolabelled for different organelle markers, including mitochondria, the Golgi apparatus, the endoplasmic reticulum, the active zone, and synaptic vesicles. To obtain a good correlation between microscopic and SIMS images, several marks on the embedding resin around the imaged area were created using the high intensity STED laser. The results showed that turnover rate of proteins was different in different organelles, particularly the turnover rate at the synapse was stronger than at other areas of the axon, and newly synthesized proteins accumulated more in the Golgi compared to the endoplasmic reticulum.

The development of labelling probes has provided multimodal imaging with flexibility, higher specificity and accuracy. Isotopic and fluorescent labelling probes for specific proteins were developed that could be used for both microscopic and NanoSIMS imaging [77, 78, 90, 91]. These probes showed very good correlation between the isotopic images from NanoSIMS and fluorescence

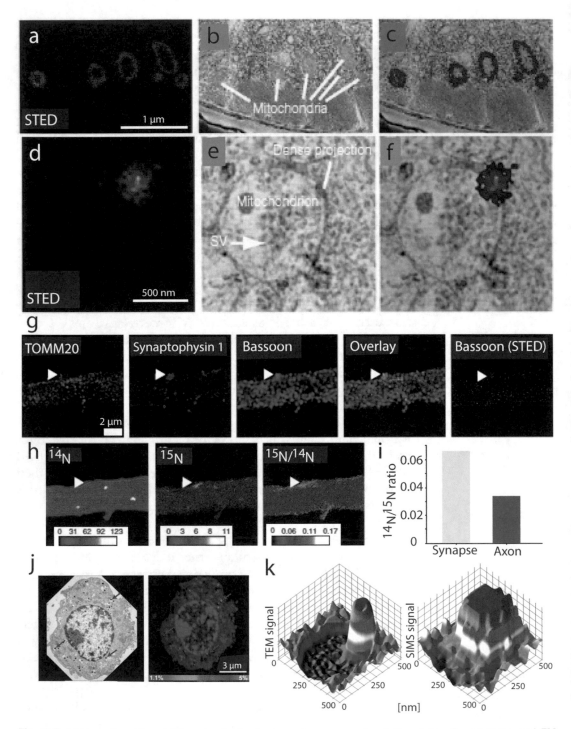

Fig. 5 Examples of multimodal imaging in the brain and synapses. (**a–c**) Correlative fluorescence and EM images of a thin section of worm expressing TOM20-citrine. (**a**) STED image, (**b**) EM image, and (**c**) overlaid image show that TOM20 localizes in the membrane of mitochondria. (**d–f**) Correlative fluorescence and EM images on a section of a worm expressing α-liprin-citrine. (**d**) STED image, (**e**) EM image, and (**f**) overlaid image show that α-liprin localizes in presynaptic dense projection area. SV: synaptic vesicles. (**g–i**) Correlative

images from STED for synaptic protein syntaxin 1, SNAP-25, and endoplasmic reticulum marker calnexin on BHK cells.

Multimodal analysis using NanoSIMS, transmission electron microscopy (TEM), and cellular electrochemistry were successfully carried out to investigate the distribution of the neurotransmitter dopamine across single vesicles of PC12 cells [92]. The cells were incubated with dopamine precursor, ^{13}C L-DOPA, which was incorporated into newly synthesized dopamine inside the vesicles. The cells were then fixed, dehydrated with ethanol, embedded in resin and cut into 70 nm thickness slices. Correlation of NanoSIMS and TEM images visualized dopamine localization between vesicle compartments, that is, the halo and the protein rich dense core. To study the dynamics of neurotransmitter transfer within single vesicles, the cells were treated with reserpine, a drug to deplete dopamine from vesicles. An electrochemistry method was used to quantify vesicle content and dopamine release during exocytosis and it showed that the dopamine transfer between vesicle compartments was kinetically limited at the time scale of hours.

3   Summary

As highlighted in this review, analysis of the localization, chemical nature, and biological turnover of targets in brain and synapses at nanoscale resolution has been possible using the current state-of-the-art imaging technologies. Super-resolution microscopy allows for measuring the functional organization of fluorescently labeled molecules and their dynamics. On the other hand, SIMS and NanoSIMS have been used to visualize the chemical structure of brain and synapses, and their molecular turnovers using the isotopic pulse and chase strategy. Moreover, combination of optical and secondary ion mass spectrometry imaging has been enabled

◄

Fig. 5 (continued) STED and NanoSIMS imaging of axons of hippocampal neurons. (**g**) Confocal microscopic images of a neuronal axon immunostained for mitochondrial marker TOMM20, synaptic vesicle marker synaptophysin 1, active zone marker bassoon, their overlaid image, and STED image of bassoon. Arrowhead indicates a synapse where all three labels localize. (**h**) NanoSIMS images of the same axons for ^{14}N, ^{15}N, and ^{15}N/^{14}N ratio, respectively. (**i**) The bar chart of the ^{15}N/^{14}N ratio shows high protein turnover at the synapse compared to the axonal area. (**j–k**) Multimodal NanoSIMS and TEM imaging to study dopamine distribution inside single synaptic vesicles. (**j**) Correlated TEM and NanoSIMS images (^{13}C^{14}N^{-}/^{12}C^{14}N^{-}) of a PC12 cell, respectively, to observe dopamine enrichment. The cell was previously incubated with precursor ^{13}C-L-DOPA. Vesicles are highlighted by red arrows. (**k**) 3D surface plots of TEM (left) and ^{13}C^{14}N^{-} NanoSIMS ion imaging (right) for incubation of ^{13}C-L-DOPA for 12 h followed by reserpine treatment indicate that the dopamine distribution inside a single vesicle is uneven. The yellow and red signals in the TEM plot show the protein dense core, the dark blue shows the halo. The red and dark blue in the NanoSIMS plot show higher and lower signals of the ^{13}C enrichment, respectively. Panels **a–f** are reproduced with permission from Ref. 88. Panels **g–i** are reproduced with permission from Ref. 89. Panels **j** and **k** are reproduced with permission from Ref. 92

owing to the development of isotopic and fluorescence labeling probes, which broadens the possible scope for measurement and correlation of various parameters such as function, morphological structure, chemical structure and molecular turnover of the brain. The future perspective in brain and synaptic research would heavily involve multimodal imaging as it provides substantially more information than any of these technologies used separately. Strategies of this technology combination, however, need to be further improved including instrumental modification, development of labeling probes, and sample preparation, which are compatible to multimodal imaging techniques, and data analysis protocols for interpretation of very complex data.

Acknowledgments

The authors acknowledge the support of grants from the Swedish Research Council (International Postdoc Grant) and the German Research Foundation (SFB 1286/B1) to N.T.N.P., and from the European Research Council (ERC Consolidator Grant NeuroMolAnatomy, 614765) to S.O.R.

References

1. Kliman M, Vijayakrishnan N, Wang L, Tapp JT, Broadie K, McLean JA (2010) Structural mass spectrometry analysis of lipid changes in a Drosophila epilepsy model brain. Mol BioSyst 6(6):958–966

2. Xun Z, Sowell RA, Kaufman TC, Clemmer DE (2007) Lifetime proteomic profiling of an A30P r-synuclein of Parkinson's disease. J Proteome Res 6:3729–3738

3. Han X, Holtzman D, McKeel D Jr, Kelley J, Morris J (2002) Substantial sulfatide deficiency and ceramide elevation in very early Alzheimer's disease: potential role in disease pathogenesis. J Neurochem 82:809–818

4. Farooqui AA, Horrocks LA, Farooqui T (2000) Glycerophospholipids in brain: their metabolism, incorporation into membranes, functions, and involvement in neurological disorders. Chem Phys Lipids 106:1–29

5. Matsumoto J, Sugiura Y, Yuki D, Hayasaka T, Goto-Inoue N, Zaima N et al (2011) Abnormal phospholipids distribution in the prefrontal cortex from a patient with schizophrenia revealed by matrix-assisted laser desorption/ionization imaging mass spectrometry. Anal Bioanal Chem 400(7):1933–1943

6. Wilhelm BG, Mandad S, Truckenbrodt S, Kröhnert K, Schäfer C, Rammner B et al

(2014) Composition of isolated synaptic boutons reveals the amounts of vesicle trafficking proteins. Science 344(6187):1023–1028

7. Dani A, Huang B, Bergan J, Dulac C, Zhuang X (2010) Superresolution imaging of chemical synapses in the brain. Neuron 68(5):843–856

8. Willig KI, Rizzoli SO, Westphal V, Jahn R, Hell SW (2006) STED microscopy reveals that synaptotagmin remains clustered after synaptic vesicle exocytosis. Nature 440(7086):935–939

9. Phan NT, Fletcher JS, Ewing AG (2015) Lipid structural effects of oral administration of methylphenidate in Drosophila brain by secondary ion mass spectrometry imaging. Anal Chem 87(8):4063–4071

10. Phan NT, Mohammadi AS, Dowlatshahi Pour M, Ewing AG (2016) Laser desorption ionization mass spectrometry imaging of drosophila brain using matrix sublimation versus modification with nanoparticles. Anal Chem 88(3):1734–1741

11. Mikula S, Denk W (2015) High-resolution whole-brain staining for electron microscopic circuit reconstruction. Nat Methods 12 (6):541–546

12. Revelo NH, Rizzoli SO (2015) Application of STED microscopy to cell biology questions. In: Verveer PJ (ed) Advanced fluorescence

microscopy: methods and protocols. Springer, New York, NY, pp 213–230

13. Neupane B, Ligler FS, Wang G (2014) Review of recent developments in stimulated emission depletion microscopy: applications on cell imaging. J Biomed Opt 19(8):080901-080901-080901-080909

14. Westphal V, Rizzoli SO, Lauterbach MA, Kamin D, Jahn R, Hell SW (2008) Video-rate far-field optical nanoscopy dissects synaptic vesicle movement. Science 320(5873):246–249

15. Fornasiero EF, Opazo F (2015) Super-resolution imaging for cell biologists: concepts, applications, current challenges and developments. BioEssays 37(4):436–451

16. Balzarotti F, Eilers Y, Gwosch KC, Gynnå AH, Westphal V, Stefani FD et al (2017) Nanometer resolution imaging and tracking of fluorescent molecules with minimal photon fluxes. Science 355(6325):606–612

17. Xu K, Babcock HP, Zhuang X (2012) Dual-objective STORM reveals three-dimensional filament organization in the actin cytoskeleton. Nat Methods 9(2):185–188

18. Huang B, Wang W, Bates M, Zhuang X (2008) Three-dimensional super-resolution imaging by stochastic optical reconstruction microscopy. Science 319(5864):810–813

19. Curdt F, Herr SJ, Lutz T, Schmidt R, Engelhardt J, Sahl SJ et al (2015) isoSTED nanoscopy with intrinsic beam alignment. Opt Express 23(24):30891–30903

20. Rönnlund D, Xu L, Perols A, Gad AKB, Karlstro AE, Widengren J (2014) Multicolor fluorescence nanoscopy by photobleaching: concept, verification, and its application to resolve selective storage of proteins in platelets. ACS Nano 8(5):4358–4365

21. Tonnesen J, Nadrigny F, Willig KI, Wedlich-Soldner R, Nagerl UV (2011) Two-color STED microscopy of living synapses using a single laser-beam pair. Biophys J 101 (10):2545–2552

22. Hell SW (2007) Far-field optical nanoscopy. Science 316:1153–1158

23. Hell SW, Wichmann J (1994) Breaking the diffraction resolution limit by stimulated emission: stimulated-emission-depletion fluorescence microscopy. Opt Lett 19(11):780–782

24. Gottfert F, Wurm CA, Mueller V, Berning S, Cordes VC, Honigmann A et al (2013) Coaligned dual-channel STED nanoscopy and molecular diffusion analysis at 20 nm resolution. Biophys J 105(1):L01–L03

25. Schmidt R, Wurm CA, Jakobs S, Engelhardt J, Egner A, Hell SW (2008) Spherical nanosized focal spot unravels the interior of cells. Nat Methods 5(6):539–544

26. Rust MJ, Bates M, Zhuang X (2006) Sub-diffraction-limit imaging by stochastic optical reconstruction microscopy (STORM). Nat Methods 3(10):793–795

27. Tam J, Merino D (2015) Stochastic optical reconstruction microscopy (STORM) in comparison with stimulated emission depletion (STED) and other imaging methods. J Neurochem 135(4):643–658

28. Vicidomini G, Moneron G, Han KY, Westphal V, Ta H, Reuss M et al (2011) Sharper low-power STED nanoscopy by time gating. Nat Methods 8(7):571–573

29. Westphal V, Blanca CM, Dyba M, Kastrup L, Hell SW (2003) Laser-diode-stimulated emission depletion microscopy. Appl Phys Lett 82 (18):3125–3127

30. Willig KI, Harke B, Medda R, Hell SW (2007) STED microscopy with continuous wave beams. Nat Methods 4(11):915–918

31. van de Linde S, Loschberger A, Klein T, Heidbreder M, Wolter S, Heilemann M et al (2011) Direct stochastic optical reconstruction microscopy with standard fluorescent probes. Nat Protoc 6(7):991–1009

32. Chen F, Tillberg PW, Boyden ES (2015) Expansion microscopy. Science 347 (6221):543–548

33. Chen F, Wassie AT, Cote AJ, Sinha A, Alon S, Asano S et al (2016) Nanoscale imaging of RNA with expansion microscopy. Nat Methods 13(8):679–684

34. Tillberg PW, Chen F, Piatkevich KD, Zhao Y, Yu CC, English BP et al (2016) Protein-retention expansion microscopy of cells and tissues labeled using standard fluorescent proteins and antibodies. Nat Biotechnol 34 (9):987–992

35. Chozinski TJ, Halpern AR, Okawa H, Kim HJ, Tremel GJ, Wong RO et al (2016) Expansion microscopy with conventional antibodies and fluorescent proteins. Nat Methods 13 (6):485–488

36. Chang JB, Chen F, Yoon YG, Jung EE, Babcock H, Kang JS et al (2017) Iterative expansion microscopy. Nat Methods 14 (6):593–599

37. Ries J, Kaplan C, Platonova E, Eghlidi H, Ewers H (2012) A simple, versatile method for GFP-based super-resolution microscopy via nanobodies. Nat Methods 9(6):582–584

38. Opazo F, Levy M, Byrom M, Schafer C, Geisler C, Groemer TW et al (2012) Aptamers as potential tools for super-resolution microscopy. Nat Methods 9(10):938–939

39. Fernandez-Suarez M, Ting AY (2008) Fluorescent probes for super-resolution imaging in

living cells. Nat Rev Mol Cell Biol 9 (12):929–943

40. Dean KM, Palmer AE (2014) Advances in fluorescence labeling strategies for dynamic cellular imaging. Nat Chem Biol 10(7):512–523

41. Grimm JB, English BP, Chen J, Slaughter JP, Zhang Z, Revyakin A et al (2015) A general method to improve fluorophores for live-cell and single-molecule microscopy. Nat Methods 12(3):244–250. p 243 following 250

42. Vreja IC, Nikic I, Göttfert F, Bates M, Kröhnert K, Outeiro TF et al (2015) Super-resolution microscopy of clickable amino acids reveals the effects of fluorescent protein tagging on protein assemblies. ACS Nano 9 (11):11034–11041

43. Sun X, Zhang A, Baker B, Sun L, Howard A, Buswell J et al (2011) Development of SNAP-tag fluorogenic probes for wash-free fluorescence imaging. Chembiochem 12 (14):2217–2226

44. Gautier A, Juillerat A, Heinis C, Correa IR Jr, Kindermann M, Beaufils F et al (2008) An engineered protein tag for multiprotein labeling in living cells. Chem Biol 15(2):128–136

45. Revelo NH, Kamin D, Truckenbrodt S, Wong AB, Reuter-Jessen K, Reisinger E et al (2014) A new probe for super-resolution imaging of membranes elucidates trafficking pathways. J Cell Biol 205(4):591–606

46. Lukinavicius G, Reymond L, D'Este E, Masharina A, Gottfert F, Ta H et al (2014) Fluorogenic probes for live-cell imaging of the cytoskeleton. Nat Methods 11(7):731–733

47. Phan NTN, Li X, Ewing AG (2017) Measuring synaptic vesicles using cellular electrochemistry and nanoscale molecular imaging. Nat Chem Rev 1:0048

48. Bates M, Huang B, Dempsey GT, Zhuang X (2007) Multicolor super-resolution imaging with photo-switchable fluorescent probes. Science 317(5845):1749–1753

49. Chozinski TJ, Gagnon LA, Vaughan JC (2014) Twinkle, twinkle little star: photoswitchable fluorophores for super-resolution imaging. FEBS Lett 588(19):3603–3612

50. Passarelli MK, Pirkl A, Moellers R, Grinfeld D, Kollmer F, Havelund R et al (2017) The 3D OrbiSIMS-label-free metabolic imaging with subcellular lateral resolution and high mass-resolving power. Nat Methods 14 (12):1175–1183

51. Fisher GL, Bruinen AL, Ogrinc Potocnik N, Hammond JS, Bryan SR, Larson PE et al (2016) A new method and mass spectrometer design for TOF-SIMS parallel imaging MS/MS. Anal Chem 88(12):6433–6440

52. Chandra S, Smith DR, Morrison GH (2000) Subcellular imaging by dynamic SIMS ion microscopy. Anal Chem 72(3):104A–114A

53. Fletcher JF, Rabbani S, Henderson A, Blenkinsopp P, Thompson SP, Lockyer P et al (2008) A new dynamic in mass spectral imaging of single biological cells. Anal Chem 80:9058–9064

54. Benninghoven A (1994) Chemical analysis of inorganic and organic surfaces and thin films by static time-of-flight secondary ion mass spectrometry (TOF-SIMS). Angew Chem Int Ed Engl 33:1023–1043

55. Kollmer F (2004) Cluster primary ion bombardment of organic materials. Appl Surf Sci 231–232:153–158

56. Touboul D, Kollmer F, Niehuis E, Brunelle A, Laprevote O (2005) Improvement of biological time-of-flight-secondary ion mass spectrometry imaging with a bismuth cluster ion source. J Am Soc Mass Spectrom 16 (10):1608–1618

57. Weibel D, Wong S, Lockyer N, Blenkinsopp P, Hill R, Vickerman JC (2003) AC60 primary ion beam system for time of flight secondary ion mass spectrometry: its development and secondary ion yield characteristics. Anal Chem 75:1754–1764

58. Phan NTN, Fletcher JS, Sjövall P, Ewing AG (2014) ToF-SIMS imaging of lipids and lipid related compounds in Drosophila brain. Surf Interface Anal 46(S1):123–126

59. Sjovall P, Lausmaa J, Johansson B (2004) Mass spectrometric imaging of lipids in brain tissue. Anal Chem 76:4271–4278

60. Rabbani SN, Barber A, Fletcher JS, Lockyer NP, Vickerman JC (2013) Enhancing secondary ion yields in time of flight-secondary ion mass spectrometry using water cluster primary beams. Anal Chem 85(12):5654–5658

61. Angerer TB, Blenkinsopp P, Fletcher JS (2015) High energy gas cluster ions for organic and biological analysis by time-of-flight secondary ion mass spectrometry. Int J Mass Spectrom 377:591–598

62. Brison J, Robinson MA, Benoit DS, Muramoto S, Stayton PS, Castner DG (2013) TOF-SIMS 3D imaging of native and non-native species within HeLa cells. Anal Chem 85(22):10869–10877

63. Angerer TB, Fletcher JS (2014) 3D imaging of TiO$_2$ nanoparticle accumulation in *Tetrahymena pyriformis*. Surf Interface Anal 46:198–203

64. Dowlatshahi Pour M, Jennische E, Lange S, Ewing AG, Malmberg P (2016) Food-induced

changes of lipids in rat neuronal tissue visualized by ToF-SIMS imaging. Sci Rep 6:32797

65. Lanekoff I, Phan NT, Van Bell CT, Winograd N, Sjovall P, Ewing AG (2013) Mass spectrometry imaging of freeze-dried membrane phospholipids of dividing *Tetrahymena pyriformis*. Surf Interface Anal 45 (1):211–214

66. Nagata Y, Ishizaki I, Waki M, Ide Y, Hossen MA, Ohnishi K et al (2014) Glutaraldehyde fixation method for single-cell lipid analysis by time-of-flight secondary ion-mass spectrometry. Surf Interface Anal 46:185–188

67. Bradford A, Atkinson J, Fuller N, Rand RP (2003) The effect of vitamin E on the structure of membrane lipid assemblies. J Lipid Res 44 (10):1940–1945

68. Monroe EB, Jurchen JC, Lee J, Rubakhin SS, Sweedler JV (2005) Vitamin E imaging and localization in the neuronal membrane. JACS Commun 127:12152–12153

69. Tucker KR, Li Z, Rubakhin SS, Sweedler JV (2012) Secondary ion mass spectrometry imaging of molecular distributions in cultured neurons and their processes: comparative analysis of sample preparation. J Am Soc Mass Spectrom 23(11):1931–1938

70. Roddy TP, Cannon DM, Ostrowski SG, Ewing AG, Winograd N (2003) Proton transfer in time-of-flight secondary ion mass spectrometry studies of frozen-hydrated dipalmitoylphosphatidylcholine. Anal Chem 75:4087–4094

71. Kurczy ME, Piehowski PD, Van Bell CT, Heien ML, Winograd N, Ewing AG (2010) Mass spectrometry imaging of mating Tetrahymena show that changes in cell morphology regulate lipid domain formation. Proc Natl Acad Sci U S A 107(7):2751–2756

72. Lanekoff I, Sjovall P, Ewing AG (2011) Relative quantification of phospholipid accumulation in the PC12 cell plasma membrane following phospholipid incubation using TOF-SIMS imaging. Anal Chem 83 (13):5337–5343

73. Boxer SG, Kraft ML, Weber PK (2009) Advances in imaging secondary ion mass spectrometry for biological samples. Annu Rev Biophys 38:53–74

74. Zhang DS, Piazza V, Perrin BJ, Rzadzinska AK, Poczatek JC, Wang M et al (2012) Multi-isotope imaging mass spectrometry reveals slow protein turnover in hair-cell stereocilia. Nature 481(7382):520–524

75. Steinhauser ML, Bailey AP, Senyo SE, Guillermier C, Perlstein TS, Gould AP et al (2012) Multi-isotope imaging mass spectrometry quantifies stem cell division and metabolism. Nature 481(7382):516–519

76. Frisz JF, Lou K, Klitzing HA, Hanafin WP, Lizunov V, Wilson RL et al (2013) Direct chemical evidence for sphingolipid domains in the plasma membranes of fibroblasts. Proc Natl Acad Sci U S A 110(8):E613–E622

77. Vreja IC, Kabatas S, Saka SK, Krohnert K, Hoschen C, Opazo F et al (2015) Secondary-ion mass spectrometry of genetically encoded targets. Angew Chem Int Ed 54 (19):5784–5788

78. Kabatas S, Vreja IC, Saka SK, Hoschen C, Krohnert K, Opazo F et al (2015) A contamination-insensitive probe for imaging specific biomolecules by secondary ion mass spectrometry. Chem Commun 51 (67):13221–13224

79. Jiang H, Goulbourne CN, Tatar A, Turlo K, Wu D, Beigneux AP et al (2014) High-resolution imaging of dietary lipids in cells and tissues by NanoSIMS analysis. J Lipid Res 55(10):2156–2166

80. Rakowska PD, Jiang H, Ray S, Pyne A, Lamarre B, Carr M et al (2013) Nanoscale imaging reveals laterally expanding antimicrobial pores in lipid bilayers. Proc Natl Acad Sci U S A 110(22):8918–8923

81. Peteranderl R, Lechene C (2004) Measure of carbon and nitrogen stable isotope ratios in cultured cells. J Am Soc Mass Spectrom 15 (4):478–485

82. He C, Hu X, Jung RS, Weston TA, Sandoval NP, Tontonoz P et al (2017) High-resolution imaging and quantification of plasma membrane cholesterol by NanoSIMS. Proc Natl Acad Sci U S A 114(8):2000–2005

83. Sosinsky GE, Crum J, Jones YZ, Lanman J, Smarr B, Terada M et al (2008) The combination of chemical fixation procedures with high pressure freezing and freeze substitution preserves highly labile tissue ultrastructure for electron tomography applications. J Struct Biol 161(3):359–371

84. McDonald KL, Webb RI (2011) Freeze substitution in 3 hours or less. J Microsc 243 (3):227–233

85. Grovenor CRM, Smart KE, Kilburn MR, Shore B, Dilworth JR, Martin B et al (2006) Specimen preparation for NanoSIMS analysis of biological materials. Appl Surf Sci 252 (19):6917–6924

86. Korogod N, Petersen CC, Knott GW (2015) Ultrastructural analysis of adult mouse neocortex comparing aldehyde perfusion with cryo fixation. elife 4. https://doi.org/10.7554/eLife.05793

87. Lee RFS, Riedel T, Escrig S, Maclachlan C, Knott GW, Davey CA et al (2017) Differences in cisplatin distribution in sensitive and

resistant ovarian cancer cells: a TEM/Nano-SIMS study. Metallomics 9(10):1413–1420

88. Watanabe S, Punge A, Hollopeter G, Willig KI, Hobson RJ, Davis MW et al (2011) Protein localization in electron micrographs using fluorescence nanoscopy. Nat Methods 8(1):80–84

89. Saka SK, Vogts A, Krohnert K, Hillion F, Rizzoli SO, Wessels JT (2014) Correlated optical and isotopic nanoscopy. Nat Commun 5:3664

90. Kabatas S, Agui-Gonzalez P, Saal KA, Jahne S, Opazo F, Rizzoli SO et al (2019) Boron-containing probes for non-optical high-resolution imaging of biological samples. Angew Chem Int Ed Engl 58(11):3438–3443

91. Kabatas S, Agüi-Gonzalez P, Hinrichs R, Jähne S, Opazo F, Diederichsen U et al (2019) Fluorinated nanobodies for targeted molecular imaging of biological samples using nanoscale secondary ion mass spectrometry. J Anal At Spectrom 34(6):1083–1087

92. Lovric J, Dunevall J, Larsson A, Ren L, Andersson S, Meibom A et al (2017) Nano secondary ion mass spectrometry imaging of dopamine distribution across nanometer vesicles. ACS Nano 11(4):3446–3455

Chapter 4

Advancing Array Tomography to Study the Fine Ultrastructure of Identified Neurons in Zebrafish (*Danio rerio*)

Marlene Strobel, Frederik Helmprobst, Martin Pauli, Manfred Heckmann, Christina Lillesaar, and Christian Stigloher

Abstract

Array tomography (AT) provides a versatile workflow for correlated light and electron microscopy (CLEM). In short, biological tissues are embedded in EM-resins for immunolabeling, cut in ultrathin section arrays, which are mounted on glass slides, labeled and imaged for immunofluorescence at the light microscope and then prepared for scanning electron microscopy (SEM) imaging. Light- and electron micrographs obtained from the identical regions of interest of the same sections are then correlated to an aligned composite image series. We adapted this protocol to identify and image the Mauthner neuron of the developing zebrafish embryo. The Mauthner neuron is an identifiable neuron, which can be easily labeled by retrograde tracing with for example rhodamine dextran. We take advantage of the fact that the fluorescence of rhodamine is retained after embedding in the LR White resin. Furthermore, we expanded the workflow to reach a near-to-native ultrastructural preservation and good antigenicity of the nervous tissue, by applying high pressure freezing and freeze substitution. Moreover, we add structured illumination microcopy (SIM) as imaging modality to allow tracing of fine neuronal projections and increase correlation accuracy.

Key words Array tomography, Zebrafish, *Danio rerio*, Mauthner neuron, Reticulospinal neuron, Identified neuron

1 Introduction

An important research topic for cellular neurobiology is the analysis of identified neurons with high resolution imaging techniques to get an extended picture of the connections, the cellular architecture as a whole and synaptic architecture in particular. This is technically challenging, as intact neurons in the central nervous system are typically embedded in a dense network within the nervous tissue. The technique we present here solves the problem to identify a specific neuron by application of direct labeling of the neuron of interest with dyes that allow fluorescent imaging followed by

Irene Wacker et al. (eds.), *Volume Microscopy: Multiscale Imaging with Photons, Electrons, and Ions*, Neuromethods, vol. 155, https://doi.org/10.1007/978-1-0716-0691-9_4, © Springer Science+Business Media, LLC, part of Springer Nature 2020

electron microscopic analysis of the full ultrastructural context. Our approach builds on the technical principle to combine fluorescent labeling on EM resin embedded samples with electron microscopic analysis of the very same section (*see* for example refs. 1, 2). This principal approach was thoroughly described for application in mammalian nervous tissue and coined *Array Tomography* (AT) by Micheva and Smith [3]. More topics relating to AT approaches can be found in Chapters 5–8 of this volume. We recently advanced the AT-approach and replaced classical aldehyde fixation with high-pressure freezing and freeze substitution (HPF/FS) for near-to-native sample preservation, and added super-resolution fluorescence microscopy for the light imaging step in this correlative light and electron microscopy (CLEM) workflow [4, 5]. Moreover, we recently showed that the principles of AT can be applied for RNA in situ hybridization to detect RNA localization in the full ultrastructural context [6]. As a next step, to make AT even more relevant for neurobiological questions, we now describe a workflow to directly label identified neurons. Notably, the rhodamine coupled tracer we use retains fluorescence in the immunocompetent EM-resin LR White, therefore allowing direct superresolved imaging of the labeled neurons without any further antibody labeling steps, which are typically necessary in the standard AT protocol.

As a starting point for establishment of the technique we decided to use the experimentally tractable Mauthner neuron of the developing zebrafish, which has served as a model neuron in neurobiology for decades. The Mauthner neuron is a classical model for synaptic transmission where pioneering work on quantal release in central synapses was performed, and has since then stayed an attractive model system for neurobiological research [7]. Further, the Mauthner neuron has been suggested to adhere to the identifiable neuron concept initially described for neurons in insect nervous systems [8]. This means that a specific neuron can be reproducibly found at the same location, is exhibiting the same properties and is executing the same function in all individuals of the same species. The Mauthner neuron fulfills at least some of these criteria making it one of few identifiable neurons among vertebrate species.

The Mauthner neuron is a sensory integrator that initiates the escape response in fish and amphibians [9], and has been suggested to be evolutionary related to the mammalian reticular neurons within the nucleus gigantocellularis [10]. The Mauthner belongs to a group of bilaterally paired segmentally organized homologous neurons situated in the hindbrain [11]. The Mauthner neuron is the largest and most anterior of these cells and is found in rhombomere 4 both in developing and adult zebrafish [11, 12]. It has one thick dendrite projecting laterally and receiving input from the auditory system [9, 12]. In addition, it has one main and several smaller ventral dendrites, which are targeted by multiple sensory systems. Also its axon is extreme with a diameter reaching up to about 15 μm in the adult zebrafish [12]. The axon is myelinated and is crossing the midline to project posteriorly all along the spinal cord where it contacts motor neurons and interneurons.

The Mauthner neuron receives afferent input via several distinct types of synapses distributed over the dendrites and the soma. The morphological character of this input appears stable during post-hatching stages in zebrafish, and exhibits a high degree of similarity between the developing zebrafish and adult goldfish [13, 14]. The described types of synapses include myelinated club endings located on the distal part of the lateral dendrite, terminal boutons found on the dendrites and soma, unmyelinated club endings present on the dorsomedial portion of the perikaryon adjacent to the axon cap, and spiral fiber terminals within the axon cap [14]. Further, gap junctions were reported to frequently be found on the initial segment of the axon, on the ventral dendrite and ventral soma as well as on the distal lateral dendrite [14]. In contrast, gap junctions were only rarely seen on the dorsal surface of the cell. Also, mixed synapses with both chemical and electrical transmission contact the Mauthner neuron [13].

In addition to the morphological properties of the afferent synapses on the Mauthner neuron, detailed neurochemical investigations have identified a number of different neurotransmitters present in the boutons targeting the Mauthner neuron [7]. Interestingly, terminals containing different transmitters are spatially differentially located at the Mauthner neuron, with glutamatergic and GABAergic terminals dominating at the lateral dendrite, glycinergic and dopaminergic at the axon cap, and serotonergic and glutamatergic terminals at the ventral dendrite. Moreover, auditory responses of the Mauthner neuron have been extensively investigated, and furthermore, this circuit is modulated by visual and lateral line inputs [9].

The properties of the Mauthner neuron, that is, extremely large size and well-defined location, allows performing morphological investigations and electrophysiological recordings of a vertebrate identifiable neuron, which is contributing significantly to the beauty of its circuitry.

2   Materials

2.1   Rhodamine Dextran Backlabeling of Reticulospinal Neurons

1. Danieau's solution: 17.4 ml (1 M) NaCl, 210 μl (1 M) KCl, 120 μl (1 M) $MgSO_4$, 180 μl (1 M) $Ca(NO_3)_2$, 1.5 ml (1 M) HEPES, optional 1 ml methylene blue (final 0.1%) and adjusted to 1 l with ddH_2O.

2. 0.4% Tricaine (Western Chemical Inc.) dissolved in ddH_2O, pH 7.4 with NaOH.

3. 0.7% Low melting agarose (Sigma-Aldrich) dissolved in Danieau's solution and stored in aliquots of 1.5 ml at RT.

4. Tetramethylrhodamine, anionic, lysine fixable, MW 3000, stored at −20 °C protected from light (Invitrogen, Molecular Probes, cat. no D3308).

5. Petri dish with bottom covered with 1% agarose (*see* Fig. 1).

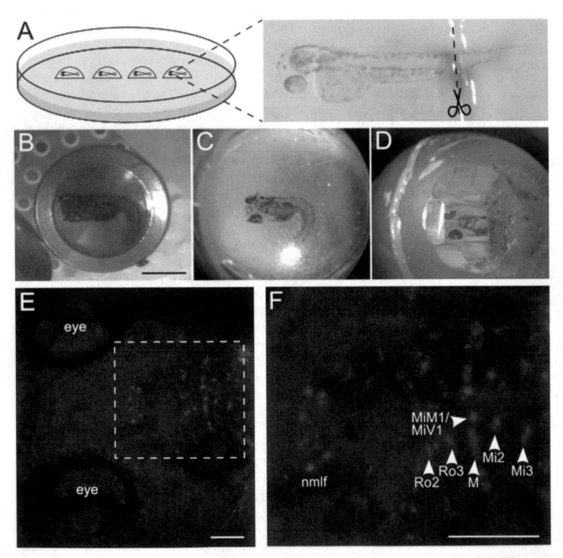

Fig. 1 Illustrations demonstrating the specimen at different steps of preparation. (**a**) 3 dpf zebrafish mounted in a drop of low melting agarose on a plate of standard agarose. To backfill the reticulospinal neurons a sharp razorblade is used for cutting the tail and a crystal of rhodamine dextran is directly applied at the incision site. (**b**) Larva in freezing chamber during the freeze substitution process. Dorsal view, anterior to the left. (**c**) The same larva shown after embedding in LR White resin. (**d**) After trimming of the LR White block the same larva is seen. (**e**) The same larvae shown with epifluorescence highlighting the reticulospinal neurons. Dorsal view, anterior to the left. (**f**) High magnification of boxed area in (**e**). Reticulospinal neurons are marked (arrow heads) and annotated (Ro2, Ro3, MiM1, MiV1, Mi2, Mi3) according to ref. 11. Mauthner neuron (M), nucleus of the medial longitudinal fascicle (nmlf). Scale bar in **b** = 1 mm, in **e** = 100 μm and in **f** = 125 μm

6. Zebrafish (*Danio rerio*) embryos (4 dpf). All fish are kept on a 14:10-h light/dark cycle at 28 °C in Danieau's solution. Animals are staged according to Kimmel et al. [15]. All experiments are performed according to the animal welfare regulations of the District Government of Lower Franconia.

2.2 High Pressure Freezing

1. High pressure freezer: protocols are conducted with the EM HPM100 system (Leica Microsystems, Vienna, Austria) and accompanying accessories.

2. Sample carrier sandwich are composed of Type A (recesses 100 and 200 μm) (Leica Microsystems) and Type B (recesses 150 and 150 μm) (BALTIC Preparation, Art. Nr. 665).

3. Lecithin solution (about 3 mg lecithin is dissolved in 1 ml chloroform).

4. Liquid nitrogen.

5. 0.4% Tricaine (Western Chemical Inc.) dissolved in ddH$_2$O, pH 7.4 adjusted with NaOH.

6. 0.7% Low melting agarose (Sigma) dissolved in Danieau's solution

7. Animals to be frozen.

2.3 Freeze Substitution

1. Freeze substitution apparatus: protocols are performed with the EM AFS2 system (Leica Microsystems) and accompanying accessories (metal washing containers with bottom plates and liquid permeable plastic baskets).

2. Liquid nitrogen, anhydrous acetone and absolute ethanol.

3. Freeze substitution solution: 0.1% KMnO$_4$ in anhydrous acetone, freshly prepared before use.

4. LR White Medium Grade Acrylic Resin (London Resin Company Ltd.)

5. Small glass vials with lid.

2.4 Embedding

1. LR White Medium Grade Acrylic Resin (London Resin Company Ltd.)

2. Gelatin embedding capsules.

2.5 Ultramicrotomy

1. Ultramicrotome EM UC7 (Leica Microsystems).

2. Histo Jumbo diamond knife (DiATOME, Biel, Switzerland).

3. Poly-L-lysine coated slides (Polysine, Thermo Scientific).

4. Eyelash mounted with super glue on a toothpick or small syringe.

5. Syringe.

6. Glue (Pattex Gel Compact).

7. Spinell black 47400 (Kremer Pigmente, Aichstetten, Germany).

8. Xylene.

2.6 Light Microscopy

1. Hydrophobic pen (Immunopen, Wako).

2. Staining chamber (*see* **Note 10**).

3. Tris-buffer (50 mM Tris in ddH$_2$O, pH 7.6).

4. Blocking solution consisting of 0.1% BSA and 0.05% Tween 20 in 50 mM Tris buffer, pH 7.6, Hoechst 33342 (Invitrogen).

5. Primary AB against GFP (polyclonal chicken anti-GFP, Abcam; ab13970).

6. Secondary AB antibody goat anti-chicken IgG (H + L) Alexa Fluor 488 conjugate (Thermo Fisher Scientific).

7. Mowiol.

8. High Precision Microscope Cover Glasses (Carl Roth).

9. ELYRA S.1 super-resolution structured illumination microscope (Zeiss).

2.7 Contrasting and Carbon Coating

1. Decarbonized ddH$_2$O.

2. 2.5% uranyl acetate in ethanol.

3. 50% Reynolds's lead citrate [16] in decarbonized ddH$_2$O.

4. Conductive silver paint.

5. SEM specimen holder stubs with carbon stickers.

6. Carbon coating is conducted using the carbon coater Med 010 (Balzers Union).

2.8 Scanning Electron Microscopy

Field emission scanning electron microscope JSM-7500F (JEOL, Japan).

2.9 Image Processing

Fiji software package (https://imagej.net/Fiji/Downloads) [17].

2.10 Alignment of Serial Electron Micrographs

IMOD software package (version 4.7. http://bio3d.colorado.edu/imod) [18].

2.11 Correlation with Icy

eC-CLEM plugin for the Icy software package (http://www.icy.bioimageanalysis.org/plugin/ec-CLEM#documentation) [19, 20].

3 Methods

3.1 Rhodamine Dextran Backlabeling of Reticulospinal Neurons

Our procedure for retrograde labeling of the Mauthner neuron is based on McLean and Fetcho [21]. For an in vivo backlabeling of reticulospinal neurons 3 dpf zebrafish embryos are first sedated in 0.4% tricaine, and then transferred into an aliquot with melted 0.7% low melting agarose (max. 30 °C). The embryo is transferred together with a drop of agarose to the petri dish covered with 1% standard agarose, and carefully positioned on the lateral side at the interface between the low melting agarose and the standard agarose plate (*see* Fig. 1). The low melting agarose with the embryo is left to set for a few minutes. The tails of the fish are then cut off between somite 22 and 24 with a quick and distinct movement using a sharp razor blade. Immediately thereafter a few rhodamine dextran crystals are put at the lesion site. After 5 min the fish embryos are carefully freed from the low melting agarose using forceps, transferred into Danieau's solution and incubated for at least 16 h to allow the tracer to migrate intra-axonally in a retrograde direction. For the next steps the fish are sorted for the brightest signal using a fluorescence stereomicroscope.

3.2 High Pressure Freezing

The freezing platelets are coated with lecithin for a smooth removal of the sample after HPF/FS (about 3 mg Lecithin is dissolved in 1 ml chloroform, a drop filling the platelet depression is pipetted and dried). For high pressure freezing 4 dpf zebrafish embryos are first completely sedated in 0.4% tricaine and then transferred into freezing chambers (recesses 150 and 100 μm) containing 0.7% low melting agarose (max. 30 °C) as filler and freeze protectant according to previously published HPF protocols [22, 23]. The fish are subsequently cryoimmobilized with an EM HPM100 high pressure freezing machine at >20,000 K/s freezing speed and >2100 bar pressure and stored in liquid nitrogen until freeze substitution (*see* **Note 1**).

3.3 Freeze Substitution

The samples are processed for freeze substitution using an EM AFS2 system. The following basic freeze substitution protocol is modified from previously published protocols [4, 5, 24] and is summarized in Table 1. Metal washing containers with bottom washing rings and liquid permeable plastic baskets (Leica Microsystems) are loaded with a solution of 0.1% $KMnO_4$ in anhydrous acetone and cooled down to −90 °C. Using precooled forceps, the samples are transferred from liquid nitrogen into the plastic baskets (*see* **Note 2**). After an incubation of 16 h the freeze substitution solution is exchanged to efficiently remove residual water (*see* **Note 3**).

The samples are kept in the freeze substitution solution at −90 °C for a total of 80 h. The temperature is then gradually ramped up to −45 °C over the course of 11 h. At −45 °C the

Table 1
Freeze substitution protocol

Solution	Temperature, °C	Time span	Note
1. 0.1% KMnO$_4$ in anhydrous acetone	−90	80 h	Exchange after 16 h
2. 0.1% KMnO$_4$ in anhydrous acetone	−90 → −45	11 h	Linear temp. ramp
3. Anhydrous acetone	−45	3 h	Wash 4×
4. Two-thirds anhydrous acetone/one-third ethanol	−45	30 min	Wash 1×
5. One-third anhydrous acetone/two-thirds ethanol	−45	30 min	Wash 1×
6. Absolute ethanol	−45	1 h	Wash 2×
7. Absolute ethanol	−45 → 4	16 h	Linear temp. ramp
8. Absolute ethanol	4	1 h	Wash 2×

samples are washed four times within 3 h with anhydrous acetone. Acetone is then exchanged by ethanol, because acetone may inhibit polymerization of LR White resin. Anhydrous acetone is exchanged with one-third ethanol in acetone and incubated for 30 min, then 30 min with two-thirds ethanol in acetone, and finally two times (30 min each) with pure ethanol. Then a linear temperature rise from −45 to 4 °C over the course of 16 h follows. Afterward, the samples are washed two times with pure ethanol over the time course of 1 h. Then the samples are transferred into small glass vials with 50% LR White resin in ethanol. The samples need to be kept at 4 °C during the infiltration steps (*see* **Note 4**). By pipetting up and down with a glass pipette with a widened opening the samples are removed from the freezing platelets. If pipetting is not sufficient the tip of a needle can be used carefully to facilitate the removal. The glass vials are then covered with a lid and incubated for 16 h at 4 °C. The next infiltration step is performed with pure LR White resin with three washing intervals for 1, 4 and 16 h to allow a complete infiltration of the tissue with LR White resin.

3.4 Embedding

The samples are finally transferred into gelatin capsules containing pure LR White resin using a glass pipet. The sample should sink to the bottom of the gelatin capsule and can be oriented for ultramicrotomy. The gelatin capsule is filled with pure LR White up to the rim (*see* **Note 5**). The capsules are thermally cured in an upright position at 48–52 °C for at least 48 h. For an alternative UV polymerization of the samples at 4 °C that we applied to other specimens such as the roundworm *C. elegans see* [4, 5].

3.5 Ultramicrotomy

Prior to ultramicrotomy the gelatin layer has to be removed from the tip of the capsule using a razor blade. Starting with a coarse trimming with the razor blade, the embedded fish larva (Fig. 1c) is approached. Once the tissue of the fish is reached, the block is trimmed to achieve a trapezoid block face (Fig. 1d). The rhodamine fluorescence (Fig. 1e, f) can be used for targeted cutting [25, 26]. Two opposite sides of the block framing the fish need to be trimmed in a parallel manner to obtain a ribbon of consecutive sections (*see* **Note 6**). The region containing the Mauthner neuron is located roughly at the height of the otic vesicle. To check for the right position some sections with a thickness of 250–500 nm can be stained with methylene blue and analyzed with a standard wide-field light microscope.

Ribbons cut for AT are collected on glass slides. We use Poly-L-Lysine coated slides which offer a very reliable adhesion of the ribbons. For sectioning of ribbons we use the Histo Jumbo diamond knife from DiATOME, which provides a large boat where a glass slide can be submerged (*see* **Note 7**). For AT we use a section thickness of 100 nm. The length of the ribbon is limited by the dimensions of the glass slide, but it is possible to put several ribbons in parallel on one glass slide. A ribbon of consecutive sections can be detached from the knife's edge by using a mounted eyelash and guided toward the glass slide. The first section should touch the glass-water-interface. By reducing the water level, the intact ribbon sticks to the glass. The glass slide can then be carefully lifted from the boat and dried at room temperature (*see* **Note 8**). The attached ribbons on the glass slide can be stored for several days to weeks (*see* **Note 9**).

3.6 Light Microscopy Preparations

For the following labeling steps we use a modified version of the AT protocol published by Micheva and Smith [3, 27]. To keep all solutions within a restricted area, and to prevent drying out of the sections, the area surrounding the sections is framed by a hydrophobic pen. A simple humid and dark glass chamber is prepared and glass slides are placed inside (*see* **Note 10**). Our here presented protocol concentrates on the labeling of identified neurons by rhodamine staining. For many neurobiological questions it might be interesting to additionally label specific epitopes using antibodies. As an example we provide as reference the broadly applicable staining of GFP epitopes in **Note 11** using a well-established line labeling serotonergic neurons [28], but do not further discuss the immunofluorescence antibody staining in the main frame of the protocol as these steps have been described comprehensively before (*see* for example refs. 3–5, 27, 29). For correlation we use a nuclear counterstaining with Hoechst 33342 diluted 1:10,000 in Tris-buffer. Finally, one washing step with ddH_2O is carried out to remove salt residuals. The water is removed as far as possible. The sections are mounted with mounting medium, such as Mowiol (*see* **Note 12**) and covered with High Precision Microscope Cover Glasses. The slides can be stored at 4 °C in the dark. The light microscopic image acquisition (*see* **Note 13**) should be carried out as soon as possible as fluorescent signals tend to degrade quickly. This is particularly relevant if epitopes are labeled with immuno-fluorescence steps in addition. To find the sections under the light microscope the glue mixture with the black pigment powder helps to retrieve the transparent sections (*see* **Note 6**).

3.7 Contrasting and Carbon Coating

After the acquisition of the light microscopic images is complete, the sections are processed for SEM imaging. The cover slips are carefully removed from the glass slides and the Mowiol mounting medium is removed by rinsing it off with water. The sections may now dry out and can be stored for a few months. To reduce the size of the glass slides they are cracked around the sections using a diamond pen so that only the part containing the sections remains. Our protocol for contrasting includes an incubation in 2.5% uranyl acetate in ethanol for 15 min and then in 50% Reynolds's lead citrate [16] in water for 10 min (*see* **Note 14**). After contrasting the glass slide pieces are mounted to SEM specimen holders. In order to reduce charging under the electron beam, the glass slide pieces are surrounded with a contact adhesive, such as silver paint. Additional carbon coating is also essential for good SEM imaging results, as it reduces charging effects efficiently (*see* **Note 15**).

3.8 Scanning Electron Microscopy

For SEM image acquisition we use the field emission scanning electron microscope JSM-7500F (JEOL, Japan) with a LABE detector (for backscattered electron imaging at extremely low acceleration voltage). We achieve the best results with our machine using an acceleration voltage of 5 kV, a probe current of 0.3 nA and a working distance of 6–8 mm (*see* **Note 16**). For a discussion on some other SEM imaging possibilities *see* [5]. An excessive exposure to the electron beam should be avoided to protect the sections (*see* **Note 17**).

3.9 Image Processing

Before starting with the correlation, the light microscopic images have to be adjusted. Image acquisition with SIM results in z-stacks. However, the 100 nm thick sections would fit into a single layer each, z-stacks are required for proper image processing and to generate the super resolved images (*see* also [5]). For correlation, we produce a maximum intensity projection using ImageJ for all channels (*see* **Note 18**). If an area of interest is larger than the field of view of a single acquisition the images have to be stitched together. For this we use the Stitching tool of ImageJ/FIJI [17].

3.10 Alignment of Serial Electron Micrographs

When SEM image acquisition of the serial sections is complete the images or mosaics can be put together as a stack in a topological order. For image stack alignment we use eTomo which is included in the software package IMOD [18].

After starting eTomo, the function "Align Serial Sections" should be selected and the stack loaded into the software. Default settings should be used if nothing else is specified. By ticking the box "Search for" in the tab "Align" different options can be chosen (*see* **Note 19**). If all images were taken with the same magnification the option "Rotation/Translation" is sufficient. The option "Full linear transformation" will transform the images for a smooth alignment result, but it causes distortion of the images. The alignment is started with "Initial Auto Alignment." Select "Midas" to correct mistakes by shifting the images and save the corrections. By selecting "Revert Auto Alignment to Midas" recent transformations with Midas are taken into account. To finally create the aligned stack switch to the tab "Make Stack," choose the option "Global alignments (remove all trends)" and click on "Make Aligned Stack" (*see* **Note 20**). The aligned stack can be processed for segmentation with the 3dmod software, which is included in the IMOD software package as well.

3.11 Correlation with Icy Plugin eC-CLEM

The aligned stack obtained by the eTomo software is saved in ".mrc" file format and needs to be converted into ".tiff" format to extract the image sequence.

The corresponding light microscopy composite and electron micrograph are opened in the Icy software platform [30] with the eC-CLEM plugin [19, 20] (*see* Fig. 2). At first, we select a computation of the transformation by choosing the option "2D (X, Y, [T])". In the eC-CLEM plugin panel it is possible to determine which image should be transformed. As we aligned the serial SEM images, we choose the SEM image as not to be modified and use it as the target image. The light microscopy image will then be selected to be transformed and resized according to the SEM image. For an unbiased correlation we switch off the channels of interest, that is, in this case the rhodamine channel. The correlation performed by matching the intense heterochromatin staining of Hoechst 33342 with the corresponding electron dense heterochromatin patterns of nuclei in the SEM images. The brightness of all channels can be increased or decreased in the color display. The correlation is initiated by pressing the start button. The first landmark point is seeded on a distinct place on the SEM target image and adjusted on the source image. In 2D registration three initial homology points are necessary to compute the initial transformation (*see* **Note 21**). By seeding more homology points on the correspondent images the correlation can be further improved. While high magnification correlations work well with rigid transformations, correlation of lower magnification images gains precision from non-rigid transformations (*see* Figs. 2 and 3).

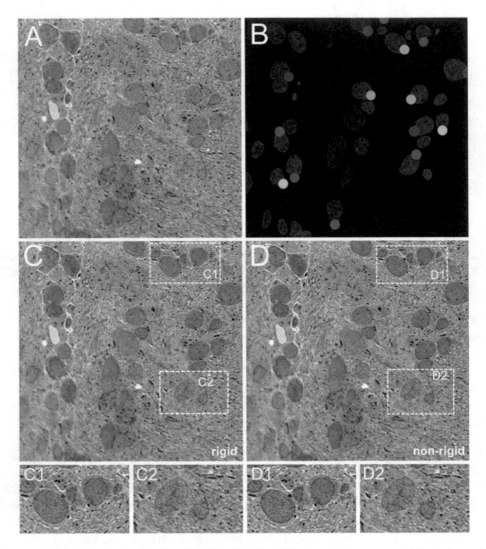

Fig. 2 Comparison of correlation quality with eC-CLEM [19, 20] using rigid and non-rigid transformations at low magnifications. (**A**) Setting of reference points (random colors) on SEM image. (**B**) Definition of reference points (random colors) on SIM image. Note that corresponding points on SEM and SIM images show same color. (**C**) Correlation with rigid transformation. Two higher magnified examples are shown in **C1** and **C2**. Corresponding regions in (**C**) are boxed. (**D**) Correlation with rigid transformation. Two higher magnified examples are shown in **D1** and **D2**. Corresponding regions in **D** are boxed. Note the more accurate correlation in **D, D1,** and **D2** versus **C, C1,** and **C2**, respectively

3.12 3D
Reconstruction

For 3D reconstruction of the Mauthner neuron, the light microscopic signal of rhodamine is used. We combine the FIJI [17] and the IMOD/3dmod Software packages (Version 4.8.37) [18] in the following processing workflow. The approximated time needed for each step including the computation time on a current standard desktop computer for ~100 array sections is estimated and is indicated in brackets:

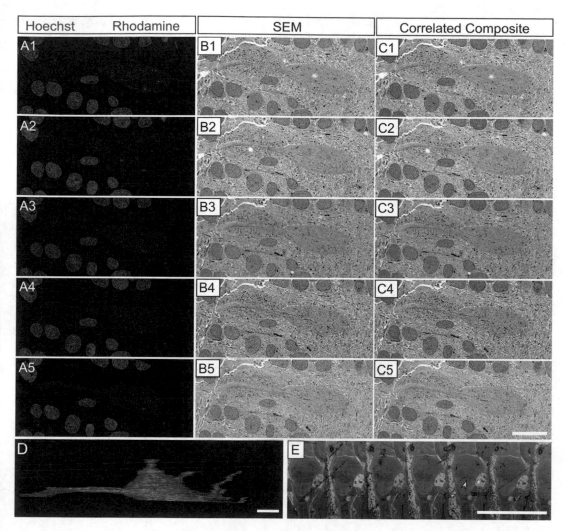

Fig. 3 Exemplary results of the array tomography approach to identify and image the Mauthner neuron. **(A1–A5)** Five 100 nm array sections imaged at the SIM microscope. Blue signal represents nuclear stain (Hoechst 33342) and red signal the endogenous rhodamine signal resulting from retrograde labeling. **(B1–B5)** Same sections as in **A1–A5**, respectively, imaged with SEM after contrasting and carbon coating. **(C1–C5)** Correlated composite images resulting from overlay of **A1–A5** and **B1–B5**, respectively. **(D)** IMOD 3D model of part of the Mauthner neuron generated after thresholding and curation from the rhodamine signal of a stack of 96 array sections. The model shows part of the soma region on the right with dendritic arms projecting outward. The long broad arm to the left is the axon projecting toward the midline. **(E)** Series of consecutive array sections imaged at very low magnification at the SEM for orientation. On the fifth section from the left the region of the Mauthner somata is indicated by an arrowhead and the otic vesicle by an asterisk. Scale bar for **A1–C5** in **C5** = 10 μm, in **D** = 5 μm and in **E** = 1 mm

1. The red rhodamine signal is separated from the combined and aligned image stack in Fiji with the "Split Channels" function (~5 min).

2. The whole stack is pre-filtered with the Gaussian Blur Filter (Sigma 3–5) (~2 min).

3. On each image, depending on the signal quality, a separate threshold with dark background is generated with the "Threshold" function in Fiji (~45 min).

4. This new stack is used to generate an auto contour model with imod and the "imodauto" function in the command window (Windows) or the terminal (macOS and Linux) (~15 min).

   imodauto –h 1 stack.tif output_model.mod

5. The output_model.mod is meshed and the surfaces are split into different objects (in command window or terminal).

   Meshing: imodmesh output_model.mod (~10 min).

   Surface Splitting: imodsortsurf –s output_model.mod surface_model.mod (~5 min).

6. The model is opened in 3dmod as model file. A map of the different generated objects is opened. Select the Mauthner neuron object using the "Sculpt" tool of the "Drawing Tools" and note the object number. The Mauthner neuron is separated from the other rhodamine labeled neurons by generating a new model with the "imodextract" command. The Mauthner neuron can be identified in 3dmod by its object number. Again, this function has to be entered in the command window or terminal. (~5 min).

   imodextract "Mauthner Neuron object number" surface_model.mod Mauthner.mod.

7. The model of the Mauthner neuron (*see* Fig. 3) is now carefully reviewed on the original SEM stack with the correlated rhodamine signal and contouring mistakes are corrected by hand with the Drawing tools in 3dmod which are part of the IMOD software package. (~30 min).

3.13 Conclusion

This principle approach should be broadly applicable to trace Mauthner neurons as well as other reticulospinal neurons present in embryos of other fish species and tadpoles. Furthermore, other types of neurons or neurites labeled via retrograde or anterograde tracing can be analyzed this way. Moreover, this approach should be well suited for experimental situations where neurons, or other cells of interest, are labeled with rhodamine derivates by other means, such as electroporation (*see* **Note 22**) or microinjection. This is particularly interesting for an experimental situation, where a neuron or neuronal connection is characterized by electrophysiology first. The very same cell can then be labeled with rhodamine derivates and its ultrastructural context can be analyzed. Furthermore, due to the flexibility of the underlying array tomography approach, several epitopes of interest can be localized in addition using immunolabeling. In addition, it is possible to stain for RNAs in the principally same framework as described here, by switching to fluorescent RNA in situ hybridization steps on the LR White sections [6]. With the here presented protocol to label identified cells in the full ultrastructural context in a near-native state we add another level of information to this very versatile group of approaches subsumed under the name "array tomography."

4 Notes

1. It is recommended to make the sample as thin as possible without damaging the tissue to prevent ice crystal formation. A freezing chamber depth of 250 μm fits very well for 4 dpf zebrafish larvae.

2. To prevent a mixup of samples, each substitution metal and plastic container has its own individual number, symbol (metal containers) or number of notches (plastic container) carved into its frame. The samples must not deviate from −90 °C (± max. 2 °C) to prevent ice crystal formation. Therefore, watch the temperature in the chamber carefully, especially when opening the chamber and during manipulations in the AFS2, such as sample transfer and solution exchanges. The metal containers are covered, for example with simple flat round Teflon disks completely covering the chamber, to prevent evaporation of the solution.

3. To reduce unwanted temperature gradients, all solutions need to be carefully precooled before putting them into the AFS2. Furthermore, we equilibrate the fresh solutions in an additional metal container inside the AFS2 before applying the solution to the samples.

4. We prefer to use glass pipettes because LR White might negatively interact with some kinds of plastic before polymerization. To make a glass pipette with a wide opening, we shorten conventional Pasteur pipets to provide a sufficiently wide opening to pipet the sample without breaking it. Before use, the edges of the breaking point are smoothened by melting them with a Bunsen burner to protect the sample from sharp edges resulting from broken glass.

5. A small paper strip with a label is added to the capsule for simple sample identification and resin block archiving. Since oxygen inhibits polymerization of LR White resin try to minimize air bubbles in the gelatin capsules and close them tightly. A little air bubble may remain at the tip of the capsule distal to the sample. This will not prohibit polymerization.

6. With LR White being less hydrophobic than epoxy resins like Epon, the consecutive sections do not usually form stable ribbons. To stabilize the ribbon, we use a glue mixture

consisting of ordinary contact adhesive glue (Pattex Gel Compact), which is diluted with xylene in a ratio of roughly 1:1. The glue mixture can be mixed in an Eppendorf tube using a toothpick. We add Spinel Black 47400 (deepest black) pigment powder to the glue mixture to aid localizing the ribbon during light microscopy [5]. The glue mixture is applied to one edge of the block face using a very thin needle. It is important to add the glue mixture only to the cutting edge forming an attachment zone for the consecutive section. If some glue drops onto the block face it will be removed with the first section.

7. Prior to ultrathin sectioning, the glass slide should already be submerged in the boat and covered with water.

8. Because the sections are almost transparent it is advisable to encircle the region of the arrays on the bottom side of the objective slide with a water permanent pen.

9. If immunolabeling is included we recommend to proceed as soon as possible as the sections begin to show reduced quality of the staining for certain epitopes after a few days. To protect the embedded rhodamine signal, keep sections in the dark as much as possible.

10. A few layers of water soaked tissue paper are placed in a large glass petri-dish, which is covered with a nontransparent lid, sealing the humidity chamber. Soaking of the slides with excess water from the bottom of the humidity chamber is prohibited by putting them on ridges such as a pair of reused 50 ml reaction tube caps.

11. The anti-GFP-staining procedure starts by rehydrating and blocking the sections by applying a blocking solution consisting of 0.1% BSA and 0.05% Tween-20 in 50 mM Tris buffer, pH 7.6) for 5 min. The primary antibody, in the current protocol against GFP (polyclonal chicken anti-GFP, Abcam; ab13970), is diluted 1:500 in the blocking solution and centrifuged at maximum speed in a table top centrifuge ($13,000–16,000 \times g$) for 2 min to pellet debris and conglomerates. 200 μl of antibody solution is sufficient to cover an array region of 2 cm × 1 cm. Only the supernatant is used. The sections are incubated in the primary antibody for 1 h at room temperature. Afterward, the sections are washed five times in 5 min intervals with Tris-buffer. To prevent drying of the sections during the exchange of solutions, we use a flow-through method. Two pipets are used, one to remove a solution from the sections and one to simultaneously add new solution. The secondary antibody, here goat anti-chicken IgG (H + L) Alexa Fluor 488 conjugate (Thermo Fisher Scientific) diluted 1:1000 in blocking solution, is also centrifuged, and the supernatant is applied to the sections followed by

incubation for 30 min at RT in darkness. After incubation with the secondary antibody, the sections are washed as before with Tris buffer. For a comprehensive description of immunolabeling, SIM, and dSTORM applications see [4, 5].

12. Alternatively, it is possible to use glycerol or other water soluble mounting media suited for immunofluorescence.

13. In principle any type of fluorescent light microscope can be used to image arrays. We prefer a SIM microscope, as it does not require any additional sample preparation steps for super-resolution imaging and is rather quick for multi-channel acquisitions. Furthermore, the SIM technique [31] allows for imaging reaching about 120 nm in lateral resolution and very reliable channel alignments. Therefore, it allows very precise correlation of the fluorescent and SEM signals. For a comparison of SIM versus dSTORM applications for array tomography see [4].

14. The lead citrate solution is mixed with preboiled ddH_2O (boil for at least 10 min and cool down), as the lead citrate reacts with dissolved CO_2 in carbonated water leading to an electron dense precipitate on the sections. Each contrasting solution is centrifuged before use at $13,000$–$16,000 \times g$ for 5 min to pellet debris and precipitates. For contrasting we place the resized slides on a piece of Parafilm and cover them with a glass lid. To remove CO_2 while incubating with lead citrate, some NaOH pellets should be placed around the slides. Additionally, you should avoid exposing the lead citrate to your breath as it contains CO_2.

15. An even carbon layer of approx. 10 nm is sufficient for our setup.

16. We recommend to first take an overview of the desired position of the sections at about 700–1000× magnification. Then a higher magnification image acquisition can be carried out. If a larger area is to be recorded with tiling an overlap of at least 20% is advisable. For a proper alignment of the sections, we recommend to image the same region of the consecutive sections.

17. As we experienced an increasing damaging and blistering of the brain tissue during SEM, we first scan the region of interest at low magnification (<1000×) with a decreased emission current of 2 µA for at least 1 min. We then adjust the emission current to 10 µA and conduct the image acquisition. We repeat this preparation before image acquisition at any desired magnification.

18. As an alternative to maximum projection it is possible to choose the slice in the SIM z-stack with the highest signal intensity. All channels have to be flipped horizontally if an inverted fluorescence microscope is used.

19. If all images were taken with the same magnification the option "Rotation/Translation" is sufficient.

20. If the aligned stack is not satisfactory it is possible go back to "Midas" and try to improve the alignment manually.

21. By locking images together (select the lock in the upper left corner in the two images), Icy allows to synchronize the navigation while zooming in and out of the images.

22. An elegant possibility to fill single neurons is electroporation of fluorescent dyes as described in [32, 33]. We use an upright fixed stage microscope with motorized z drive (Zeiss) on a *xy*-translation table (Linos) and a vibration isolation table (Newport). Zebrafish larvae are immobilized by tricaine and mounted with 0.7% low-melting agarose (*see* Subheading 3.1) in a custom made plexiglas bath chamber with cover slip bottom. For electroporation bath chambers are filled with Danieau's solution and put on the stage of the microscope. Microelectrodes with a resistance of 7–11 MΩ were made from 2 mm borosilicate glass capillaries with filament (GBF 200F 10, Science Products) using a DMZ Universal Puller (Zeitz instruments) at least 1 day before electroporation to reduce adhesion between electrode tip and cell surfaces. Electrodes are mounted on a micromanipulator (Scientifica). Serotonergic cells, on which we tested the technique, were identified by endogenous GFP expression [28]. The skin of the larvae is penetrated with an electrode filled with extracellular saline. After opening the skin, the electrode is replaced with an electrode filled with biotinylated and TMR conjugated Dextran (micro-ruby, Thermo Fisher, 10 mg/ml). Gentle pressure is applied to avoid clogging of the electrode during the approach to the target cell. Three second trains of rectangular voltage pulses (width 1ms, frequency 100 Hz) with an amplitude of up to several Volts are applied using a Axoporator 800 A (molecular devices). Polarization of the rectangular pulses is chosen to drive the fluorescent dye out of the electrode due to electrostatic forces. After applying the pulse train the electrode is retracted and the electroporation result is controlled with epifluorescence.

Acknowledgments

We want to thank Manfred Schartl for generously providing us access to the fish facility. Furthermore, we thank Georg Krohne, Markus Engstler, Sebastian Markert, Markus Sauer, Marcus Behringer, Jean-Louis Bessereau, Camilla Luccardini, and Hong Zhan for many supportive discussions throughout the different stages of the project. We thank Claudia Gehrig, Daniela Bunsen, and Brigitte

Trost for excellent technical support. We would like to thank Swarnima Joshi and Felix Erwin for their contribution during early stages of this project. This project was supported by the Universitätsbund Würzburg (AZ14-48). C.L. was funded by the Bayerische Gleichstellungsförderung.

References

1. Albrecht U, Seulberger H, Schwarz H et al (1990) Correlation of blood-brain barrier function and HT7 protein distribution in chick brain circumventricular organs. Brain Res 535:49–61

2. Schwarz H, Humbel B (2014) Correlative light and electron microscopy using immunolabeled sections. In: Kuo J (ed) Electron microscopy. Humana, New York, NY, pp 559–592

3. Micheva KD, Smith SJ (2007) Array tomography: a new tool for imaging the molecular architecture and ultrastructure of neural circuits. Neuron 55:25–36

4. Markert SM, Britz S, Proppert S et al (2016) Filling the gap: adding super-resolution to array tomography for correlated ultrastructural and molecular identification of electrical synapses at the C. elegans connectome. Neurophotonics 3:041802–041802

5. Markert SM, Bauer V, Muenz TS et al (2017) 3D subcellular localization with superresolution array tomography on ultrathin sections of various species. In: Verkade TM-R (ed) Methods in cell biology. Academic, New York, NY, pp 21–47

6. Jahn MT, Markert SM, Ryu T et al (2016) Shedding light on cell compartmentation in the candidate phylum Poribacteria by high resolution visualisation and transcriptional profiling. Sci Rep 6:35860

7. Korn H, Faber DS (2005) The Mauthner cell half a century later: a neurobiological model for decision-making? Neuron 47:13–28

8. Eaton RC, Lee RKK, Foreman MB (2001) The Mauthner cell and other identified neurons of the brainstem escape network of fish. Prog Neurobiol 63:467–485

9. Medan V, Preuss T (2014) The Mauthner-cell circuit of fish as a model system for startle plasticity. J Physiol Paris 108:129–140

10. Pfaff DW, Martin EM, Faber D (2012) Origins of arousal: roles for medullary reticular neurons. Trends Neurosci 35:468–476

11. Metcalfe WK, Mendelson B, Kimmel CB (1986) Segmental homologies among reticulospinal neurons in the hindbrain of the zebrafish larva. J Comp Neurol 251:147–159

12. Lee RKK, Eaton RC (1991) Identifiable reticulospinal neurons of the adult zebrafish, Brachydanio rerio. J Comp Neurol 304:34–52

13. Nakajima Y (1974) Fine structure of the synaptic endings on the Mauthner cell of the goldfish. J Comp Neurol 156:375–402

14. Kimmel CB, Sessions SK, Kimmel RJ (1981) Morphogenesis and synaptogenesis of the zebrafish Mauthner neuron. J Comp Neurol 198:101–120

15. Kimmel CB, Ballard WW, Kimmel SR et al (1995) Stages of embryonic development of the zebrafish. Dev Dyn 203:253–310

16. Reynolds ES (1963) The use of lead citrate at high pH as an electron-opaque stain in electron microscopy. J Cell Biol 17:208–212

17. Schindelin J, Arganda-Carreras I, Frise E et al (2012) Fiji: an open-source platform for biological-image analysis. Nat Methods 9:676–682

18. Kremer JR, Mastronarde DN, McIntosh JR (1996) Computer visualization of three-dimensional image data using IMOD. J Struct Biol 116:71–76

19. Paul-Gilloteaux P, Heiligenstein X, Belle M et al (2017) eC-CLEM: flexible multidimensional registration software for correlative microscopies. Nat Methods 14:102–103

20. Heiligenstein X, Paul-Gilloteaux P, Raposo G et al (2017) eC-CLEM: a multidimension, multimodel software to correlate intermodal images with a focus on light and electron microscopy. Methods Cell Biol 140:335–352

21. McLean DL, Fetcho JR (2004) Relationship of tyrosine hydroxylase and serotonin immunoreactivity to sensorimotor circuitry in larval zebrafish. J Comp Neurol 480:57–71

22. Nixon SJ, Webb RI, Floetenmeyer M et al (2009) A single method for cryofixation and correlative light, electron microscopy and tomography of zebrafish embryos. Traffic 10:131–136

23. Schieber NL, Nixon SJ, Webb RI et al (2010) Modern approaches for ultrastructural analysis of the zebrafish embryo. In: Electron microscopy of model systems. Academic, New York, NY, pp 425–442

24. Weimer RM (2006) Preservation of *C. elegans* tissue via high-pressure freezing and freeze-substitution for ultrastructural analysis and immunocytochemistry. Methods Mol Biol 351:203–221

25. Kolotuev I, Schwab Y, Labouesse M (2010) A precise and rapid mapping protocol for correlative light and electron microscopy of small invertebrate organisms. Biol Cell 102:121–132

26. Kolotuev I, Bumbarger DJ, Labouesse M et al (2012) Targeted ultramicrotomy: a valuable tool for correlated light and electron microscopy of small model organisms. Methods Cell Biol 111:203–222

27. Micheva KD, O'Rourke N, Busse B et al (2010) Array tomography: immunostaining and antibody elution. Cold Spring Harb Protoc 2010:pdb.prot5525

28. Lillesaar C, Stigloher C, Tannhäuser B et al (2009) Axonal projections originating from raphe serotonergic neurons in the developing and adult zebrafish, *Danio rerio*, using transgenics to visualize raphe-specific pet1 expression. J Comp Neurol 512:158–182

29. Micheva KD, Busse B, Weiler NC et al (2010) Single-synapse analysis of a diverse synapse population: proteomic imaging methods and markers. Neuron 68:639–653

30. de Chaumont F, Dallongeville S, Chenouard N et al (2012) Icy: an open bioimage informatics platform for extended reproducible research. Nat Methods 9:690–696

31. Gustafsson MGL (2000) Surpassing the lateral resolution limit by a factor of two using structured illumination microscopy. J Microsc 198:82–87

32. Haas K, Sin W-C, Javaherian A et al (2001) Single-cell electroporationfor gene transfer in vivo. Neuron 29:583–591

33. Ruthazer ES, Schohl A, Schwartz N et al (2013) Labeling individual neurons in the brains of live xenopus tadpoles by electroporation of dyes or DNA. Cold Spring Harbor Protoc 2013:pdb.prot077149

Chapter 5

A Low-Tech Approach to Serial Section Arrays

Waldemar Spomer, Andreas Hofmann, Lisa Veith, and Ulrich Gengenbach

Abstract

Three-dimensional reconstructions based on ultrathin serial sections have long been used in traditional transmission electron microscopy (TEM). In modern field emission scanning electron microscopes (SEM), it is also possible to image such sections with a resolution comparable to that obtained in a standard TEM, using secondary or backscattered electrons at relatively low landing energies. A far greater number of sections can be observed without changing the support as sections are not placed on a typical TEM grid but on much larger substrates, such as a piece of silicon wafer or a conductively coated glass coverslip. In this chapter, we describe a workflow for reliably creating sections and how to place them on the substrate as an array of long ribbons. We discuss sample block trimming to obtain straight ribbons of sections, how to prepare and handle the substrate, and how to approach and align the knife to the block face. Regarding substrate handling in the knife boat, we introduce a combination of micromanipulators based on a "supporting hand" concept. These also help with smooth retrieval of the section array from the water onto the substrate without damaging the order of the sections.

Key words Array tomography (AT), Serial sections, Substrate holder, Correlative AT, CLEM

1 Introduction

Analyzing biological samples using three-dimensional (3D) reconstructions has been an objective in electron microscopy for more than half a century. Initially, scientists produced very basic 3D models by slicing biological samples into few (10–25) sections, transferring them onto copper grids and analyzing them in the transmission electron microscope (TEM) [1]. They produced hand-drawn 3D sketches of individual cellular organelles or of larger subcellular domains such as parts of neuromuscular junctions based on the cross-section images they acquired with the TEM [2]. The number of obtainable sections increased continuously, and up to 2000 were mentioned in a 1972 Nature paper describing computer-assisted 3D reconstruction of the *Daphnia magna* neuropil [3]. As microscopy has evolved and computers have become more powerful and less expensive, an increasing interest in analyzing bigger sample volumes at ultrastructural resolution ("volume

Irene Wacker et al. (eds.), *Volume Microscopy: Multiscale Imaging with Photons, Electrons, and Ions*, Neuromethods, vol. 155, https://doi.org/10.1007/978-1-0716-0691-9_5, © Springer Science+Business Media, LLC, part of Springer Nature 2020

electron microscopy" [4]) is now prevalent. For TEM-based approaches, ribbons of serial sections have to be carefully maneuvered onto tiny (3 mm diameter) slot grids [5], covered with relatively thin films that may rupture easily. Depending on the size of the block face, or in other words the target region, only a few sections fit onto one grid, thus requiring large numbers of delicate grids to be handled when larger volumes are to be reconstructed. For imaging in a scanning electron microscope (SEM) with its large chamber, long ribbons of sections may be placed directly on silicon wafers or other substrates. Since ideally these sections are arranged on the substrate as a neat array, Micheva and Smith [6] coined the term array tomography (AT) for this method (cf. also [7] for a general review of volume EM methods). Silicon wafers are much larger than copper grids, perfectly flat and conductive, making them an ideal substrate for SEM imaging. With the introduction of correlative light and electron microscopy (CLEM) [8] and correlative array tomography (CAT) [9] new substrates became necessary. These need to be both conductive for electron microscopy (EM) and transparent for light microscopy (LM). Glass coverslips with a conductive but transparent layer of carbon or indium tin oxide (ITO) therefore became the quasi-standard substrates for SEM-based correlative imaging methods.

As individual section handling can be somewhat cumbersome (except when using an automated device such as the ATUMtome, cf. Chapter 7), a modified cutting process is proposed here. This generates long ribbons, which are significantly easier to handle than individual sections and already provide a well-defined order of the sections on the substrate. This is a definitive advantage over a recently proposed method—"MagC"—where sections are collected from the water surface in an unordered manner, requiring their order to be reestablished subsequently using additional microscopic and computational postprocessing [10]. This chapter will deal with the preparation steps leading to well-ordered ribbons, their handling with a custom-designed device and transfer onto a substrate. The basic steps for creating arrays of ribbons on a substrate are as follows:

- Embedding the sample in a matrix suitable for ultramicrotomy, for example, in epoxy resin following protocols as used for traditional TEM analysis.
- Trimming the sample block.
- Coating the sample block with an adhesive to ensure the sections adhere to each other and form a ribbon.
- Cutting the sample into ultrathin sections.
- Transferring the ribbons onto the substrate.

With the exception of embedding (which will be discussed in Chapters 1, 4, 6, 7, 9, 11 for different types of samples), every step will be examined as a separate topic. Likewise, the automated collection of images in the SEM from arrays will be described in Chapters 6 and 7.

2 Materials

To obtain high quality arrays of ribbons on an appropriate substrate, a skilled operator and various tools are required, as listed in Table 1. The most important tool is the ultramicrotome (Fig. 1a), basically consisting of a knife holder, a sample holder (Fig. 1b), and a microscope with appropriate illumination. The sample holder can be mechanically moved towards the knife in nanometer steps.

Table 1
Tools and materials for creating section arrays for AT

Tool/material	Example
Ultramicrotome	PowerTome Series by RMC Boeckeler Inc., USA UltraCut Series by Leica Microsystems, Austria
Trimming knife	Diamond trimming knife, (DiATOME, Switzerland) or glass knife
Diamond knife	Ultra with normal boat or Jumbo, (DiATOME, Switzerland)
Manipulator	Eyelash Manipulator (Science Services, Germany) or self-made tool (eyelash or very fine hair from a cat's fur glued to a toothpick)
Brush	Self-made brush to apply glue (few hairs from a watercolor brush glued to a toothpick)
Silicon wafer	e.g., Si-Mat, http://si-mat.com/silicon-wafers.html Doping: P/Bor, orientation: <100>, thickness: 525 ± 25 μm, resistivity: 1–30 Ω cm
ITO coverslips	e.g., CorrSlide™ (very thin coating plus fiducials), Optics Balzers (Liechtenstein), or custom-made coverslips from Diamond Coatings (UK), coating of different thicknesses, no fiducials, cf. **Note 1**
Substrate holder	detailed drawings can be found in supplementary material of Ref. 16
Distilled water	
Adhesive for sample block coating	Mixture of contact cement and diluting agent, e.g., Pattex classic + Roti-Histol (or xylene)
Adhesive for attaching substrate	Fixogum (Marabu, Germany)
Optional	
Antistatic device	Static Line II (DiATOME, Switzerland)
Wafer cutter	RV 125 (AVT, Germany)

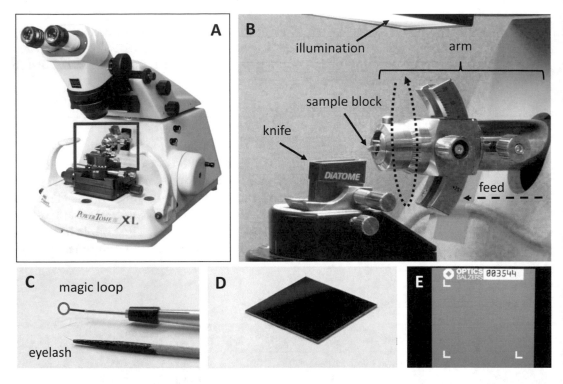

Fig. 1 Ultramicrotome and tools. Commercially available ultramicrotomes (**a**) consist chiefly of a moving arm holding the sample block in the block holder and a knife sitting in the knife holder (**b**). To cut a section, the arm is moved up and down on an elliptical path. On every downward movement, a slice is removed from the sample block. Moving the sample towards the knife—the feed—is also performed by the arm. Useful tools are an eyelash attached to a toothpick for directing sections on the water surface and a metal loop for picking up a few sections for quick examination (**c**). Typical conductive substrates for looking at sections in an SEM are silicon wafers (**d**) or ITO-coated glass coverslips (**e**) when prior observation in a light microscope is desired

Cutting ultrathin sections requires a sharp knife such as a glass or diamond knife, both equipped with a boat filled with distilled water (see Fig. 6a). The water ensures that the sections do not stick to the knife but swim on the water surface straight after cutting.

To handle the sections that swim on the water surface in the knife boat an eyelash attached to a toothpick (Fig. 1c) can be used. This tool enables the operator to release the sections from the knife edge and to move them around on the water surface. The magic loop (Fig. 1c) is useful if only a few sections have to be retrieved from the knife boat, for example for a quick check to identify the target region. Appropriate substrates for AT are silicon wafers (Fig. 1d), coverslips coated with a conductive ITO layer (Fig. 1e, see also **Note 1**) or glass slides covered with a thin carbon layer, produced, for example by sputter coating. Conductive—and at the same time transparent—glass is necessary for correlative array tomography (CAT) where both LM and SEM imaging of the same sample are intended.

3   Methods

3.1   Sample Block Preparation

3.1.1   Trimming

In AT based on section ribbons, sample block trimming serves two purposes:

1. Exposing the sample on the front face (block face, Fig. 2b bottom) of the sample block. After embedding, the sample is usually not perfectly oriented in the resin block nor exposed at

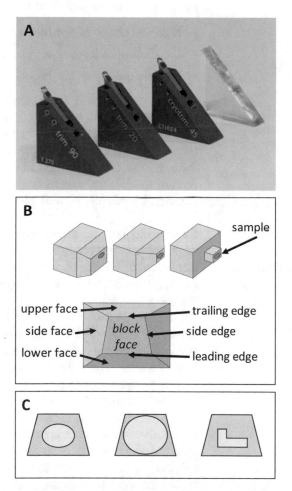

Fig. 2 Trimming knives and block shapes. (**a**) Glass knife (right) and diamond trim knives for different applications: 45° trim knife (second from right) for a small number of sections producing a pyramid with 45° angled sides (cf. middle block in **b**), 20° trim knife (middle diamond knife) for larger numbers of sections, producing a pyramid with steeper sides and 90° knife (left) for hundreds of sections, producing a cuboid block (cf. right block in **b**). (**b**) Different trimming geometries: Untrimmed sample block (left), pyramid (middle) and cuboid (right). (**c**) Block face of embedded symmetrical sample with sufficient embedding material surrounding it (left), block trimmed too close to the sample (middle), asymmetric sample in an orientation that may cause curved ribbons (right)

the surface. Before creating sections, it is therefore necessary to trim the resin block in order to remove excess material.

2. Preparing the side faces of the sample block in the right manner is crucial for successful ribbon formation: The upper and lower faces (leading and trailing edge when inserted into the ultramicrotome's sample holder) must be trimmed exactly parallel to each other.

The sample block can be trimmed manually using a razor blade or the ultramicrotome. The most important parameters here are:

- Size and shape of the block face.
- Parallelism of leading and trailing edge.
- Surface quality of the faces.

Manual trimming with a razor blade is suitable for coarsely removing the bulk material. The resulting surface quality is poor and it is not possible to achieve parallel edges. After coarse trimming with a razor blade, metal flakes could stick to the trimmed faces of the block. To avoid damaging the diamond knife, it is recommended to retrim the sides using a glass or diamond trimming knife (check also **Note 2**).

Selecting the Knife for Trimming with the Ultramicrotome

Depending on the amount of excess embedding material to be removed and the preferred surface quality, different types of knives can be used. Precise trimming is possible with glass knives (Fig. 2a right), because the knife is fixed in the knife holder and the ultramicrotome ensures a precise trimming movement. The same applies for diamond trimming knives (Fig. 2a left), but here the surface quality is better than when trimming with a glass knife. A smooth surface quality increases the tendency of the sections to form ribbons when they are cut. Glass knives can produce high-quality surfaces but only as long as the edge is new and sharp. This quickly becomes blunt compared with a diamond trimming knife, which lasts for much longer if treated properly. On the other hand, glass knives are much less expensive and can be self-made. Diamond trim knives are available in different shapes: Old models come with a flat edge, and basically look like a glass knife (cf. Fig. 2a left and right). In order to trim the sides of the block the knife holder has to be rotated. Newer versions have 20° or 45° (Fig. 2a, center two) angled sides, producing 20° inclined or 45° inclined pyramids (Fig. 2b top, block in center), respectively if the knife is simply moved from the left to the right side of the sample block to trim opposite side faces (left/right and top/bottom). To keep the area of the block face constant over a large number of sections (hundreds) it may be advisable to trim a "cuboid" (Fig. 2b top, right block), instead of a pyramid—a 90° trim knife (Fig. 2a left) is available from Diatome on request for this purpose.

Target Size and Shape of the Sample Block

When trimming the sample block, various aspects should be considered, depending on sample type and embedding. The embedding material may support the sample during the cutting process, especially if the sample is not infiltrated (well) by the embedding resin. In such cases, it helps to leave a certain amount of embedding material all around the sample (Fig. 2c left). For homogeneous material that is well infiltrated, trimming as close to the sample as possible is more suitable (Fig. 2c middle). As compression is always an issue, even with 35° cutting knives, major differences in the mechanical properties of the sample itself, between embedding material and sample, as well as an asymmetrical sample (Fig. 2c right), increase the risk of uneven section compression. This in turn will lead to curved ribbons instead of straight ones.

As regards sample block geometry, the block face has a 2D shape, while the sample block has a 3D shape. The most common shapes for blocks are the pyramid (Fig. 2b top, left and middle) and the cuboid (Fig. 2b top, right). Sections that are cut from a cuboid do not increase in size when large numbers of sections are cut (see above). With respect to the shape of the block face it helps to make the lower edge (slightly) longer than the upper edge (Fig. 2c). This ensures that the following section is able to release the previous section completely from the knife edge during cutting. If the upper edge of the block face is larger than the lower edge, a part of the previous section might stick partially to the knife edge while the next section is shifting it forward. This may result in a cluster of sections instead of a proper ribbon, potentially leading to loss of section order and/or section damage.

Trimming Parallel Edges for Straight Ribbons

In AT, where a large number of sections are to be collected, ribbons should be as straight as possible. Curved ribbons have several disadvantages compared with straight ones:

- More (and in the case of ITO-coated coverslips, also expensive) substrates are necessary due to poor usage of substrate space.

- More time-consum ption because more substrates need to be changed.

- Higher probability of ribbon separation, because curved ribbons are exposed to a momentum when manipulated in the knife boat (Fig. 6e).

Three main factors have an impact on the straightness of the ribbons:

1. Compression of the sample and the embedding material (depending on their mechanical properties): This factor is not easy to control. The precise mechanical properties (such as hardness) of the sample and the embedding material would need to be known, which is rarely the case. If hardness of

sample and embedding matrix are not properly matched, sectioning of an asymmetric sample (cf. Fig. 2c right—L-shape) would lead to asymmetric compression and thus to curved ribbons. If possible, the sample should be retrimmed to give a more symmetrical distribution of sample vs embedding matrix.

2. Parallel leading and trailing edges of the block face are of utmost importance. In order to achieve this, the following rules must be complied with:
   - A sample block mounted in the sample holder must not be removed during the entire trimming process.
   - The angle of the arc segment holder must not be changed during the whole process.
   - Check parallelism of leading and trailing edge several times during the trimming process by removing the sample block holder from the arm and viewing it from above (i.e., inserted into the trimming holder placed in the knife stage).
   - After trimming the block sides it is advisable to cut the block face to ensure that it is orthogonal, but see also **Note 3**.

3. Even distribution of the applied adhesive: Irregularly curved ribbons are usually due to nonuniformly and/or too thickly applied adhesive (see also Subheading 3.1.2 below and Fig. 3).

3.1.2 Coating the Sample Block with Adhesive

Sections from sample blocks trimmed right before cutting tend to adhere to each other well when cutting a series. They form ribbons that are stable enough for a certain number of sections and thus for a certain length of ribbons. The ribbon length achievable based on adhesion between sections alone cannot be reliably predicted in advance. It depends on a number of parameters such as the embedding material, trimming quality, and cutting parameters. To additionally increase the adhesion between sections it is recommended to homogeneously apply a small amount of adhesive on the upper and/or lower side of the block (Fig. 3).

Selecting of Adhesive and Application Tool

The adhesive consists of a commercial glue and a thinner. An established commercial glue is contact cement (such as Pattex classic) diluted with xylene or a limonene-based thinner (such as Roti-Histol). The literature is not consistent regarding the mixing ratio, varying from 3:1 to 1:500. We use a glue–thinner mixing ratio of 1:1–1:3. After mixing the components thoroughly, a very thin film should be carefully applied to the upper and/or lower face of the trimmed block (Fig. 3b). It is important not to coat the block face, as this may interact chemically with the embedded sample (Fig. 3d, right panel). The literature is also inconsistent with regard to the point of application of the adhesive. We recommend a first trial applying it only to the lower side of the block. If the adhesive is applied to the upper face, we occasionally observed a bead of

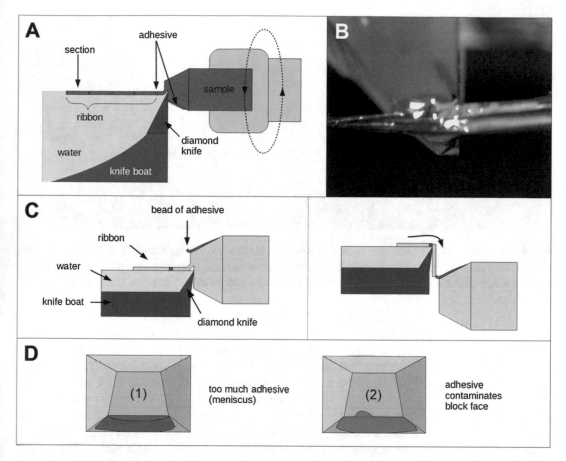

Fig. 3 Applying adhesive for producing section ribbons. (**a**) Schematic side view of sections glued together (glue shown in green) to improve ribbon stability when swimming in the knife boat. (**b**) Application of adhesive on the sample block with a thin brush. (**c**) Bead of adhesive accumulating on top of the sample block that may occur after a few hundred sections and may pull cut sections over the knife edge and damage them (right). (**d**) Most common issues when coating the sample block with adhesive: Too much adhesive (1) causes a meniscus which may result in curved ribbons. Careless coating of adhesive (2) may contaminate the sample

adhesive accumulating at this point when cutting large numbers of sections. This bead may pull the section back over the knife edge directly after cutting (Fig. 3c right panel). After application the adhesive should dry for 5–10 min.

3.2 Substrate Preparation

While the adhesive is drying on the sample block, the substrate can be prepared. This should be handled with flat tweezers (special tweezers exist for wafers) and the operator should wear (nitrile) gloves to avoid contamination such as fingerprints when handling the substrate. The first step is to wipe the substrate with a tissue soaked in ethanol or isopropanol—no streaks or particles should be visible after this. To ensure proper cleanliness, it may be necessary to work in a cleanroom-like environment.

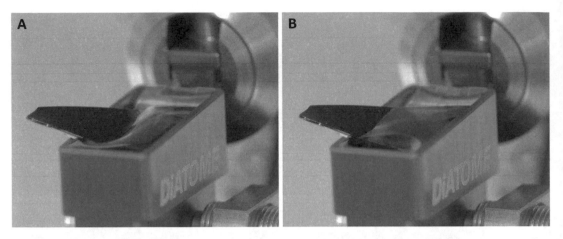

Fig. 4 Plasma treatment of substrate. Untreated silicon wafer inserted into knife boat (**a**) with water forming a steep meniscus. Treatment of the wafer with a plasma cleaner using air as process gas produces a hydrophilic surface which allows a low contact angle between substrate and water and thus improves wetting (**b**)

The second step is to render the surface hydrophilic by treating the substrate (silicon, glass) with a plasma cleaner using air (nitrogen/oxygen plasma). This decreases the contact angle, which enables wetting of the substrate (Fig. 4) and eventually collection of the ribbons. Plasma treatment remains effective for a few hours at the most and heavily depends on the environment (humidity). The substrate should therefore be plasma-treated immediately before use. Settings (power, time) of a particular plasma cleaner can easily be ascertained by observing the behavior of a water droplet dispensed from a pipette onto the wafer. When the droplet does not form a hemisphere but runs out into a tiny flat puddle (i.e., has a very low contact angle) hydrophilicity should be sufficiently high.

3.3 Substrate Handling

Without a dedicated substrate holder, collecting section ribbons onto a substrate by hand was almost impossible without section damage or partial reorientation of sections (cf. Chapter 6). When collecting a large number of sections for 3D reconstructions—where every section matters—this is not acceptable. In order to improve this situation, various instruments have been devised to support the operator [11, 12]. Based on the "supporting hand concept" we developed an assembly of micromanipulators [13–17] consisting of several axes (Fig. 5a) for the following reasons:

- To enable adjustment on different microtome setups (axes 1–3), which is only necessary for the initial setup once per sample block.

- To realize smooth lift-up trajectories for the substrate (axes 4–6, Fig. 5b).

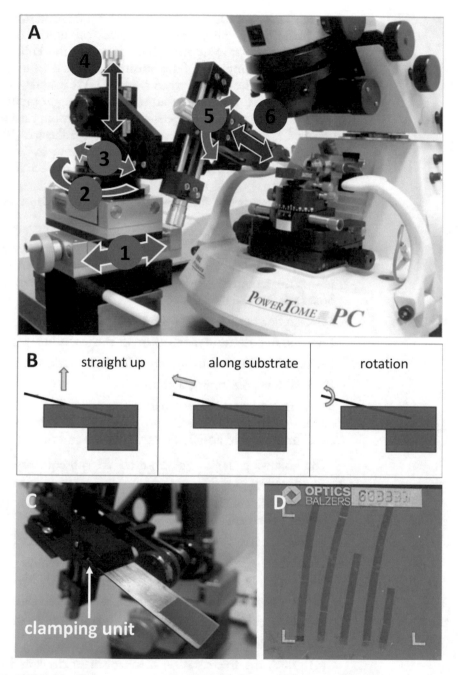

Fig. 5 Substrate holder. This consists of a combination of micromanipulators for six axes of movement and is adjustable to different microtome setups (**a**). Schematic side view of knife illustrating different lift-out trajectories (**b**) of the substrate carrier (**c**). The very smooth movement of the micromanipulators allows several ribbons to be collected on a single substrate (**d**)

In addition, the rotational axis 2 offers a free rotation mechanism which enables the operator to rotate the holder out of the knife workspace, for example, when changing the knife or after work has been completed. The substrate is fixed in a clamping unit mounted at the end of axis 6 (Fig. 5c). We usually work with a Jumbo knife and mount the substrate on an appropriate carrier plate, such as a clean microscope slide or an aluminum carrier cut to the size of a slide, with an easily removable adhesive (Fixogum Marabu, Germany) at least 10 min before using it. We usually plasma-treat the entire assembly to guarantee perfect wetting in the knife boat.

The basic steps for adjusting the micromanipulator assembly are as follows:

- Insert knife into its holder.
- Move z-axis (4) to the upper end position.
- Clamp the mounted substrate, position it roughly over the knife boat (axis 1 and 2).
- Perform fine positioning of the substrate.

  Lock the free-rotation mechanism of the rotation table (axis 2).

  Perform fine rotation using screw for fine rotation (axis 2).

  Perform fine positioning with axis 3.

- Move the substrate down using the z-axis (4) and check if substrate has contact with the walls of the knife boat. If it does, improve fine positioning with axes 3 and 6.

The following steps are required to subsequently lift and change the substrate:

1. Lift up with axis 4 or 5 (recommended) until the water no longer has contact with the substrate or carrier plate.
2. Unlock the free-rotation mechanism of the rotation Table (2).
3. Turn the holder towards the operator to enable substrate handling.
4. Remove the carrier plate with the substrate from the clamp.
5. If desired, clamp a new substrate mounted on a carrier plate.
6. Turn it to the knife.
    (a) Lock the free-rotation mechanism of the rotation table (axis 2).
    (b) Perform fine rotation using the screw for fine rotation (axis 2).
7. Move the substrate down (axis as in **step 1**) until it touches the bottom of the knife boat then move a little bit up, ensuring that the substrate/carrier plate does not touch the knife boat.
8. To ensure optimal wetting of the substrate, it may be necessary to adjust axis 5.

3.4   Cutting

3.4.1   Knife Setup

See **Note 4** to select the correct knife for sectioning.

The knife first needs to be aligned to the block face. Several adjustment options exist for this purpose. Figure 6b shows the options for aligning orientation between knife and block face. If the trimming rules for obtaining parallel edges (Subheading 2.3) have been properly applied, there is little need to adjust the angle on the arc segment holder in most cases (green double arrow in Fig. 6b). To verify proper alignment, the back/bottom light below the knife can be used. This enables a projection of the knife edge on the block face (Fig. 6c) to be seen, which needs to be positioned just a few micrometers away from the knife edge. On a perfectly aligned knife, this projection appears as a stripe perfectly in parallel with the knife edge. If the knife is not properly aligned to the block face, the light stripe is not in parallel with the knife edge (as shown in Fig. 6c). Alignment should be performed in the following sequence:

1. Align by rotating the knife (blue double arrow in Fig. 6b) and check if the light stripe is in parallel with the knife edge.

2. Adjust the sample rotation (green single arrow in Fig. 6b) until the lower edge of the block face is perfectly in parallel with the knife edge. Knife and sample rotation mutually influence each other and may need to be corrected iteratively.

3. Align by adjusting the angle on the arc segment holder (green double arrow in Fig. 6b) until the light stripe keeps the distance to the knife edge constant when moving the sample up and down using the hand wheel.

Finally, the knife boat has to be filled with water. The perfect water level is achieved when the complete surface reflects the top light.

3.4.2   Initial Cut

Sectioning can start when the knife is aligned. The first sections will not be the full size of the block face but rather wedge-shaped. The number of sections required to attain the first complete section depends mainly on the quality of the alignment. When the first complete section is completed and the desired starting point is reached—usually when sample material appears in the sections—sectioning should be stopped and the waste sections removed. The substrate then has to be lowered into the water. It is important not to completely submerge it but to retain a small part of the substrate, kept dry, out of the water, to which the section ribbons will be attached. It may be necessary to readjust the water level at this point before ribbon sectioning can start.

3.4.3   Best Practice for Large Volume Arrays

- Do not touch anything, keep away from the ultramicrotome, and use camera-based monitoring. Potential problems arising during sectioning are illustrated in Fig. 6d–f and will be discussed in **Note 5**.

Fig. 6 Sectioning. Diamond sectioning knives for different applications (**a**): standard knife for ultrathin sections (right) for a small number of sections or a Jumbo knife (left) for a very large number of sections. Alignment of knife edge to block face (**b**, **c**) illustrating the possible movements of knife holder (blue arrows) and block (green arrows) in (**b**); red square indicates field of view shown in (**c**). Reflection of knife edge on block face (**c**) helps with alignment, see text. Possible problems occurring while sectioning: Top views of knife boat with straight ribbon (**d**) or curved ribbon (**e**) with cracks between sections (insert), induced by a transverse momentum, for example, air flow hitting the water surface. Chatter, that is, thickness variations in a section (**f**), is usually caused by mechanical impact

- Use an antistatic device: The problem of electrostatic charging of sections may occur when a large number of sections are cut. This may lead to wrinkles and/or folding of a given section over its successor. An antistatic device may help to avoid these adverse effects. This produces ions that neutralize static charges on the sections. It is important to point the electrode of the antistatic device at the area around the knife edge and the block face. Details on how to position the electrode can be found in the device manual. Such a device is listed as an optional component in Table 1.

3.5   Section Handling

After the first ribbon has been cut it needs to be collected on the substrate. This is achieved in the following handling steps:

1. Releasing the ribbon from the knife edge.

2. Guiding the ribbon to the substrate.

3. Attaching the ribbon to the substrate.

The most important tool for ribbon manipulation is the eyelash tool (*see* Subheading 2). A very soft hair from a cat's fur is even better—sections tend to stick less to this than to an eyelash.

3.5.1   Releasing the Ribbon from the Knife Edge

To release the ribbon from the knife edge: One method is to gently stroke the eyelash across the knife edge until the ribbon releases. This is critical because the knife edge and the ribbon may be damaged. To support this task, Fahrenbach's paper [18] suggests to raise the water level in the boat.

3.5.2   Guiding Ribbons to the Substrate

To transport the ribbon to the substrate the eyelash tool can be used to push or pull the ribbon. If pulled, too much pressure on a section may damage it. If pushed, it may stick to the eyelash and may be damaged or even lost. Another possibility, more secure but also time-consuming, would be to generate a slight water flow with the eyelash tool to move the ribbon.

3.5.3   Attaching the Ribbon to the Substrate

For large volume arrays it is more advantageous to collect several ribbons on one substrate. In this case, the ribbons previously cut need to be attached to the substrate to make sure that they do not drift away when the next ribbons are cut. This type of mutual influence occurs (cf. also **Note 6**) in particular with densely arranged ribbons. To attach a ribbon, it needs to be moved towards the substrate until one end becomes pinned to that part of the substrate protruding from the water. The free end of the ribbon is still swimming on the water surface. The next ribbon can now be produced and attached to the substrate close to the previous ribbon in the same manner.

3.6   Ribbon Transfer to the Substrate Surface

After all ribbons have been sectioned and pinned with one end to the substrate, their free ends still float on the water. The task is now to fully transfer the ribbons to the substrate, so that they become attached over their entire length without damage and wrinkles. To this end, the substrate should be gently lifted out of the water avoiding undesired water flows or turbulences that might damage the ribbons (*see* also **Note** 7). Here, the substrate holder offers three options (Fig. 5b): The straight-up movement (left), the pull-out along the substrate axis (middle) and the rotation leading to a tilt-up movement of the substrate (right). We recommend the tilt-up adjustment or the straight-up movement. If the lift-out is successful the result should be as shown in Fig. 5d. Ribbons are finely aligned and stretched without wrinkles. We do not recommend treating the ribbons with chloroform to stretch them or to use a hotplate to dry the ribbons. Both methods stress the ribbons and may have a negative impact on their preservation. When using the substrate holder, only a very small amount of water remains on the substrate and needs only a few minutes to dry.

A short movie illustrating the most critical steps of ribbon production can be found in an open access article by Wacker et al. [17].

4   Notes

1. ITO-coated glass coverslips are available from a number of companies. There is a trade-off between conductivity and granularity of the coating: A highly conductive coating is somewhat thick (several 100 nm) and has a high surface roughness which may be detrimental for picking up long ribbons without breaking them. A very thin and smoother coating is less conductive, but this may still be sufficient when using low primary electron energies in the SEM. For ITO-coated coverslips without fiducials it is not obvious which side is conductive. Measuring conductivity using a multimeter helps to determine this.

2. Warning when selecting the appropriate trimming tool.

   When using razor blades for coarse trimming, tiny metal flakes might break off from the blade and become attached to or even inserted into the resin. These will damage the expensive sectioning knife. The knife manufacturers therefore recommend following trimming with a razor blade by smoothing all sides using a diamond trim knife, which is cheaper than a sectioning knife. A fresh glass knife will serve the same purpose.

3. Trimming parallel edges.

   The last step, cutting the block face to achieve orthogonality may not be possible under certain circumstances, for example if the target volume is directly exposed at the surface of the block. One such example would be adherent cell monolayers.

4. Selecting the appropriate sectioning knife.

Depending on the number of sections required, various different sectioning knives may be used. The boat of a standard ultramicrotome knife for ultrathin sections (Fig. 6a right) can take a substrate of about 7–8 mm width and is therefore good for up to about 100 sections (see also Chapter 6). If more sections are needed, a knife with a "Jumbo" boat—originally developed for histology work—may be equipped with a high-quality diamond knife for ultrathin sections (Fig. 6a left). This can use an entire microscope slide and thus harbor a very large number of sections. If thicker sections (500–2000 nm) for LM analysis only are desired, the original "histo Jumbo" is a less expensive option (cf. [6]).

5. Breaking of ribbons.

The main problem encountered during sectioning is ribbon breaking (Fig. 6e), caused by various issues. Curved ribbons tend to break more easily than straight ones, because they experience a transverse momentum when being moved. Depending on the reason for curving, different remedies can be used:

(a) Too much adhesive applied—trim away old adhesive and apply a thin film of new adhesive.

(b) Too little adhesive applied—add a bit more adhesive to lower and/or upper face.

(c) Sample compression.

(d) Leading and trailing edge not parallel—try to rotate the knife, not the sample, when you trim and check parallelism several times during the trimming process.

Even ribbons cut from perfectly trimmed blocks may break as shown in Fig. 6e—in this case probably an airflow (breathing, opening a door) hit the water surface, leading to tiny cracks between sections and thus the bending of the ribbon. To avoid that, the microtome can be encased (e.g., as shown in [5]).

Another problem that is common to all ultramicrotomy work is chatter (Fig. 6f)—thickness variations within one section, usually caused by mechanical impact such as by the user touching the microscope while monitoring the sectioning process. Placing the microtome on a vibration damping table and/or staying away from it during sectioning will alleviate this problem.

6. Detaching of ribbons.

To avoid ribbons already attached to the substrate from swimming away when the next ribbon approaches, move the ribbons only longitudinally, not laterally, especially when close to the previously cut ribbons. Water surface flow is induced by moving ribbons. This effect increases when the water is not clean (e.g., layer of thinner from glue mixture or particles on the water surface). Vibrations caused by the user accidentally hitting the setup may also induce attached ribbons to be released or even break.

7. Flushing of ribbons.

Sometimes ribbons that are attached on the far left or right side of the substrate are flushed away when lifting the substrate out of the boat. This is due to the fact that a small gap between knife boat wall and substrate accelerates the water flowing from the substrate down into the boat. If possible, increase the gap between knife boat and substrate or use narrower substrate.

References

1. Gay H, Anderson TF (1954) Serial sections for electron microscopy. Science 120 (3130):1071–1073. https://doi.org/10.1126/science.120.3130.1071

2. Andersson-Cedergren E (1959) Ultrastructure of motor endplate and sarcoplasmic components of mouse skeletal muscle fibers as revealed by three dimensional reconstructions from serial sections. J Ultrastruct Res Suppl 1:5–181

3. Levinthal C, Ware R (1972) Three dimensional reconstruction from serial sections. Nature 236:207–210. https://doi.org/10.1038/236207a0

4. Peddie CJ, Collinson LM (2014) Exploring the third dimension: volume electron microscopy comes of age. Micron 61:9–19

5. Harris KM, Perry E, Bourne J et al (2006) Uniform serial sectioning for transmission electron microscopy. J Neurosci 26:12101–12103

6. Micheva KD, Smith SJ (2007) Array tomography: a new tool for imaging the molecular architecture and ultrastructure of neural circuits. Neuron 55:25–36

7. Wacker I, Schröder RR (2013) Array tomography. J Microsc 252:93–99

8. Polishchuk RS, Polishchuk EV, Marra P et al (2000) Correlative light-electron microscopy reveals the tubular-saccular ultrastructure of carriers operating between golgi apparatus and plasma membrane. J Cell Biol 148:45–58

9. Oberti D, Kirschmann MA, Hahnloser RHR (2011) Projection neuron circuits resolved using correlative array tomography. Front Neurosci 5:50

10. Templier T (2019) MagC, magnetic collection of ultrathin sections for volumetric correlative light and electron microscopy. eLife 8:e45696. https://doi.org/10.7554/eLife.45696

11. Horstmann H, Körber C, Sätzler K et al (2012) Serial section scanning electron microscopy (S3EM) on silicon wafers for ultra-structural volume imaging of cells and tissues. PLoS One 7(4). https://doi.org/10.1371/journal.pone.0035172

12. Hanssen E (2016) A third hand for array tomography. Microsc Microanal 22:1152–1153. https://doi.org/10.1017/s1431927616006607

13. Wacker I, Spomer W, Hofmann A et al (2015) On the road to large volumes in LM and SEM: new tools for array tomography. Microsc Microanal 21:539–540. https://doi.org/10.1017/S1431927615003499

14. Spomer W, Hofmann A, Wacker I et al (2015) Advanced substrate holder and multi-axis manipulation tool for ultramicrotomy. Microsc Microanal 21:1277–1278. https://doi.org/10.1017/S1431927615007175

15. Wacker I, Chockley P, Bartels C et al (2015) Array tomography: characterizing FAC-sorted populations of zebrafish immune cells by their 3D ultrastructure. J Microsc 259:105–113

16. Wacker I, Spomer W, Hofmann A et al (2016) Hierarchical imaging: a new concept for targeted imaging of large volumes from cells to tissues. BMC Cell Biol 17:38

17. Wacker IU, Veith L, Spomer W et al (2018) Multimodal hierarchical imaging of serial sections for finding specific cellular targets within large volumes. J Vis Exp 133:e57059. https://doi.org/10.3791/57059

18. Fahrenbach WH (1984) Continuous serial thin sectioning for electron microscopy. J Electron Microsc Tech 1:387–398. https://doi.org/10.1002/jemt.1060010407

Chapter 6

Large Volumes in Ultrastructural Neuropathology Imaged by Array Tomography of Routine Diagnostic Samples

Irene Wacker, Carsten Dittmayer, Marlene Thaler, and Rasmus Schröder

Abstract

Routine samples in pathology and neuropathology are usually prepared according to certified standard sample preparation protocols that do not necessarily introduce the large amounts of heavy metals required to generate optimized contrast and to render the final resin block conductive. Imaging of such samples by volume electron microscopy (EM) methods such as serial block face scanning electron microscopy (SBFSEM) or focused ion beam scanning electron microscopy (FIB-SEM) can thus be challenging due to both contrast and charging issues. Array tomography on the other hand, where hundreds of ultrathin serial sections are deposited on conductive substrates and imaged in a modern field emission scanning electron microscope (FESEM) does not encounter such problems: Section arrays may be poststained with heavy metals leading to superior imaging contrast even from weakly stained blocks. Using a sample from a patient with a congenital myopathy (nebulin-related myopathy) characterized by the so-called electron-dense nemaline rods in muscle fibers we describe preparation of section arrays and how they are imaged in a FESEM in an automated way using a typical, commercially available software platform. We further demonstrate how we can target individual cells by hierarchical imaging cascades. Alignment/registration of image stacks using freeware packages such as Fiji and its TrakEM2 plugin and semiautomated single plane-based segmentation of the nemaline rods using IMOD are also explained.

Key words Volume electron microscopy, Array tomography (AT), Serial sections, FESEM, 3D pathology, Muscle pathology, Myopathy, Nemaline

1 Introduction

Array tomography (AT) was introduced more than 10 years ago to help visualize the molecular architecture and ultrastructure of neural circuits [1]. In that work, brain tissue was chemically fixed and embedded in a hydrophilic resin (LR white). Hundreds of ultrathin sections were cut using a stand-alone sample preparation tool

Electronic supplementary material The online version of this chapter (https://doi.org/10.1007/978-1-0716-0691-9_6) contains supplementary material, which is available to authorized users.

Irene Wacker et al. (eds.), *Volume Microscopy: Multiscale Imaging with Photons, Electrons, and Ions*, Neuromethods, vol. 155, https://doi.org/10.1007/978-1-0716-0691-9_6, © Springer Science+Business Media, LLC, part of Springer Nature 2020

(ultramicrotome) and placed on coated glass slides, forming ordered arrays of serial sections. By repeated cycles of antibody staining, imaging, elution, and restaining neurons and diverse synapse populations could be characterized in 3D in a considerable volume. After examination in the light microscope, arrays were stained with heavy metals and imaged in a FESEM (field emission SEM), providing ultrastructural context for the entities labelled in the preceding imaging cycles.

In the meantime, a number of similar approaches—with or without characterization of the material by light microscopy—have been introduced and reviewed [2]. Although AT lends itself very well to correlative light and electron microscopy (CLEM) approaches ([3], *see* also Chapters 4 and 7), it is also useful for 3D reconstructions and analysis of cells and tissues based entirely on ultrastructural information [4, 5]. The option to poststain section arrays with heavy metal salts is a decided advantage over the block face-based methods SBFSEM and FIB-SEM (*see* also Chapters 9–12) where contrast and conductivity have to be generated by large amounts of heavy metals introduced into the sample block during sample preparation or by charge compensation methods during imaging in an SEM (cf. Chapter 9). Since pathological samples are prepared according to standardized protocols [6], usually not optimized for maximum heavy metal content, block face imaging of such samples may encounter serious contrast and charging problems. In contrast, imaging of weakly metalized routine samples from pathological archives by AT works very well as the samples are cut up into ultrathin (up to 100 nm) slices which are placed on a conductive support. The charges produced in such a thin slice by a low voltage incident electron beam (1–3 keV primary electron energy) are usually well dissipated by the conductive support.

Another advantage of AT and of importance for pathological samples is the fact that the section arrays can be kept for an "indefinite" period of time—albeit in a dust-free environment without extremes of temperature and humidity. Furthermore, due to the stable adherence of the ultrathin sections to the conductive substrate, they will not distort during long-time storage. They may be imaged repeatedly, using either different equipment or another scanning resolution while SBFSEM and FIB-SEM destroy the sample during the slicing/imaging cycle.

As an example from pathology for a typical AT-based workflow for 3D ultrastructural reconstruction of a considerable volume we chose nebulin-related nemaline myopathy. This is one of the most common congenital myopathies, caused by mutations in the gene of the giant protein nebulin. The clinical appearance is dominated by muscle weakness in a wide variety of degrees [7]. Its histopathological hallmarks are rod-shaped structures (nemaline bodies), which are electron dense inclusions variable in shape and size (1–7 μm in length, 0.3–2 μm in width). They show ultrastructural and immunocytochemical signs of z-discs and therefore are thought to originate from them [8]. Their structure and distribution is rather complex and a 3D analysis instead of the conventional representative 2D projection imaging using a transmission electron microscope (TEM) might lead to new insights regarding the formation of nemaline bodies, their connectivity, and exact 3D structure.

2 Materials

2.1 Fixation and Embedding

– 2.5% glutaraldehyde in 0.1 M sodium cacodylate.

– 0.1 M sodium cacodylate.

– 1% osmium tetroxide in 0.05 M sodium cacodylate.

– 30%, 50%, 80%, 96%, 100% acetone.

– 1% uranyl acetate and 0.1% phosphotungstic acid in 70% acetone.

– Resin mix consisting of Renlam® M-1 (SERVA 13825), an araldite replacement epoxy resin; 2-dodecenylsuccinic acid anhydride (DDSA; SERVA 20755), a hardener; and DMP30 (SERVA 36975), an accelerator.

  For infiltration mix 25 g Renlam and 26 g DDSA, stir for 30 min. This mixture may be stored frozen and thawed just before use. Add 4% DMP30, mix well, and use one part of that mixture plus one part of dry 100% acetone to infiltrate the tissue.

  For final embedding stir 25 g Renlam and 26 g DDSA for 30 min, add 2.5% DMP30 and mix well again. This mixture may also be stored at −20 °C in 15 ml or 50 ml plastic tubes or hypodermic syringes (without needles).

– Silicone embedding molds.

2.2 Making Arrays

– Microtome blades (N35, FEATHER) for coarse trimming.

– Ultramicrotome.

– Diamond knives for trimming and sectioning (*see* also **Note 4.2.1** and Chapter 5).

– Eyelash glued to toothpick or thin silicone tubing cut open and shaped into something resembling an eyelash.

– Glue mixture to stabilize ribbons: Pattex Gel Compact (Henkel), diluted with xylene, from 1:1 to 1:10 depending on sample block characteristics.

– Silicon wafer (e.g., from http://si-mat.com/silicon-wafers.html, doping: P/Bor, orientation: ⟨100⟩, thickness: 525 ± 25 μm, resistivity: 1–30 Ω cm).

- Diamond glass cutter pen.
- Instrumentation for glow discharge/plasma cleaning.
- Richardson's stain [9]: Mix equal amounts of stock solutions A and B.
  Stock solution A: 1% Azure II in aqua dest.
  Stock solution B: 1% Methylene blue and 1% Sodium Borate in aqua dest.

2.3 Poststaining

- 1% ethylenediaminetetraacetic acid (EDTA) in aqua dest.
- 5% uranyl acetate in 50% ethanol.
- Reynolds' lead citrate (adapted from [10]).
  Dissolve 1.33 g lead (II) nitrate $(Pb(NO_3)_2)$ in 10 ml destH$_2$O.
  Dissolve 1.76 g trisodium citrate dihydrate $(Na_3(C_6H_5O_7) \cdot 2H_2O)$ in 10 ml dH$_2$O.
  Mix both and add 1 M sodium hydroxide (NaOH) until the solution is clear.
  Fill up with dH$_2$O to 50 ml.
- Aqua dest and ultrapure water (e.g., aqua ad iniectabilia) for washing of sections (*see* also **Note 4.3.1**).

2.4 SEM Imaging

- Aluminum stubs for SEM.
- Conductive silver paint.
- Field emission scanning electron microscope (FESEM), here ZEISS Merlin with a BSD4-detector (detector for back-scattered electrons, BSE).
- Image automation and microscopy control software package, here ZEISS Atlas 5 solution with Atlas AT software module.
- Sample holder (e.g., carousel for nine stubs).

2.5 Image Processing

- Trak EM2 [11], a plugin for the Fiji open source software [12] for registration of image data stacks.
- The open source software package IMOD [13] for segmentation and visualization.

3   Methods

3.1 Fixation and Embedding

Fix fresh human muscle tissue (open biopsy) in 2% glutaraldehyde in 0.1 M sodium cacodylate buffer overnight at 4 °C (*see* **Note 4.1.1**).

Cut into small blocks with an edge length of about 1 mm using a razor blade.

Incubate in 1% osmium tetroxide in 0.05% sodium cacodylate buffer for a minimum of four hours (to overnight) at room temperature, wash 3× for 10 min in 0,1 M sodium cacodylate buffer. Dehydrate for 20 min each in 30% and 50% acetone, block-stain for 60 min in 1% uranyl acetate and 0.1% phosphotungstic acid in 70% acetone. Continue dehydration with 25 min in 80% acetone, 20 min 96% acetone, and 3× 20 min in 100% acetone. Infiltrate over night with Renlam-mix (with 4% accelerator) in 100% dry acetone (1:1, v/v), leaving the lids of the vessels open so that the acetone may evaporate. Incubate with fresh Renlam (final embedding mix with 2.5% accelerator) for 4 h and embed in flat silicone molds, using fresh Renlam final embedding mix. Polymerize for 48 h at 70 °C (*see* **Note 4.1.2**).

3.2   Making Section Arrays

For quality control of the biopsy and to define the target region in the block there are two options (*see* also Fig. 1, flow chart: "Targeting 1"): Either place a few semithin (500 nm) sections on a slide and stain them for light microscopy (LM) or place several ultrathin (50–100 nm) sections on a piece of silicon wafer and stain them for electron microscopy (*see* Subheading 3.3 below). A classical LM stain is Richardson's stain [9]: Incubate sections with a droplet of stain on a hot plate for 1 min at 80 °C, then rinse with distilled water and let dry.

In our example, the tissue was not equally well infiltrated across the entire biopsy (Fig. 2a), leading to many ring folds in peripheral areas (Fig. 2b). However, a central region with many cells sectioned in transverse orientation and containing nemaline rods (Fig. 2c) was well preserved and selected for further analysis. With the 50× objective of an epi-illumination LM it is possible to unequivocally identify rod-positive cells in sections on Si wafers (Fig. 2c).

To produce arrays of large numbers of ultrathin sections, trim the block to a trapezoid shape around the selected region of interest (ROI) with leading and trailing edge exactly parallel using a diamond trim tool (*see* **Note 4.2.1**).

If necessary, coat the leading and trailing edge of the block (*see* **Note 4.2.2**) with a very thin layer of a mixture made from Pattex Gel Compact and xylene [14].

Cut silicon wafer pieces using a diamond glass cutter pen with a ruler to a size of about 25 × 10 mm to fit into a small knife boat. Clean the reflecting silicon surface using acetone and a lintfree wipe, rinse with 70% ethanol, then aqua dest. and hydrophilize by glow discharging in a MED020 sputter coater for 90 s. Insert the freshly hydrophilized silicon substrate into the filled boat of a semi-ultra diamond knife and lower the water level until the surface close to the knife's edge appears silvery.

Cut ribbons of 60 nm thin sections (*see* **Note 4.2.3**), arrange them with an eyelash and take images repeatedly (for example with a smartphone camera or optical camera via the stereo microscope of

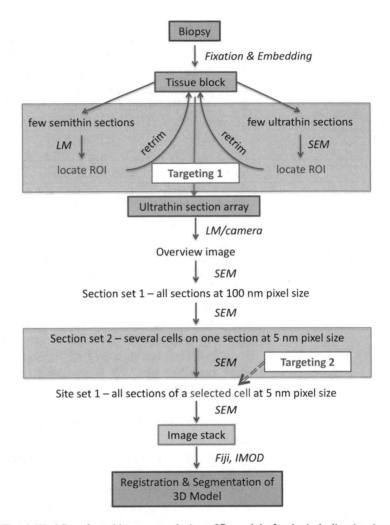

Fig. 1 Workflow from biopsy sample to a 3D model of selected ultrastructural features: an important issue here is targeting, that is, selecting a suitable region of interest (ROI). In our example there are two levels of targeting, first at low resolution, potentially using a light microscope (LM). Here the aim is to select a region suitable for electron microscopy, for example, devoid of embedding artifacts (cf. Fig. 2b). Retrimming of the block may be necessary as a result. The second targeting step, selecting a cell exhibiting the desired phenotype is done on medium resolution scanning electron microscope (SEM) images. Once all images of this cell are recorded at high—ultrastructural—resolution the resulting image stack has to be registered before the features of interest can be analyzed by segmentation

the ultramicrotome) to document how the ribbons are arranged in the boat (Fig. 2d).

Deposit the ribbons onto the silicon wafer by lowering the water level: Aspirate with a syringe, always a small volume at a time and correct drifting ribbons with the eyelash.

Let the arrays dry thoroughly before continuing with the next step.

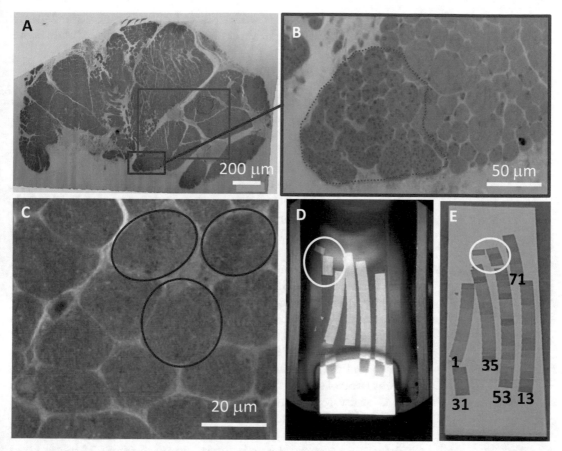

Fig. 2 Preparation of serial section arrays: Identifying a suitable region in the biopsy by epi-illumination light microscopy (**a–c**) of a few sections placed on a piece of silicon wafer and poststained with uranyl acetate and lead citrate. Peripheral regions of the entire biopsy (**a**) are often covered with numerous little folds or wrinkles, one example surrounded by the dashed yellow line in (**b**). There is however a central region (e.g., blue box) without such artifacts. It also contains many cells (red circles) harboring groups of nemaline rods (**c**), detectable with a 50× objective (region corresponding to frame **c** is indicated by green circle in (**a**)). After retrimming, arrays of serial sections are prepared—shown in (**d**) floating in the knife boat. The sections on the dried arrays are labeled to indicate the order in which they were cut (**e**). Stray sections 13 and 50–52 (in white circles) moved during aspiration of water from the boat

3.3 Poststaining

Filter all solutions directly (through a 0.22 μm syringe filter) onto the ribbons. Use enough liquid to cover all sections. For each of the three wash steps prepare three 25 ml glass beakers with distilled water plus one additional 25 ml beaker with ultrapure water (e.g., water for injection) for the final wash before drying (ten beakers total, *see* **Note 4.3.1**).

Incubate arrays for 5 min with 1% ethylenediaminetetraacetic acid (EDTA) in distilled water to reduce later precipitate formation [15].

Wash extensively in distilled water by moving the wafer 30× up and down in each of the three beakers for the first wash (*see* **Note 4.3.2**).

Incubate arrays 5 min with 5% uranyl acetate in 50% ethanol, wash as above in the beakers for the second wash.

Incubate arrays 4 min with Reynolds' lead citrate, wash as above in the beakers for the third wash (including the additional wash with ultrapure water, *see* **Note 4.3.3**).

Dry arrays overnight and mount on an aluminum stub using conductive silver paint.

Take an overview image (*see* **Note 4.3.4**) of the array using a digital consumer camera (e.g., a Smartphone camera) (Fig. 2e). This image will later be used for initial navigation in the SEM (cf. Subheading 3.4.1 below).

3.4 FESEM Imaging

3.4.1 Initial Steps (See Note 4.4.1.1)

Mount the sample on an appropriate sample holder (e.g., Fig. 5a) and load it into the SEM. Set up with appropriate sample positioning (working distance), accelerating voltage (electrical high tension) and electron beam current for optimum imaging with the detector to be used and the imaging resolution required. After initial adjustment of stage, column and imaging parameters (e.g., focus and astigmatism) on the sample using conventional instrument control software (SmartSEM) open a new project in the Atlas 5 AT software (for more monitor screenshots of the Atlas software *see* also Fig. 6, Chapter 7).

For efficient navigation align the overview image (Fig. 3a) of the section array to the coordinate system of the SEM stage: Load the digital image of your array (cf. Subheading 3.3) using the "`Data Import`" tab in the sample setup menu. Grab images from the three outermost sections of your array (arrowheads in Fig. 3a) using the SE detector and a short dwell time. Adjust brightness, contrast and dwell time in such a way that you clearly see the corners of your sections (*see* **Note 4.4.1.2**).

Open the "`Align Data`" module by right click into the overview image, adjust its transparency so that you can recognize the section edges on the SEM images, choose translate, rotate, shear, scale in the menu below the image (Fig. 3b).

Arrange the first corner of your color overview image on top of the corresponding corner in the SEM image and confirm this in the software by placing a pin (right click on the chosen position).

Pull the second corner of the color image on top of the corresponding corner in the SEM image and place the second pin by right clicking in the corner.

Adjust the third corner, place your pin and click "`Finish alignment`" (Fig. 3c).

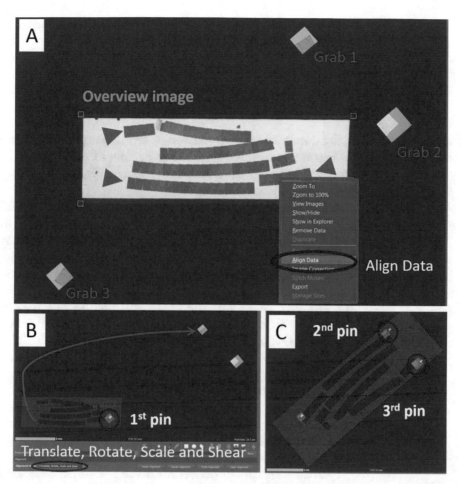

Fig. 3 Importing data into Atlas software and alignment with the SEM stage coordinate system: overview image from LM or consumer camera imported and placed between three SEM images (Grabs 1–3) taken from three corners (green arrowheads) of the array (**a**). Using the "Align Data" functionality first one section of this image is overlaid onto the corresponding SEM image (here Grab 3) by translational movement (**b**) and attached there (first pin). Then the overview image is rotated to overlay the next SEM image (here Grab 1) with the corresponding section (second pin) and finally by slight scaling and shearing the last two corresponding images are aligned (third pin)

3.4.2 Hierarchical Imaging (See Note 4.4.2.1)

The first step in the hierarchical imaging cascade is defining the section set: Outline the first section using the "region definition tool," then create the section set by cloning this ROI to all of the sections using the "stamp tool." Holding down the right mouse button while placing your ROI allows you to rotate the ROI to accommodate curved ribbons.

At this step you should make sure that you define your section ROIs in the right order, because this is the order the software will record the images. While the images can be reordered before export for later analysis, this is an extra step that is unnecessary if the section ROIs are defined in the correct order initially. It helps to be extra careful at this step (*see* **Note 4.4.2.2**).

Now design an image recording protocol appropriate for what you want to achieve in this first SEM image acquisition step. In our example, at this stage we want to identify individual muscle cells, so we choose intermediate resolution, meaning a pixel size of 100 nm, and a moderate dwell time of 4.5 μsec using the BSD4 detector for backscattered electrons.

3.4.3 Controlling Focus and Stigmation (See Note 4.4.3.2)

A software tool helps to test and optimize parameters for auto focus and auto stigmation (AFAS): The autofocus routine tests a limited focus range in a limited number of steps (Fig. 5c), starting from a manually optimized setting.

The range has to be matched with the expected focus difference ΔZ between two consecutive sections (cf. Fig. 5b and *see* **Note 4.4.3.1**). To get a rough estimate for a proper focus range measure the working distance (= optimal focus) in exact values (at least four decimal digits) for corresponding positions on two consecutive sections. Do that just outside your target ROI using SmartSEM (cf. Fig. 6b). Make sure the range you then set for the autofocus routine is a bit larger than the difference you measured. The routine divides this range into an uneven number of steps (default is nine, in Fig. 5c we use 11), ideally starting from a slightly blurry image (Fig. 5d), going through ideal focus (Fig. 5e) and ending with a blurry image again (not shown). The algorithm processes all images and extracts a "sharpness metric" for each that is then normalized to the sharpest image. If all parameters are well adjusted a peak—corresponding to the sharpest image—is found (Fig. 5c), *see* **Notes 4.4.3.2–4.4.3.4**).

Once your auto-function routines' performance is satisfactory you have to define the focus values for the starting section of each ribbon: In our example, using the `"check protocol"` function set focus values for sections 1, 14, 35, and 53, which are the starting sections of the four long ribbons (cf. Fig. 2e). For the "stray sections" number 13, 31–34, and 50–52 focus values have also to be determined manually, because the stage has to travel a fair distance to record those in the right order—meaning potentially big focus differences as when a new ribbon starts, which cannot be handled by the AFAS routine.

Now start automatic image acquisition of the section set over the entire array from the `"acquire"` tab.

3.4.4 Finding (Targeting 2) and Imaging the "Right Cell" at High Resolution

In the 100 nm pixel size images of the section set you can clearly see different cell types and even distinguish between muscle fibers with and without nemaline rods (Fig. 8a).

But it is not possible to discern finer cellular detail, such as changes in the organization of the microfilament system (Fig. 8b). In order to select cells exhibiting such changes in the areas occupied by the nemaline rods we did a screening on a "guide wafer." This was prepared directly after the 77-section-array intended for 3D analysis and contained just a few sections to help orienting within the sample. The section set feature was used in this case for high resolution automated imaging of many cells in the same section. Then these cells were ranked according to certain attributes (number of nemaline rods, filaments present in the area occupied by rods, cell size). The "interesting" cells were then marked and this "quality map" (Fig. 8c) of the muscle part was printed out and used for the identification of target cells during the 3D recording session (see also flow chart: Targeting 2).

To generate the sites (*see* **Note 4.4.4.1**) for high-resolution imaging right click into the first section of the section set to open the "`manage sites`" feature and place "site ROIs" over selected target cells. Check that the ROI indeed includes the selected cell in its entirety by going through all sections of a given site, and adjust ROI placement before starting automatic imaging (*see* **Note 4.4.4.2**).

Create a high-resolution image recording protocol for your site set: Here our aim is to resolve membranes and the microfilament system, so we chose an image pixel size of 5 nm and a dwell time of 6.4 μs using the BSE detector, in our case denoted as BSD4. AFAS parameters were AF range 8 μm, AS range 1%, AFAS dwell time 5 μs, and pixel size ratio 300% (15 nm AFAS pixel size).

Again, use the "`check protocol`" option to set focus values for sections 1, 14, 35, 53—the starting sections of the long ribbons and for the "stray sections," as above (*see* **Note 4.4.4.3**).

Now you can start automatic image acquisition of the site set.

3.4.5 Exporting Data, Documentation

Recorded image stacks of sets may be checked by the "`view images`" option. You can toggle through all the sections to check for focus and whether your cell is completely within the ROI. If an image did not turn out well there is a reshoot option.

For documentation of the project setup (overviews, section and site sets and their respective positions) it can help to take screenshots of the Atlas main correlative view. When you activate a certain ROI, image and microscope parameters will be displayed in the selection details window.

To export a single image use a right click into the ROI in the main user interface view, choose `export`, define the export pixel size (e.g., 5 nm), crop close to your structure of interest and click export. For a batch export of an image stack right click on its name in the project tab in the project browser window.

Another way to communicate, show, and discuss data is the generation of a movie using a number of keyframes. The Atlas 5 software will interpolate and create a fly-in movie (*see* Supplementary Movie S1) demonstrating the different types of ROIs and the type of resolution obtainable (*see* **Note 4.4.5.1**).

3.5 Registration

As we record a 3D dataset in a serial 2D manner a number of processing steps from 2D to 3D are necessary. For these tasks several software packages (both freeware and commercial solutions) are available. As starting point we discuss here one possible, easily accessible freeware based solution: First, align the acquired image stack using the TrakEM2 plugin [11] in the Fiji open source software package [12]: follow an initial least-square alignment (rigid, default values) by a second, elastic alignment (rigid) with a resolution parameterization in TrackEM2 of 16–50, testing maximally five layers. Export these alignments as two separate tif-stacks (`make flat image`) and rename the individual tif-images according to your favorite display software needs. Here bulk renaming tools may be used to replace parts of the name strings or to add an ordering string (001, 002, 003, and so on) to the file names according to, for example, creation time, other ordering sequences in the name string, or additional information text files (such as Excel sheets or txt files from other alignment software packages) (*see* **Notes 4.5.1– 4.5.4**).

3.6 Segmentation

For convenience we briefly sketch here the IMOD [13] visualization tools as also described in detail in http://bio3d.colorado.edu/imod/doc/man/newstack.html.

Export all tif-images of the stack to one mrc-file using Cygwin software with the IMOD-terminal command line tool "`tif2mrc`".

To compensate for uneven image brightness between the individual images, use the command line tool "`newstack`" (option "`float`" with value "2").

Import the mrc-file into the graphical user interface of 3Dmod for visualization of the aligned data.

For segmentation, use the "`contur-auto`" option and adjust the histogram for optimized auto-segmentation of the electron dense rods (*see* **Note 4.6.1**).

Alternatively, nemaline rods can be segmented using the "`iso-surface`" tool. Figure 10 illustrates the typical IMOD visualization of the reconstructed density in xyz-slices (Fig. 10a), a surface rendered object (Fig. 10b, here one branching nemaline rod), and an ensemble image of segmented objects in their 3D environment represented by a slice through the 3D density data (Fig. 10c) (*see* **Note 4.6.2**).

For reference and further visual information on the technical workflow described here we point also to the movie article [3] and its supporting information.

4   Notes

4.1   Fixation and Embedding

1. Most chemicals used here are hazardous, so make sure you work in a fume hood and dispose of your waste according to your local authorities' regulations.

   Make sure to warm all frozen resin mixes to room temperature before opening the well closed containers to prevent moistening.

2. Dehydration and resin infiltration were done in a tissue processor (EMP-5160, RMC Boeckeler, Tucson, USA), but this is not necessary. Make sure that samples are properly agitated, for example, using a rotary shaker or similar device.

4.2   Making Section Arrays

1. Which trim tool to use: For a few sections (up to 100) a classical 45° diamond trim knife as used, for example, for cryo-trimming may be sufficient. It gives a pyramidal block with the disadvantage that the block face increases in size pretty fast when cutting large numbers of sections. For that, a 20° trim tool is better suited creating steeper sides of the pyramid. Producing rectangular pillars with the help of a 90° trim tool totally avoids increase in block face size when cutting hundreds of sections (for more information about trimming and sectioning *see* also Chapter 5).

2. Renlam resin tends to form ribbons even without coating block edges with glue mixture. For other resins or samples that do not form ribbons for other reasons (*see* **Note 4.2.3**, uneven compression) Pattex–xylene mixtures from 1:1 to 1:10 can help to stabilize ribbons.

3. Asymmetrical samples or asymmetrical embedding of the sample in the cut pyramid can lead to uneven compression and thus to curved ribbons—empty resin tends to compress in a different way from resin-embedded tissue impregnated with heavy metal. Nonuniformly applied adhesive may also cause ribbons to bend. A 35° knife and cutting thinner sections can reduce general compression. A curved ribbon may be straightened by stretching with xylene vapor, but this may induce ribbon breakage. In that case a second layer of glue mixture might help, but if nothing works, curved ribbons have to be accepted. Shorter ribbons have less curvature, meaning it may still be possible to place a few of them on one wafer. Alternatively, one has to use more substrates to accommodate a desired number of sections.

4. We usually let arrays sit overnight in a dust-free environment, for example, in a glass petri dish with a silicon pad. Thicker sections (>100 nm) and sections cut from hydrophilic resin (LR White, Lowicryl) may detach from the substrate (partially or completely) when not dried properly.

4.3 Poststaining

1. Regarding the quality of the water used for washing of the substrates and, to a minor degree, the water in the knife boat: It is extremely important to avoid formation of salt deposits and dirt smears on the sections. Depending on what is available one may have to test different options: Double glass distilled water is the canonical quality for EM work, but not many labs do have such an instrument any more. In a clinical environment ultrapure water for preparation of injections (*Aqua ad iniectabilia*) may be readily available—this was used here for the final wash. Water produced in ion exchange columns can be very dirty and may not be suitable, but this again depends on the water treatment device and also how well that is maintained.

2. To avoid transfer of solution from one beaker to the next remove the wafer very slowly so that the water runs down the surface of the wafer in one sweep, caused by the water's surface tension.

3. After the last wash, remove the wafer very, very slowly—ideally without leaving minidroplets on the sections. Remaining liquid may be removed carefully with compressed air. Check that your sections are tightly attached over their entire area, otherwise they may fold over.

4. Take your overview image with the camera lens placed exactly above and parallel to the surface of the substrate. Distortions in the overview image introduced by an oblique angle impair precision when placing the ROIs for subsequent hierarchical imaging steps.

4.4 FESEM Imaging

4.4.1 Initial Steps

1. For imaging in the SEM a field emission instrument is advantageous because its small probe size allows for good resolution even at low (1–3 keV) primary electron energy. Secondary electrons (SE) are influenced more by sample charging than back-scattered electrons (BSE). When there are issues with sample conductivity—as might happen with weakly metallized samples—BSE detectors can provide more consistent images with good contrast. Surface artifacts such as wrinkles or knife marks are also less obvious when using a BSE detector.

2. If you take your overview image and mount your sample in the SEM in the same orientation, finding the corners to image in the SEM and aligning the overview image will be simpler.

4.4.2 Hierarchical Imaging

1. To avoid the generation of huge amounts of data not relevant for the question to be analysed a hierarchical imaging strategy will be applied (Fig. 4). The Atlas software introduces specialized regions of interest (ROIs) called Sections and Sites, and sets of these ROIs (Section Sets and Site Sets) to help manage setup and collection of images over large numbers

of serial sections on the array. In our example, first a section set was defined on the overview image (Fig. 4a)—each ROI enclosing an entire section (Fig. 4b). This was recorded with 100 nm image pixels, a resolution sufficient to distinguish individual cells. Within this section set, several site sets were defined, each containing a different target cell (Fig. 4c shows one of these sites). The site sets were recorded with 5 nm pixel size—at this resolution microfilaments are visible when zooming in digitally (Fig. 4d).

2. In our example ribbons and individual sections were a bit scrambled—they had been floating and changing places in the knife boat. Keeping track of that helps establishing the right section order for later 3D reconstruction.

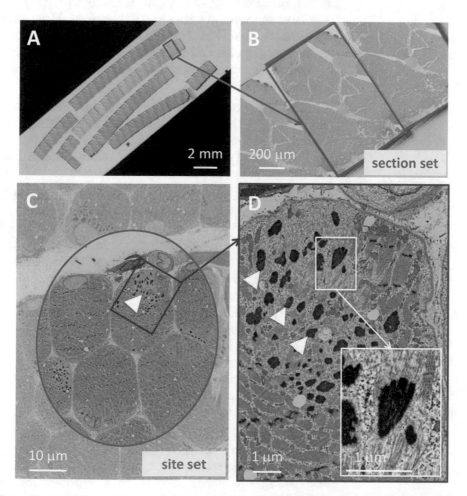

Fig. 4 Hierarchical imaging: on the overview image of the array (**a**) first a section set is defined to record entire sections (**b**) at low resolution (100 nm image pixel size). Then a second set of images (site set, **c**) containing a target cell with nemaline rods (arrowheads) is recorded at high resolution (5 nm image pixels). At this resolution microfilaments are clearly resolved (inset in **d**) as shown by digital zooming into the data

1. To compensate for slight tilts of the substrate with respect to the optical axis it is important to correct for that by applying software routines that keep the right focus and stigmation values when recording large numbers of sections—or in other words, when traveling large distances across a substrate such as a 4 in. wafer (Fig. 5a). Even on the small wafer fragment used in our example (Fig. 5b) the focus difference ΔZ was 78 μm across the entire array in y-direction ($\Delta y = 18$ mm) and 14 μm in x-direction (for $\Delta x = 5$ mm). However, the focus difference between two adjacent sections ($\Delta y = 0.7$ mm) was only 3 μm and this is a value that can be handled by the software routines implemented in Atlas 5. Although the algorithms for Auto Focus and Auto Stigmation (AFAS) are not designed to bring an image into focus and stigmation starting from a poor initial setting they do help to maintain focus and proper stigmation over large numbers of adjacent serial sections. How often these algorithms have to be called and corrections of electron optical settings are then applied depends on the sample and can be set in the image recording protocol—a good starting point is to do it on every section.

2. It is absolutely essential to have a suitable, fine enough sampling of the AFAS settings as illustrated by the fitted "sharpness metric" in Fig. 5c vs. 6a. Testing a range too large may not result in finding a proper focus and stigmation setting (well-defined maximum of the "sharpness metric").

3. Since the algorithm requires resolvable sample features to measure the sharpness it is important to optimize signal to noise by choosing the right detector and/or increase dwell time (Fig. 7). Autostigmation testing is done according to similar principles—the result is a 2D representation because two parameters—stigmator values in both x and y—are tested.

   Because by default AFAS is currently executed in the centre of each tile, it is important that this centre is not devoid of structural information (as would be the case, e.g., for empty resin or lumina in cells or tissue). When using a secondary electron detector (InLens or Everhart Thornley) the centre should be free from artifacts such as folds or dirt specks. Their usually high contrast would dominate the image and lead to an inappropriate focus.

4. For high resolution imaging of large ROIs over many sections (potentially taking hours) it is recommended to check performance of the auto-routines in a small test ROI beside the target ROI along the entire length of a representative ribbon (Fig. 6b).

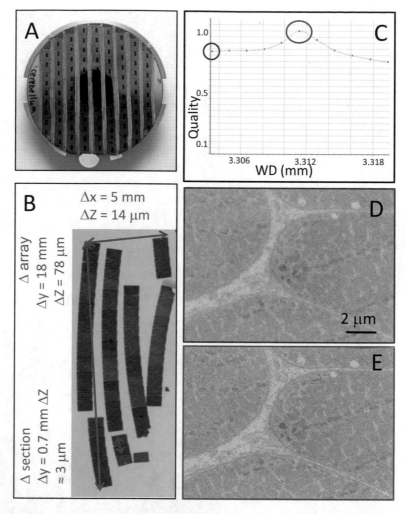

Fig. 5 Necessity of autofocus: large substrates such as a 4 in. wafer (**a**) or even a small fragment of wafer such as the example described here (**b**) are never exactly perpendicular to the optical axis, therefore the focus/working distance (WD) can be very different across the entire array (Δ array). Current Autofocus routines test only a limited focus range, in a limited number of steps (**c**). The range has to be adjusted to accommodate the focus difference between two consecutive sections (Δ section). The autofocus routine starts with a slightly blurry image (**d**) and goes through optimal focus to another blurry image. If everything is well adjusted a peak is found which produces a sharp image as a result (**e**)

4.4.4 Finding (Targeting 2) and Imaging the "Right Cell" at High Resolution

1. On a given section several smaller "sites," in our case that would be different cells (Fig. 8c) can be defined for automated imaging, either over the complete range of sections or for a defined number only.

2. It is important to consider the accuracy of the mechanical SEM stage when deciding on the size of a given ROI. Stage accuracy

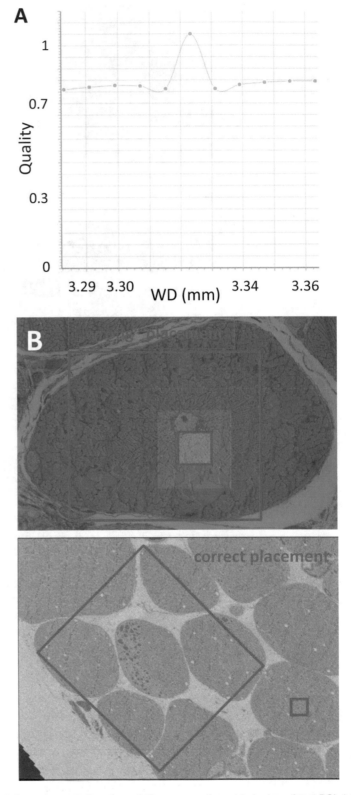

Fig. 6 Correct parameters for autofocus-sampling and placing of test ROI: (**a**) The algorithm finding the AF peak for different microscope settings relies on a

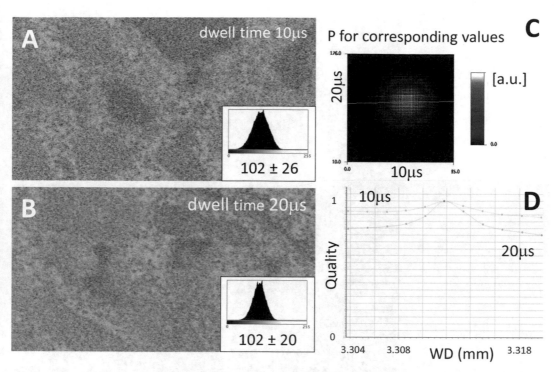

Fig. 7 Correct parameters for autofocus-dwell time and noise dependence: The dwell time (scan speed) for the AF imaging has a direct influence on the noise level of the image. (**a**, **b**) illustrate this for two dwell time settings. The images were recorded such, that min/max and mean are comparable. Note the very different standard deviation, which is much higher for short dwell times. (**c**) shows a 2D noise correlation plot (software package Fiji) which is another way of illustrating the higher noise level for shorter dwell times (10 μs vs. 20 μs). (**d**) shows the AF peaks for two different dwell times. As expected, the AF routine for longer dwell time (aka less noise) produces the better AF peak

Fig. 6 (continued) correct sampling of the working distance (WD). Depending on software it may happen that—as in this case—the sampling has not enough bins to describe a possible peak correctly. Here only one point would suggest a focus peak (hit by chance), the plotted quality curve is therefore not significant. Cf. Figure 5 where three points define the AF maximum. (**b**) The ROI (small squares) for testing the performance of the AF routine should be placed in such a way (e.g., beside the target cell as shown in the lower panel, small green frame), that contamination gathered during the AF routine is not accumulated on the area later to be imaged at high resolution (large square). Note the contamination in the upper panel within and around the small red frame. It should also be noted, that in the case of using the AF routine *after* the high-resolution imaging (AFAS performed on "previous tile") the test ROI may deliberately be placed in the imaged region—as shown here as "wrong" placement (cf. text)

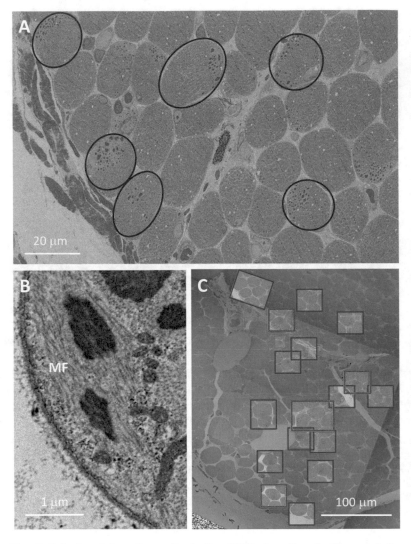

Fig. 8 Selecting interesting cells in the SEM—targeting 2. The resolution (100 nm image pixels) of an image from the section set (**a**) is sufficient to identify cells containing many nemaline rods (red circles). Details of the microfilament (MF) system, however, are only resolved at 5 nm image pixels (**b**). To identify cells with interesting constellations of MF and rods about 20 candidate cells (blue squares) on a guide wafer were screened using a section set with 5 nm image pixels. After evaluation they were marked to create a score/quality map of the section (**c**, yellow = highest score)

on the instrument we used in this example is in the order of 5 μm. That means a frame around the selected cell of at least this value was added to compensate for the mechanical inaccuracies of the SEM stage.

3. It is possible to run AFAS as **"Perform "on previous tile"** that means after high-resolution image acquisition of a given ROI. This avoids AFAS-related staining/contamination/bleaching in the acquired high-quality images when samples are prone to contamination or when using the InLens detector (cf. Fig. 6b).

4.4.5 Exporting Data, Documentation

1. Take into account that the export function will work on what you see in the "main view." So if you have changed anything regarding display (brightness, contrast, transparency) make sure that you do that in a consistent manner.

4.5 Registration

1. We want to point out, that rigid registration does usually more faithfully represent the underlying 3D structure, as rigid registration does not artificially morph structures (Fig. 9a, c). However, section compression and also local stretching (e.g., by local folds releasing stress in the sections) may be compensated by the morphing of elastic registration (Fig. 9b, d). In general, elastically registered image stacks can therefore—artificially— look smoother in their 3D visualization (cf. Fig. 9a, b).

   It is good scientific practice to always look at both the rigid and elastic registration data. Only by comparing the two one can obtain a measure for the artifacts introduced and the true underlying structure (cf. Fig. 9a, b and also Fig. 9c, d).

2. If quantitative volume results (e.g., quantitative topology/surface area/volume-to-surface ratio) is needed elastic registration must not be used. Uncontrolled morphing will introduce uncontrollable errors in volumetric data.

3. It is also very important to adapt default parameters of any used registration software to the actual picture characteristics. Here picture size, resolution, fast or slow variation of structural details according to section thickness (and thus the image sampling in z), or the ratio of area size of the ROI compared to that of the whole image need to be tested and parameters adapted accordingly.

4. Most of the time it is advantageous to test and adjust data processing on a smaller subset of images first, before running larger batch jobs with the full data set.

4.6 Segmentation

1. Semiautomated segmentation was carried out by shortcuts "a" (interpolation tool) and "b" (addition of filled material to object).

2. Depending on the software packages used multicore compute power and parallel computing (even on GPUs) can be advantageous. This has to be considered depending on data size, software used and typical visualization parameter settings.

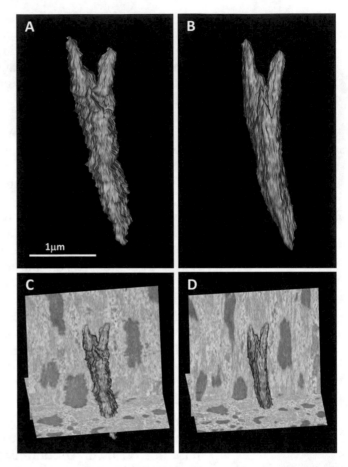

Fig. 9 Consequences of different alignment procedures for segmentation and surrounding area of an identified object in 3D. The same object rendered after least squares rigid alignment (**a**) or after an additional elastic alignment step (**b**). (**c, d**): Objects from (**a, b**) together with their embedding 3D surrounding areas represented here as two orthogonal planes cutting through the reconstructed 3D density. In (**c, d**) comparable sections are shown. Note the dissimilar appearance of object densities and surrounding; both are affected by the different alignment procedures. The multistep alignment and rendering (**b, d**) gives a smoother—presumably less artifact-prone—3D object. Note the "jitter" in the least squares rigid alignment, which is clearly visible when comparing (**a, c**) with (**b, d**)

However, for most steps during registration and visualization no overly specialized computers or high-end computing power is needed.

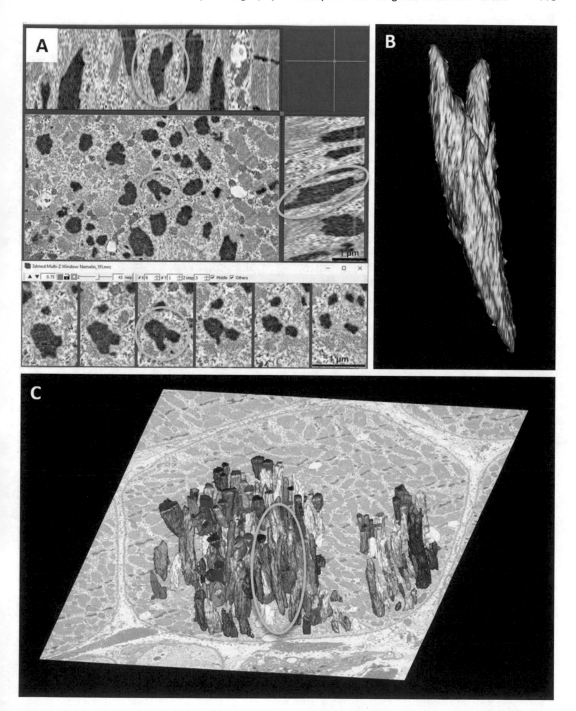

Fig. 10 Visualization of segmented objects in their 3D environment. Using the software package IMOD to highlight a branched nemaline rod: Sections through the reconstructed 3D density (**a**) showing XYZ orthoslice views (top three panels) and "Multi-Z" view (lower row). Segmented nemaline rod as isolated 3D rendered object (**b**) and within the entire population of rods (**c**)

References

1. Micheva KD, Smith SJ (2007) Array tomography: a new tool for imaging the molecular architecture and ultrastructure of neural circuits. Neuron 55:25–36

2. Wacker I, Schröder RR (2013) Array tomography. J Microsc 252:93–99

3. Wacker IU, Veith L, Spomer W et al (2018) Multimodal hierarchical imaging of serial sections for finding specific cellular targets within large volumes. J Vis Exp 133:e57059. https://doi.org/10.3791/57059

4. Wacker I, Chockley P, Bartels C et al (2015) Array tomography: characterizing FAC-sorted populations of zebrafish immune cells by their 3D ultrastructure. J Microsc 259:105–113

5. Wacker I, Spomer W, Hofmann A et al (2016) Hierarchical imaging: a new concept for targeted imaging of large volumes from cells to tissues. BMC Cell Biol 17:38

6. Stirling JW, Curry A (2009) Quality standards for diagnostic electron microscopy. Ultrastruct Pathol 31:365–367. https://doi.org/10.1080/01913120701638660

7. Malfatti E, Lehtokari V-L, Böhm J et al (2014) Muscle histopathology in nebulin-related nemaline myopathy: ultrastructural findings correlated to disease severity and genotype. Acta Neuropathol Commun 2:44

8. Malfatti E, Romero NB (2016) Nemaline myopathies: State of the art. Rev Neurol 172:614–619

9. Richardson KC, Jarett L, Finke EH (1960) Embedding in epoxy resins for ultrathin sectioning in electron microscopy. Stain Technol 35:313–325

10. Reynolds ES (1963) The use of lead citrate at high pH as an electron opaque stain in electron microscopy. J Cell Biol 16:208–212

11. Cardona A, Saalfeld S, Schindelin J et al (2012) TrakEM2 software for neural circuit reconstruction. PLoS One 7:e38011

12. Schindelin J, Arganda-Carreras I, Frise E et al (2012) Fiji: an open-source platform for biological-image analysis. Nat Methods 9:676–682

13. Kremer JR, Mastronarde DN, McIntosh JR (1996) Computer visualization of three-dimensional image data using IMOD. J Struct Biol 116:71–76

14. Blumer MJF, Gahleitner P, Narzt T et al (2002) Ribbons of semithin sections: an advanced method with a new type of diamond knife. J Neurosci Methods 120:11–16

15. Mollenhauer HH (1987) Contamination of thin sections: some observations on the cause and elimination of "embedding pepper". J Electron Microsc Tech 5:59–63

Chapter 7

Correlative Ultrastructural Analysis of Functionally Modulated Synapses Using Automated Tape-Collecting Ultramicrotome and SEM Array Tomography

Ye Sun, Connon Thomas, Takayasu Mikuni, Debbie Guerrero-Given, Ryohei Yasuda, and Naomi Kamasawa

Abstract

Live imaging of dendritic spines using advanced light microscopy (LM) provides insight into how the brain processes information to learn and form memories. As a complementary approach, electron microscopy (EM) offers a complete view of the ultrastructural characteristics of synapses, such as the size of postsynaptic density, as well as the distribution and number of synaptic vesicles in the presynaptic terminal. By bridging these two different visualization platforms, function and ultrastructure can be directly linked at the level of individual synapses. The technical challenge is how to examine the same spines in reliable and reproducible ways using two imaging modalities with completely different spatial scales. Here, we describe our detailed workflow to combine light and electron microscopy for efficient correlative analysis of spines of interest. As an example, we show how to find a dendritic spine that is stimulated with 2-photon glutamate uncaging on a CA1 pyramidal neuron expressing green fluorescent protein (GFP) in organotypic hippocampal slices. Following fluorescence observation under a 2-photon fluorescence microscope, the tissue is processed for EM using pre-embedding immunogold-labeling of GFP to locate the cell of interest. It is then sectioned with the Automated Tape Collecting Ultramicrotome (ATUMtome) to reliably and quickly collect hundreds of serial sections from a large block face (up to 3 × 3 mm). Then using a scanning electron microscope (SEM) in combination with array tomography software (Atlas 5 AT), we semiautomatically collect images at multiple resolutions. The obtained volumetric dataset is reconstructed and analyzed in a 3D manner. This workflow allows us to collect data for quantitative analysis faster than conventional serial sectioning followed by transmission electron microscopy (TEM) imaging.

Key words Correlative light and electron microscopy (CLEM), ATUMtome, Array tomography, Scanning electron microcopy (SEM), Atlas 5 AT, 2-Photon microscopy, Glutamate uncaging, Synapses

1 Introduction

A synapse is the smallest unit of cell-to-cell signal transduction between neurons and is a key component of neuronal signal transmission and plasticity [1–3]. Studies of the functional properties of a synapse together with structural characteristics in LM are

Irene Wacker et al. (eds.), *Volume Microscopy: Multiscale Imaging with Photons, Electrons, and Ions*, Neuromethods, vol. 155, https://doi.org/10.1007/978-1-0716-0691-9_7, © Springer Science+Business Media, LLC, part of Springer Nature 2020

particularly difficult because the size of synapses is close to the limit of optical resolution. Recent advances in multiphoton microscopy techniques to image and manipulate activity of single synapses using calcium sensors and glutamate uncaging have allowed researchers to analyze functional properties of single synapses in living tissue [4, 5]. While the spatial resolution of LM has been dramatically improved recently [6–8], EM is still considered to be the gold standard for ultrastructural analysis of synapses [9, 10]. TEM with its high resolution that can visualize lipid bilayers of membranes allows us to resolve many of the structural details of the post- and presynaptic components, that is, synaptic vesicles in the presynaptic terminal and postsynaptic densities in spines, which cannot be resolved even with the most advanced LM. One drawback of EM imaging is the lack of temporal resolution because samples have to be fixed. Therefore, to reveal the dynamics of individual synapses, correlative imaging between LM and EM is desirable.

Large volumes of work have been put forth to characterize and record the widely variable functional and morphological properties of dendritic spines. Structural plasticity of this tiny protrusion, most commonly found in dendrites of excitatory neurons, is known to be strongly linked to function—or dysfunction—of the brain [1]. For example, it has been suggested that spontaneous neuronal activity modifies the morphology of dendritic spines, and activity dependent structural plasticity is considered to be central to learning and memory [11–14]. Enlargement and shrinkage of dendritic spines are also associated with long-term potentiation (LTP) and long-term depression (LTD), respectively [14–17], and formation of new spines and elimination of existing spines are known to be correlated with an animal's learning [18].

To study these dynamics of dendritic spines and combine them with EM-level ultrastructural analysis, it is necessary to find the same spines of interest under both LM for live-tissue imaging and then EM after fixation. However, the correlative LM-EM imaging of spines and synapses is not trivial. Spines, with volumes between 0.01 and 0.8 μm^3, are such tiny features that it may seem to be an insurmountable task to refind the same individual in densely packed brain structure in the EM [19]. Many clever techniques have been developed to facilitate locating dendritic spines of interest [20], including but not limited to immunohistochemistry of genetically introduced fluorescent proteins [21], photo-oxidation [22, 23], expressing genetically encoded tags [24–27], fiducial markers created by near-infrared branding (NIRB) [28], diaminobenzidine (DAB) precipitation [29, 30], and laser burning [31]. Though conventional serial sectioning and TEM imaging can be used to image considerable sample volumes (and truly offers the highest resolution), it requires a high level of skill and a large investment of time. Sectioning and handling hundreds of serial sections, for example, is challenging and can be seen as an art form in and of itself. In addition, traditional TEM grids offer a relatively small area for sections and require extensive trimming of the sample. This trade-off of section area for section quantity is often unacceptable when large-scale correlation is attempted. To make this process more user friendly and consistent, numerous techniques using scanning EM (SEM) have been developed such as the automated tape collecting ultramicrotome (ATUMtome) [32], serial block-face SEM (SBF-SEM, *see* also Chapters 9 and 10 this volume) [33], focused ion beam SEM (FIB-SEM, *see* also Chapters 11 and 12 this volume) [31, 34], and array tomography (AT, *see* also Chapters 4, 6, and 8 this volume) [35–37]. Each technique has its own advantages and limitations as discussed previously [38], and this chapter will solely describe a method utilizing the ATUMtome. We will describe in detail the workflow we developed to address how glutamate uncaging adjacent to a smooth dendritic segment affects the morphological characteristics of newly formed spines and their surroundings. We will also discuss the benefits and current limitations of this technique.

Briefly, we applied pre-embedding immunogold labeling combined with array tomography to relocate and image a specific dendritic protrusion induced previously by two-photon MNI-glutamate uncaging. This was achieved using the commercially available ATUMtome (RMC Boeckeler), together with a software platform for automated SEM imaging, Atlas 5 AT array tomography (Carl Zeiss Microscopy), and a Merlin VP compact SEM (Carl Zeiss Microscopy). GFP was expressed in CA1 pyramidal neurons in cultured organotypic hippocampal slices. An immunoreaction was then carried out against GFP to label the target neuron with nanogold particles. The position of target neurons was confirmed by comparing the GFP fluorescence map with the silver enhancement pattern of the gold particles in the embedded sample, and following this, a large area of the sample (up to 3 × 3 mm in *xy*) was trimmed out for ultrathin sectioning on the ATUMtome. During microtoming, ultrathin sections were continuously collected onto a roll of Kapton tape. This tape was then aligned on a 4-in. silicon wafer and imaged in the SEM at different magnifications with multiple detectors (SE2, BSD, InlensDuo) in steps. Final images were collected at low to high resolution, and processed for 3D reconstruction of the newly induced protrusions [39].

Flowchart of the workflow and approximate timeline.

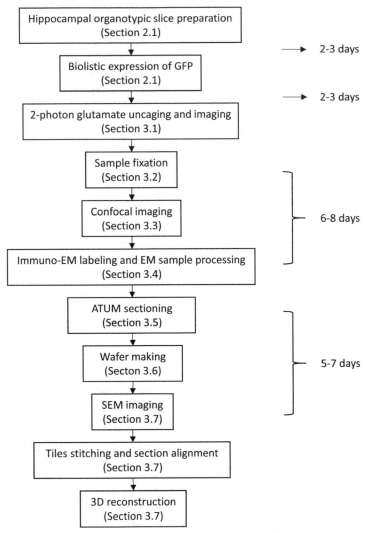

2   Materials

2.1   Hippocampal Slices and Fluorescent Protein Expression

Organotypic slice culture is a useful tool for investigating synaptic plasticity in vitro. It retains major neuronal architectures, and resembles native morphological and physiological characteristics of neurons [40, 41]. Organotypic slices can be maintained in vitro for weeks, allowing for expression of exogenous protein by various methods [42, 43]. We choose hippocampal organotypic slices to study the structural plasticity of dendritic spines on CA1 pyramidal neurons. However, samples compatible with this method could come from primary neuronal culture, acute slice, and organotypic slice culture of other brain regions, and in vivo [15, 17, 44–46].

Fluorescent protein expression is necessary to visualize neuronal dynamics in the LM. GFP is most commonly used, and other fluorescent proteins with different emission colors can also be applied [47]. There are different options to express fluorescent proteins in organotypic slices (e.g., biolistic transfection, liposome transfection, electroporation, and virus infection [42, 43]), which can be selected for a particular purpose of the experiment.

We expressed GFP with CAG promoter for strong ubiquitous expression, and identified CA1 pyramidal neurons based on neuronal morphology and location in hippocampal structure. Different promoters can be used to express fluorescent proteins in specific cell types. Additionally, temporal control of protein expression can be manipulated by inducible gene expression systems [48]. Fluorescent proteins can be fused to other functional proteins for exogenous expression, or inserted into the genome to label endogenous proteins [49, 50]. In any case, once the expression system is established, dynamic trafficking and localization of specific proteins can be correlatively studied in LM and EM.

2.2 Two-Photon Microscopy Setup and Solutions for Glutamate Uncaging

1. The two-photon microscope setup has been described previously [51]. A custom-built two-photon microscope equipped with two Ti:sapphire lasers (Coherent) was used for imaging and glutamate uncaging. One laser was tuned to 920 nm to excite GFP for structural imaging, the other laser was tuned to 720 nm for glutamate uncaging, and the intensity of each was independently controlled using Pockels cells (Conoptics). Laser power measured after the objective lens was 1–2 mW for imaging, and 3–4 mW for uncaging. The microscope was equipped with $10\times$ and $60\times$ water immersion objective lenses. The imaging chamber was connected to a circulation system, in which a solution can be circulated and bubbled with 95% O_2–5% CO_2 throughout the experiment. The temperature of the solution was maintained at 32 °C.

2. Artificial cerebrospinal fluid (ACSF): 127 mM NaCl, 25 mM $NaHCO_3$, 25 mM D-glucose, 2.5 mM KCl, 1.25 mM NaH_2PO_4. Osmolarity should be around 310 mOsm. Stored at 4 °C.

3. Glutamate uncaging solution: Bubble ACSF with 95% O_2–5% CO_2 for 10 min, then add the following components to reach the final concentration: 4 mM MNI-caged-L-glutamate (No. 1490, Tocris), 4 mM $CaCl_2$, 1 μM TTX. Prepare right before experiment. Note: $CaCl_2$ will form precipitations in ACSF without enough bubbling.

2.3 Solutions and Reagents for Pre-embedding Immuno-EM Sample Processing

1. 0.1 M Sorenson's phosphate buffer (PB), 0.1 M NaH_2PO_4, 0.1 M Na_2HPO_4, pH 7.4.

2. Fixative solution: 2% paraformaldehyde (PFA) and 2% glutaraldehyde (GA) in 0.1 M PB.

3. 50 mM glycine solution in 0.1 M PB.

4. 1% sodium borohydride in 0.1 M PB (if necessary).

5. 15% and 30% sucrose in 0.1 M PB.

6. Liquid nitrogen.

7. Tris-buffered saline (TBS), 50 mM Trizma base, 150 mM NaCl, pH 7.4.

8. Blocking buffer: 10% normal goat serum (NGS) and 1% fish skin gelatin (FSG) in TBS.

9. Primary antibody solution: Primary antibody (rabbit polyclonal anti-GFP antibody, 1/5000, ab6556, Abcam) in TBS containing 1% NGS, 0.1% FSG, and 0.05% sodium azide (NaN_3).

10. Secondary antibody solution: Nanogold conjugated secondary antibody (1/100, # 2003, Nanoprobes) in TBS containing 1% NGS, 0.1% FSG, and 0.05% NaN_3.

11. HQ silver enhancement kit (# 2012, Nanoprobes).

12. Phosphate buffered saline (PBS, GE Hyclone), pH 7.4.

13. 0.5% osmium tetroxide (OsO_4, EMS) aqueous solution.

14. 1% uranyl acetate (UA, SPI supply) aqueous solution.

15. Ethanol.

16. Acetone.

17. Propylene oxide.

18. Fluka Durcupan: component in weight A: 11.4 g, B: 10.0 g, C: 0.3 g, D: 0.05 g, mix in order.

2.4 Setups for Ultrathin Sectioning with ATUMtome and Wafer Preparation

1. Dissecting microscope.

2. ATUMtome—PowerTome equipped with ATUM and its control software (RMC Boeckeler) (Fig. 1).
   - Water level control system.
   - Antistatic ionizer.
   - Air-activated antivibration microtomy table.
   - Silent compressor.
   - Environmental chamber (Fig. 1, inset).

3. 4 mm diamond knife, ultra Maxi 35°, mounted in large-cavity blue anodized holder (DIATOME) (Fig. 2a)

4. Wafer workstation (RMC Boeckeler) (Fig. 5a).

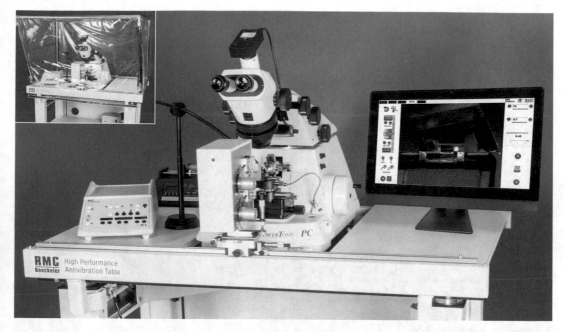

Fig. 1 Whole view of ATUMtome system. The ATUMtome is a combination of an ultramicrotome (PowerTome) and a tape collection device (ATUM), both operated by the same software. Antistatic device, water level control system, and air-activated antivibration microtomy table are included in the system. Inset shows the environmental chamber closed for more stable sectioning

5. Glow-discharged Kapton tape (RMC Boeckeler).

   *After discharging, the tape retains its hydrophilic state for 2–3 weeks. (As an easy test, drop water on the tape surface. If the water diffuses, the tape is hydrophilic and suitable for collecting sections.)

6. 4 in. diameter silicon wafer.

7. Double-sided adhesive carbon tape (50 mm width, EMS).

8. Antistatic roller.

9. Copper EM grids (100–150 mesh).

10. Carbon coater, need to have large specimen chamber for wafers (Leica EM ACE600).

11. Carbon thread.

2.5 Image Acquisition and 3D Reconstruction

1. Scanning electron microscope (SEM), Merlin VP Compact (Carl Zeiss Microscopy).

2. 4 in. Wafer holder (Carl Zeiss Microscopy).

3. Zeiss Atlas 5 AT software, V5.0.49.4 (Carl Zeiss Microscopy) for array tomography imaging.

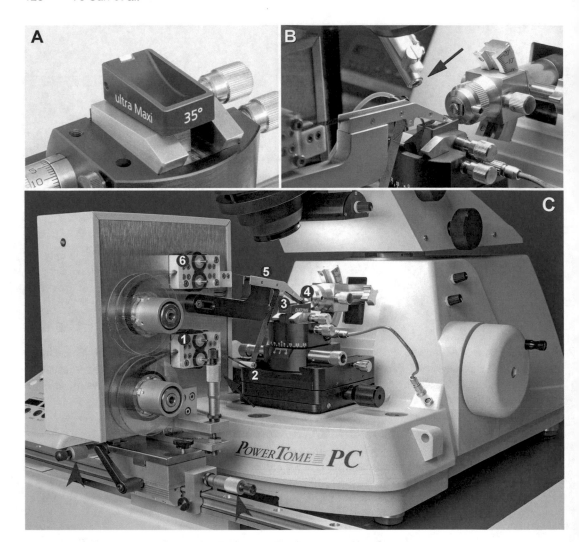

Fig. 2 ATUMtome setup. (**a**) ultra Maxi diamond knife with a 4 mm cutting edge and a 35° angle. (**b**) Close-up view of ATUMtome setup, ready for ultrathin sectioning. The tip of the antistatic device (arrow) faces the knife edge, and the nozzle of the water level control system (arrowhead) is inserted into the water of the knife boat. (**c**) Complete ATUM setup with rolls of Kapton tape. After the tape is loaded through ① bottom pinch roller, ② tape-speed potentiometer, ③ tape stabilizer ④ tape head, ⑤ tape guides, and ⑥ top pinch roller (in this order), the ATUM slides into position and is adjusted for section collection. The height of the ATUM tape head is adjusted by the Z-axis micrometer (arrow), and X- and Y-axes are adjusted by their respective micrometers (arrowheads)

4. Software for image alignment, segmentation, and 3D reconstruction TrakEM in Fiji, https://imagej.net/TrakEM2
   Microcopy Image Browser, http://mib.helsinki.fi/. Amira (Thermo Fisher Scientific).

3   Methods

This method describes the procedures to achieve correlative imaging between the 2-photon microscope and EM in a relatively fast and convenient manner for quantitative data analysis. Samples compatible with this method could come from primary neuronal culture, acute slice, organotypic slice culture, and in vivo. Manipulation of neuronal structure can be induced via different methods such as chemical stimulation, electrophysiological stimulation, photon stimulation, and animal behavior training, all while recording with the confocal or 2-photon microscope [15, 17, 44–46]. To facilitate visualization, target neurons or specific molecules are typically labeled with fluorescent proteins. Such labeling can be achieved by various techniques, such as virus infection, plasmid transfection by lipofection or electroporation, and transgenic animal production [52–55]. In this section, we describe a workflow for correlative imaging of newly formed spines induced by two-photon glutamate uncaging in biolistically transfected GFP positive CA1 pyramidal neurons in organotypic hippocampal slices [39]. It is noted that this method could be applied to a variety of different samples described above with minor adjustments.

3.1   Spinogenesis Induction by 2-Photon Glutamate Uncaging

1. Organotypic hippocampal slices can be prepared according to previous description [56], and GFP expression in a small population of CA1 neurons can be achieved by biolistic transfection as previously described [52]. In brief, prepare organotypic hippocampal slices with a thickness of 300–350 μm on Millicell membrane using postnatal day 4 (P4) to P6 C57/B6 mouse pups. After 2–3 days culture in vitro, transfect pCAG-GFP plasmid biolistically for expression in neurons. Two days after transfection, successful GFP expression can be observed with a fluorescence microscope. Cut out Millicell membrane containing the slice with an extra 2–3 mm space around the slice by a scalpel blade right before 2-photon glutamate uncaging.

2. Spinogenesis can be induced by 2-photon glutamate uncaging with a modified protocol from a previous description [57]. In brief, choose a smooth segment of secondary dendrite on a GFP-positive CA1 pyramidal neuron. Acquire z-stack images centered with the selected segment using $60\times$ objective lens at $1\times$, $5\times$, and $25\times$ zoom with 1–2 mW 920 nm 2-photon laser (Fig. 3a–c). Uncaging can be done 0.5 μm away from the side of the dendrite with 3–4 mW, 2 Hz $\times$ 40 stimulation by a 720 nm 2-photon laser (see **Note 1**).

3. Image the stimulated area at $25\times$ zoom in z-stack mode immediately after uncaging, and repeat imaging with an interval of 1 min (Fig. 3d).

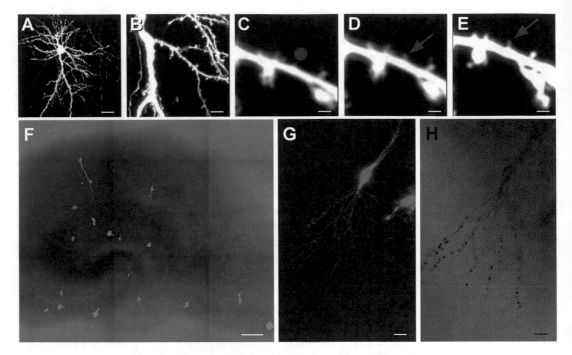

Fig. 3 Spinogenesis induced by 2-photon glutamate uncaging. The target CA1 pyramidal neuron was imaged by 2-photon microscopy with a 60× objective lens at 1× (**a**), 5× (**b**), and 25× (**c–e**) digital zooming mode. The position of glutamate uncaging is indicated by the red dot in (**c**). A small protrusion is shown immediately after uncaging (**d**, arrow) and is sustained after fixation (**e**, arrow). (**f**) Whole hippocampal slice imaged by confocal microscopy, showing GFP expressing pyramidal neurons. (**g**) Soma and dendrites of a GPF expressing neuron imaged by confocal microscopy. (**h**) Brightfield LM image of the resin embedded sample showing the same cell as captured in (**g**). Immunogold-labeled dendrites show the same structure as dendrites in the corresponding fluorescence image. Scale bars: (**a**) 25 μm; (**b**) 5 μm; (**c–e**) 1 μm; (**f**) 200 μm; (**g–h**), 20 μm

4. Image for at least 10 min to confirm the formation of a new protrusion. Remove the slice from the setup and immediately transfer into fixative solution.

5. After fixation, image the same region to confirm successful spinogenesis (Fig. 3e).

3.2   Tissue Fixation

1. Transfer the slices into the fixative solution, incubate for 1 h on ice, then wash with 0.1 M PB for 3 × 15 min on ice.

3.3   Overview Imaging of the Tissues by Confocal Laser Scanning Microscopy

1. Wrap a microscope slide with Parafilm. Cut out the center region of the Parafilm to make a window large enough to place the slice inside. Fill the window with 0.1 M PB, and submerge the slice in the solution. Metal wire can be used to hold the position of the slice (*see* **Note 2**).

2. Take z-stack confocal images using the 10× objective lens in tile mode to cover the entire slice (Fig. 3f) (*see* **Note 3**).

3.4 Pre-embedding Immunogold-EM Sample Preparation

3.4.1 Antibody Reaction and Silver Enhancement

1. Incubate the slices in 50 mM glycine for 10 min to block free aldehyde groups (*see* **Note 4**).

2. Wash with 0.1 M PB once, and incubate slices in 15% and 30% sucrose, successively, for 1–2 h each with gentle shaking at 4 °C.

3. Put the tissue slices on a folded piece of aluminum foil (approximately 4 cm × 1.5 cm, four single layers), with the slice side facing down. Use a clean disposable wipe to remove extra sucrose solution. With forceps, hold one edge of the aluminum foil and hover the slice over liquid nitrogen until slices and residual sucrose turn rigid and white, then dip the foil into liquid nitrogen, holding it for 1 min. Take the sample out, and warm it up in room temperature until slices become transparent again. Repeat this process for one more time when necessary (*see* **Note 5**).

4. Transfer the slices to a dish containing 0.1 M PB. Detach the slices from aluminum foil by gently pipetting solution over them.

5. Transfer the slices to TBS, and wash for 2 × 10 min.

6. Transfer the slices to blocking buffer, and incubate on a shaker for 1 h at 4 °C.

7. Transfer the slices to primary antibody solution, incubating for 2 days at 4 °C with gentle shaking (*See* **Note 6**).

8. Wash the slices with TBS for 4 × 15 min, followed by incubation in nanogold conjugated secondary antibody solution for 1 day at 4 °C with gentle shaking.

9. Wash the slices with PBS for 3 × 10 min, then postfix slices in 1% GA in PBS for 10 min. During the waiting time, take out HQ silver intensification kit from freezer and thaw it in room temperature water (*see* **Note 7**).

10. Wash the slices with PBS for 2 × 10 min, then wash them thoroughly with deionized water for 4 × 5 min (*see* **Note 8**).

11. Add two drops of solution A and two drops of solution B of the silver enhancement kit into a round bottom 2 ml microcentrifuge tube. Mix the solution thoroughly by vortex, then add two drops of solution C, followed by another vortex. Transfer one slice into the solution immediately after mixing, making sure the whole slice has been submerged. Keep the tube in dark for 6–8 min (*see* **Note 9**).

12. Stop the reaction by adding deionized water into the microcentrifuge tube. Pour the solution into a clean well, and add more deionized water. Briefly observe the silver enhancement conditions under a dissecting microscope. The GFP positive

neuron should be visible by brown coloration at this time point. Coloring may vary depending on sample conditions and reaction time.

13. Transfer the slices to 0.1 M PB and wash for 2×10 min to stop the reaction fully.

3.4.2 Postfixation, Dehydration, and Resin Embedding

1. Wash the slices in deionized water for 5×2 min, and transfer each slice into a separate well in a 24-well plate.

2. Remove water from the wells gradually until slices are attached to the bottom of the well (*see* **Note 10**). Add a few drops of 0.5% OsO_4 solution onto the slices. Wait for 30 s to 1 min, then add additional OsO_4 solution to make up to 0.5 ml per well (*see* **Note 11**).

3. Seal the 24-well plate with Parafilm to avoid OsO_4 evaporation or any contaminations, and keep at 4 °C for 40 min (*see* **Note 12**).

4. Wash the slices in deionized water for 5×5 min (*see* **Note 13**).

5. Immerse the slices in 1% UA solution for 35 min in dark at 4 °C.

6. Wash the slices in deionized water once for 5 min.

7. Transfer the slices to glass vials filled with 30% ethanol on a rotator for 10 min.

8. Immerse the slices successively in 50%, 70%, 90%, 100% ethanol for 10 min in each, then transfer to 100% acetone, 1:1 acetone–propylene oxide, and 100% propylene oxide (*see* **Note 14**).

9. Mix Durcupan reagents thoroughly (*see* **Note 15**).

10. Mix propylene oxide and resin at a 3:1 volume ratio, and incubate slices in the mixed solution for 1–2 h on the rotator. Then, change the solution to 1:1 for 1–2 h, and 1:3 for overnight.

11. Transfer the slices to 100% resin. Make sure they are fully submerged and keep them in the desiccator for at least 4 hours.

12. Flat embed the slices between Aclar sheets. Carefully pick up one slice sample and transfer it onto a piece of Aclar sheet with the slice side facing up (*see* **Note 16**). Slowly put another piece of Aclar sheet on the slice from one side to another to avoid inducing extra air bubbles (*see* **Note 17**).

13. Cure slices in 60 °C incubator for 2 days until resin is fully polymerized.

3.5 Trimming and Ultrathin Sectioning with ATUMtome

3.5.1 Identification of Target Region

1. Check the embedded slices under an LM. Immunostained positive neurons have darker brown color than surrounding tissue. GFP fluorescence images of the slices are helpful to identify the target neuron(s) by fitting the outer shape of the tissue slices (Fig. 3f–h).

3.5.2 Sample Trimming

1. Cut out the region of slice containing the target neuron with a scalpel blade. The size should be larger than that of the desired final block face (*see* **Note 18**).

2. Mount the piece onto a blank resin block with a small amount of superglue or resin. Make sure the slice faces up, and no bubble exists between the resin block and the sample. If the sample is mounted with resin, polymerize for a day at 60 °C.

3. Place resin block with the slice in a PowerTome specimen chuck (*see* **Note 19**).

4. Trim the tissue slice into a rectangular shape with a trimming diamond knife or the edge of a glass knife. Make sure the top and bottom of the rectangle are strictly parallel. Make the side edges of the block face 90°, so that every ultrathin section has the same size and shape. Adjust the size of the block face according to the size of the target neuron. We typically use up to 3 mm × 3 mm block face, though this is only limited by the size of the diamond knife. Larger sections are easier for automatic collection of one section at a time by the tape (*see* **Note 20**).

5. Carefully trim the surface of the slice until it is uniform and smooth with glass knife.

6. Change the knife to 35° ultra Maxi (Fig. 2a) and align the block face for cutting. Expose a small amount of tissue (~1–2 μm) before beginning section collection, and retract the sample holder for ATUM setup.

3.5.3 ATUM Setup

1. Loading tape: switch on the power button on ATUM. From the controlling software, go to "Setup"-"ATUM." Click on "Jog Top Forward" and "Jog Bottom Forward" (Fig. 4a, arrows), and both top and bottom pinch rollers should start running. Jogging speed can be adjusted from the software (Fig. 4a, arrowhead). Usually, jogging speed can be set at 2–4 mm/s (*see* **Note 21**).

2. Insert the tape reel onto the bottom tension motor, and guide the tape through the bottom pinch roller tunnel (*see* **Note 22**).

3. After the tape has been moved through the bottom pinch roller tunnel (Fig. 2c, ①) for a certain length, guide it to pass underneath the tape-speed potentiometer (Fig. 2c, ②), through the

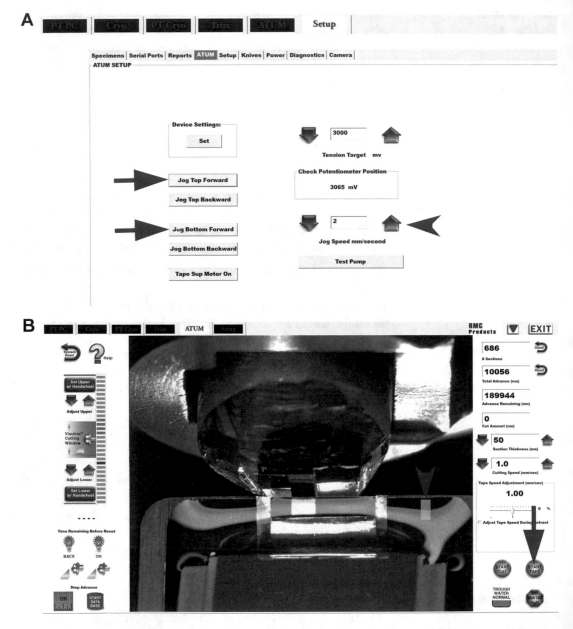

Fig. 4 ATUM control software monitor views. (**a**) ATUM setup window. Arrows point to boxes for controlling the pinch rollers' running. Jog speed is controlled by clicking the blue arrows in the right column (arrowhead). (**b**) ATUM main control window. Left panel contains buttons for PowerTome operation; middle panel shows a live camera view of sectioning. The edge of the diamond knife, ATUM tape head with Kapton tape, and a water level indicator (arrowhead) are shown. Right panel contains the PowerTome running status, section thickness and speed, and tape speed controller. A button to start/stop cutting, a button to start/stop tape running (arrow), a water level status monitor icon, and a button for water level control are found at the bottom of the right panel

tape stabilizer (Fig. 2c, ③), around the tape head (Fig. 2c, ④), between the tape guides (Fig. 2c, ⑤), and through the top pinch roller tunnel (Fig. 2c, ⑥).

4. Insert an empty tape reel to the top tension motor. Use double-sided adhesive tape to attach the Kapton tape to the empty tape reel.

5. Switch the top and bottom pinch rollers between running forward and back until potentiometer shows number around 3000 mV.

6. From the controlling software, go to the "ATUM" tab and select "start tape" (Fig. 4b, arrow). Run the tape until it makes one loop around the empty tape reel, then stop the tape.

7. Move ATUM laterally on the slide (x axis) to carefully position the tape head above the middle of the diamond knife boat, roughly opposing to the resin block. Lower the tape head until it touches the water surface by adjusting the z-axis micrometer (Fig. 2c, arrow). Adjust the precise position of the tape head by adjusting the two horizontal micrometers (x and y axis, Fig. 2c, arrowheads) until the tape head is exactly opposing the sectioning position at a distance of $1.5\times$ the section length from the knife edge (*see* **Note 23**).

8. In the software, set up section thickness, cutting speed and cutting window. Tape speed during retract can be adjusted separately to reduce the gap between sections.

9. Install the nozzle of the water level controller to the left side of the knife (Fig. 2b, arrowhead). Check the syringe has enough water with no bubbles.

10. Adjust the relative position between resin block and diamond knife as normal for ultrathin sectioning. Before cutting, start the ATUM tape, and adjust the water surface level until flat.

11. Set up water control through the software. Press "set up H_2O" and place the water adjustment rectangle over the reflection line on the side of the diamond knife (Fig. 4b, arrowhead). The reflection line of the water adjustment level must not be blocked by the shadow of the sample holder during the sectioning process, otherwise it will affect the precision of water level control. Press "maintain H_2O" (*see* **Note 24**).

12. Position the ionizer tip pointing at the cutting edge at around 3 cm away (Fig. 2b, arrow). Turn on the ionizer and adjust its strength (*see* **Note 25**).

13. Start cutting a few test sections with the tape running to make sure that the ATUM works properly. Stop cutting, carefully close the tabletop curtain to block airflow (Fig. 1 inset), and restart cutting sections.

14. Monitor sectioning process and provide necessary adjustments during sectioning. The ATUMtome computer can be monitored remotely.

15. After collecting enough sections, stop cutting, stop tape from running, and end H$_2$O maintenance. Remove the water level controller. Retract the tape head backward and lift it up from the water surface by adjusting the micrometers. Move the ATUM laterally to a safe position before disassembling the diamond knife and resin block. Clean up the diamond knife and record total cutting thickness of the sample.

16. Switch ATUM to tape loading mode. Select "Jog Bottom Forward" to reduce tension on the tape until it is loose enough to safely cut with a pair of scissors, directly under the tape head. Use clean, soft tissue paper to wipe off water from the backside of the Kapton tape.

17. Now, select "Jog Top Forward" and "Jog Bottom Back" until both tapes are released from the pinch rollers. Finish rolling the tape on the tape reels. The top tape roll has the collected sections, and is to be processed for wafer making; the bottom tape roll can be saved for further section collection.

3.6 Wafer Preparation

1. Take off the top tape reel with sections carefully, avoid touching the tape with your fingers.

2. Clean the shiny surface of a new silicon wafer with an air duster (do not use compressed air as this can leave residue), and place it in the designated area on the wafer workstation (Fig. 5a) with the shiny surface facing up. Turn on the light to facilitate visualization of the sections.

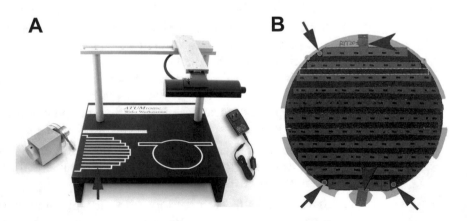

Fig. 5 Wafer preparation. (**a**) Wafer preparation workstation. Arrow points to a tape length adjusting tool, used to cut the tape fitting to the wafer dimensions. (**b**) A wafer with Kapton tape glued on and ready for imaging. The block-face was 1.5 mm × 1.7 mm for this sample. Three copper grids are placed as fiducial markers (arrows), and two pieces of copper tape are used to electrically connect the wafer to the SEM sample holder (arrowheads). Dotted lines indicate a thin segment of carbon tape connecting two adjacent lines of Kapton tape to reduce charging

3. Cut two pieces of double-sided adhesive conductive carbon tape without removing the protective covering on one side. Align and stick them carefully on the wafer surface, making sure there are no big bubbles between the wafer and the carbon tape.

4. Before exposing the carbon tape where the Kapton tape will go, use an antistatic roller to ensure the carbon tape is completely adhered to the silicon wafer. Cut the carbon tape around the edge of the wafer with a razor blade to remove the excess.

5. Cut the Kapton tape into segments with scissors or a scalpel blade. The length of each segment can be determined according to the marks on the workstation (Fig. 5a, arrow). If any large gaps exist between sections, it is useful to excise these segments of tape to maximize the number of sections on one wafer.

6. Adhere the Kapton tape onto the exposed carbon tape carefully, avoiding air bubbles underneath the Kapton tape. Align all tape segments next to each other in parallel (Fig. 5b). It is important to place the Kapton tape flat on the carbon tape. Cover the Kapton tape with the nonsticky clean paper that was removed from the carbon tape in the previous step, and roll over it with an antistatic roller.

7. Take another segment of carbon tape, and cut it into ~2 mm wide bands. Adhere these bands between every pair of Kapton tape to increase conductivity (Fig. 5b, dotted line).

8. Attach three meshed copper EM grids on the edge of the exposed carbon tape region of the wafer (Fig. 5b, arrows). These grids are used as fiducial markers for alignment in Atlas 5 AT software (*see* **Note 26**).

9. Coat the wafer with a 5 nm thick layer of carbon in a carbon coater.

3.7   SEM Imaging

1. Take an overview picture of the entire wafer with a digital camera or smart phone camera. Choose a location with a good light source, and adjust the position for tape reflection until sections can be visualized in the picture. Make sure the camera and the wafer are parallel to each other as much as possible so that less distortion will be introduced in the picture (*see* **Note 27**).

2. Insert the wafer into the SEM wafer holder and tighten it with the bottom knob and two top screws. Cut two segments of double-sided sticky copper tape, and use them to connect the holder with the first and last line of Kapton tape (Fig. 5b, arrowheads). The copper tape connection results in better conductivity. Put the holder into the SEM chamber, making sure it is installed properly.

3. Start SEM pumping, and start stage initialization (*see* **Note 28**).

4. Start Atlas 5 AT software, connect it to SEM and set the stage Z position in SmartSEM (*see* **Note 29**).

5. Start a new project, and import the overview picture as background image.

6. Create imaging protocols from "Management" > "Edit Protocol" (Fig. 6a, b, ①). Specify the parameters for each protocol that fit the purpose of each step (Fig. 6a). Protocols we used are described in the following steps.

7. With the secondary electron (SE2) detector at acceleration voltage of 3 kV, locate the three grids and take 2 μm/pixel images of the grids (with the dwell time 0.8 μs). Take images of the three grids (used as fiducials) that were placed on the wafer, and align the background image to fit to the grids' positions.

8. Select an image ROI that is slightly larger than a section and use the stamp tool to select all sections with the "section definition tool" (Fig. 6b, ② arrow). The stamping tool will create a section set. Adjust focus, brightness and contrast, so that the shape of sections can be easily distinguished (Fig. 7a). Obtain images for the whole section set (2 μm/pixel, 0.8 μs dwell time). The purpose of this step is to locate and outline all the sections, thus the dwell time can be minimized to save imaging time.

9. On top of the images captured above, use the "create polygon region" (Fig. 6b, ② arrowhead) to mark the outline of one section shape, then stamp the section outline to the other sections with "section definition tool." This will make yet another section set. The accuracy of setting the imaging region depends on the consistency of section shape and size, and the accuracy of the SE2 2 μm/pixel images (e.g., free of distortions).

10. Switch to 10 kV acceleration energy and the backscattered electron (BSE) detector for better performance in detecting gold particles. Search around the sections based on the information noted in the trimming step and find gold particle labeled profiles in the expected depth calculated from the confocal fluorescence z-stack image. The approximate depth of the target can be estimated from the thickness and numbers of the sections lined up on the wafer.

11. Right click on a section outline selection, select "Site Management" mode (Fig. 6b, ③), and make a new region that covers the targeted gold particle pattern. This process will automatically add a new imaging region set with the designated size and shape into the same relative position on each section in the series.

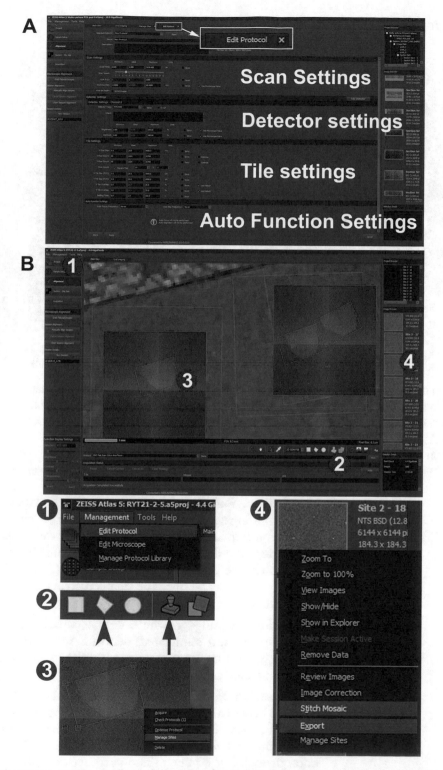

Fig. 6 Monitor views of Atlas 5 AT. (**a**) "Edit protocol" window. Parameters such as scan, detector, and tile settings can be easily modified and saved to a protocol for use in subsequent sessions. (**b**) "Main view" window. ① Management tab to edit protocols, microscope settings, and manage the protocol library; ② imaging region selection tools, such as "create polygon region" (arrowhead) and "section definition tool" (arrow); ③ right clicking the ROI provides a tab to perform site management and other functions; ④ right clicking on the gallery images, provides a tab with options to stitch the mosaic, export images, and others

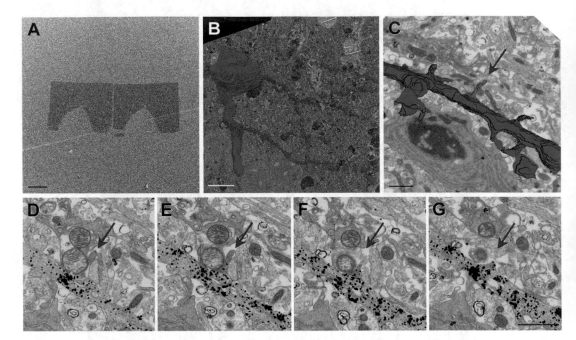

Fig. 7 Example of imaging workflow for visualization of spinogenesis. (**a**) 2 μm/pixel low magnification image taken by the SE2 detector at 3 kV shows the outline and position of two ultrathin sections. (**b**) Partial 3D reconstruction of a neuron made from a stack of 30 nm/pixel images taken by the BSE detector at 10 kV. (**c**) Dendritic segment of interest. 3D reconstruction from 4 nm/pixel images taken by InlensDuo detector at 2.5 kV, with the newly formed protrusion (yellow). (**d–g**) High resolution images of serial sections with the newly formed protrusion (yellow, arrows). Scale bar: (**a**) 200 μm; (**b**) 10 μm; (**c**) 1 μm; (**d–g**) 1 μm

12. Image the region set and follow gold particles labeled pattern in surrounding sections. Imaging is performed with a 30 nm/ pixel resolution and 10 μs dwell time at this step. The purpose of this step is to find gold-labeled profiles, thus brightness and contrast can be enhanced to make the gold particles highly visible, which may sacrifice the detailed structure of the tissue (*see* **Note 30**).

13. Determine the range of the sections containing the target neuron soma and stimulated dendrite segment. Set up an imaging region on each section with the "section definition tool," and start automatic imaging.

14. In general, autofocus range should be defined as the working distance between in focus and slight out of focus. Focusing size should be around 300–500% of imaging size, so that it gives less chance of focusing on an empty space, but provides enough focusing accuracy. The autofocusing position has to be set in the autofocus mode (*see* **Note 31**).

15. Stitch images when more than one tile is taken in a mosaic and export images (Fig. 6b, ④).

16. Alignment between images is a critical step for processing ATUM-SEM images. As section compression is an issue, sometimes distortion correction is required before alignment. After aligning all the images, make a rough 3D reconstruction model of the gold particle labeled structures with a 3–5 sections interval using a reconstruction software (Fig. 7b) (*see* **Note 32**).

17. For rough reconstruction, precise distortion correction is usually not required. By comparing the reconstructed image with the 2-photon image, an accurate location of the target dendritic segment is determined.

18. Switch to InlensDuo detector with BSE mode for higher resolution imaging. Once switching detectors, it is best to make a new session in Atlas and perform another manual alignment between sessions to achieve higher localization accuracy. Set up the final imaging regions which contain the target spine. Spine ultrastructure can be visualized at 4 nm/pixel with 75 µs dwell time. For higher resolution imaging, a tile size of 4 k × 4 k works better (Fig. 7d–g) (*see* **Note 33**).

19. Stitch final images, export and create 3D reconstruction (Fig. 7c). We used Microscopy Image Browser [58] and Amira (FEI) for alignment, segmentation, and reconstruction, though a plethora of free and commercial software exists.

4   Notes

1. Uncaging should be done after the slices have been submerged in circulating uncaging buffer for at least 10 min to achieve full penetration of MNI-caged glutamate.

2. Make sure metal wire does not touch the slices, but only the Millicell membrane.

3. Make sure the slice does not dry out during this process. Try to minimize imaging time.

4. Sodium borohydride can be added in this step to further quench free aldehydes.

5. Handle liquid nitrogen carefully.

6. Adjust incubation time according to different sample conditions if necessary.

7. Before performing the procedure, a reagent test is recommended by mixing one drop of each A, B, and C solution. When they work properly, the mixed solution should show a darker color after a few minutes.

8. It is important to wash samples thoroughly in deionized water to remove all salt of the buffer and reduce nonspecific reaction.

9. Reaction time depends on sample. Prolonged reaction time increases the background, thus it is better to stop reaction once and repeat the process if further reaction is needed.

10. Make sure slices are unfolded and flat.

11. Handle OsO_4 very carefully.

12. Incubation time can be adjusted for sample conditions.

13. It is important to wash the slices thoroughly to remove the phosphate buffer before adding UA to avoid precipitation.

14. Handle propylene oxide carefully.

15. Handle Durcupan resin carefully.

16. Pay special attention that you do not introduce air bubbles between the Millicell membrane and Aclar, and if any, remove them carefully.

17. The resin should be degassed in a desiccator to remove bubbles before it is used for embedding.

18. It should be easy to detach Aclar from the slice. However, if it does not come off easily, try to dip the sandwich sheet into liquid nitrogen shortly to facilitate the removal of Aclar.

19. The height of the resin block may need to be adjusted. The surface of the slice should be standing out from the Power-Tome specimen chuck by 1–2 mm. Taller resin blocks may create instability while sectioning.

20. Make a note of the XY position of the target neuron in the block, which will help to identify the orientation of the sections for imaging.

21. When encountering any issue with installing the tape, try to slow down the jogging speed.

22. Make sure you are wearing gloves all the time when handling the tape, use forceps instead of bare hands to avoid debris and grease attaching to the tape. Keep the environment as clean as possible.

23. Pay special attention when adjusting the tape head position of the ATUM. Make sure the tape head does not hit the knife. Make sure the tape head does not touch the bottom of the diamond knife boat when ATUM is running.

24. ATUM running will affect the water surface around the tape head, so it is better to set up water level with ATUM running.

25. Ionizer helps to maintain stable sectioning conditions for a long time by reducing electrostatic effects on the sections' surface.

26. For better alignment performance, separate these grids evenly on the edge of the wafer, or place them with a 90–90–180° separation.

27. A level helps to make sure the camera is placed horizontal.

28. It is beneficial to do stage initialization every time when the wafer holder is reinstalled for better realignment between sessions.

29. A Z-position of 44 mm corresponds to 6.7 mm working distance. The Z-height stays constant when using three different detectors for imaging.

30. We usually image sections with a 5–10 sections interval, and 10–20 imaged sections should be enough to provide a rough idea about the neuron's geometry.

31. When the autofocusing position is expected to fall into an empty space, image regions can be rotated so that the autofocusing position will fall into a tissue filled region. When sections are distributed over a large area, it is helpful to preset the focus for the first section of each row. When the wafer surface is not really flat, it is better to preset focus with a shorter interval to ensure that the autofocus function performs best. For more details on autofocus routines you may also consult Chapter 5 of this volume.

32. Various 3D reconstruction software packages are available. Depending on the final aim of the project, data set volume, and financial capability, users can choose the software that fit their situation best.

33. Small tiles will lead to inconsistent overlapping between tiles due to imperfect stage position control; tiles too big will lead to image distortions at the edge of the tiles.

5 Summary

We established a correlative workflow to study spinogenesis in organotypic slice culture using light and electron microscopy. A dendritic protrusion induced by 2-photon glutamate uncaging was relocalized and imaged in SEM with high resolution. 2-Photon live imaging recorded the functional morphological change of the protrusion via GFP fluorescence (Fig. 3c, d), and corresponding EM images of the same region revealed the fine structure of the protrusion and its surroundings (Fig. 7c–g). This correlative analysis provides direct evidence of how new spines form in response to stimulations. Advantages of this workflow are as follows: (1) a large block face size can cover the entire organotypic slice culture, and the tissue slice can be cut in its entire thickness for examination; (2) serially cut ultrathin sections are automatically collected with minimal human error; (3) semiautomatic imaging of sections is available in SEM when using the Atlas 5 AT software; (4) no tissue damage from fiducial markings gives better preservation of all surrounding structures in the area of interest; (5) imaging is nondestructive and sections can be imaged repeatedly in different sessions, at different magnifications as necessary; (6) compatibility with immunohistochemical labeling to identify specific proteins of interest; (7) large data-sets can be collected more quickly than with the traditional TEM serial section method; (8) the workflow can be modified for application to any sample having target cells expressing fluorescent proteins; (9) all correlative equipment and software are commercially available and require no further development.

Difficulties and limitations of this workflow are as follows: (1) large datasets can require considerably more time for postprocessing and analysis; (2) most software and hardware used for correlation must be purchased; (3) registration of sections is more difficult than alternative methods (e.g., SBF-SEM), as section wrinkles and compression can distort structure; (4) Pre-embedding immunohistochemistry as a method of labeling carries problems involving antibody/nanogold penetration depth, signal–noise ratio, and antibody compatibility from LM to EM.

6   Troubleshooting

Problem	Possible reason	Solution
Bad ultrastructure	Unhealthy slice	Improve slice culture condition
	Poor fixation condition	Try stronger fixative with higher percentage of GA or add supplements (e.g., glucose, $CaCl_2$) to fixative
High immuno-EM background	Imperfect immunostaining procedures	Adjust blocking condition, primary antibody and secondary antibody incubation time and concentration
	Insufficient antibody penetration	Increase number of freeze–thaw cycles Add surfactant (e.g., Triton X-100, Tween 20) into blocking and antibody solutions
	Excessive GA	Longer glycine incubation; Add sodium borohydride Reduce GA concentration in fixative Wash slices thoroughly with deionized water

(continued)

Problem	Possible reason	Solution
	Anion residues in slices before silver enhancement	before silver enhancement
	Reaction time for silver enhancement is too long	Shorten silver enhancement reaction time
		Perform multiple rounds of reaction with new reagents instead of single prolonged reaction
Sections picked up at various rotated angles	Distance between tape head and cutting edge is too large	Bigger resin block-face
		Move tape head closer to cutting edge
Excessive wrinkles on picked up sections	Kapton tape surface is hydrophobic	Discharge tape before use
		Adjust antistatic ionizer
Dust and debris on sections/tape	Unclean tape or environment	Check tape surface condition before use
		Keep ATUM working environment as clean as possible
Imaging region not following gold particle labeled pattern automatically in serial sections	Selected imaging region too small	Make imaging region bigger
	Serial sections are not consistent in size and shape	Make sure the sides of sample block are orthogonal to the sample surface
	Images do not represent accurate section position	Select smaller tile size (e.g., 1 k × 1 k) for 2 μm/pixel images to avoid image distortion
		Improve focus for images
Images out of focus	Imperfect set up of autofocus function	Adjust autofocusing range and pixel size according to sample condition
		Preset focus value for sections with a smaller interval
Tiles not overlapping for stitching	Size of imaging tile is too small	Set the tile size to a bigger dimension
Mismatched image location with different detectors	Inaccurate presentation of section position by images	Create new session and perform manual alignment once when switching between detectors
		Adjust stigmation and aperture position once when switching between detectors

Acknowledgments

The authors thank RMC Boeckeler and Carl Zeiss Microscopy for their support during startup and to keep the equipment up and running to establish the workflow. This work was supported by funding from Max Planck Society and a grant from Florida Atlantic University Pilot Graduate Research and Inquiry Program (GRIP).

References

1. Nimchinsky EA, Sabatini BL, Svoboda K (2002) Structure and function of dendritic spines. Annu Rev Physiol 64:313–353. https://doi.org/10.1146/annurev.physiol.64.081501.160008

2. Burns ME, Augustine GJ (1995) Synaptic structure and function: dynamic organization yields architectural precision. Cell 83 (2):187–194

3. Martin SJ, Grimwood PD, Morris RG (2000) Synaptic plasticity and memory: an evaluation of the hypothesis. Annu Rev Neurosci 23:649–711. https://doi.org/10.1146/annurev.neuro.23.1.649

4. Matsuzaki M, Ellis-Davies GC, Nemoto T, Miyashita Y, Iino M, Kasai H (2001) Dendritic spine geometry is critical for AMPA receptor expression in hippocampal CA1 pyramidal neurons. Nat Neurosci 4(11):1086–1092. https://doi.org/10.1038/nn736

5. Chen TW, Wardill TJ, Sun Y, Pulver SR, Renninger SL, Baohan A, Schreiter ER, Kerr RA, Orger MB, Jayaraman V, Looger LL, Svoboda K, Kim DS (2013) Ultrasensitive fluorescent proteins for imaging neuronal activity. Nature 499(7458):295–300. https://doi.org/10.1038/nature12354

6. Dani A, Huang B, Bergan J, Dulac C, Zhuang X (2010) Superresolution imaging of chemical synapses in the brain. Neuron 68(5):843–856. https://doi.org/10.1016/j.neuron.2010.11.021

7. MacGillavry HD, Song Y, Raghavachari S, Blanpied TA (2013) Nanoscale scaffolding domains within the postsynaptic density concentrate synaptic AMPA receptors. Neuron 78 (4):615–622. https://doi.org/10.1016/j.neuron.2013.03.009

8. Nagerl UV, Willig KI, Hein B, Hell SW, Bonhoeffer T (2008) Live-cell imaging of dendritic spines by STED microscopy. Proc Natl Acad Sci U S A 105(48):18982–18987. https://doi.org/10.1073/pnas.0810028105

9. Harris KM, Jensen FE, Tsao B (1992) Three-dimensional structure of dendritic spines and synapses in rat hippocampus (CA1) at postnatal day 15 and adult ages: implications for the maturation of synaptic physiology and long-term potentiation. J Neurosci 12 (7):2685–2705

10. Fifkova E, Anderson CL (1981) Stimulation-induced changes in dimensions of stalks of dendritic spines in the dentate molecular layer. Exp Neurol 74(2):621–627

11. Bosch M, Hayashi Y (2012) Structural plasticity of dendritic spines. Curr Opin Neurobiol 22 (3):383–388. https://doi.org/10.1016/j.conb.2011.09.002

12. Kasai H, Fukuda M, Watanabe S, Hayashi-Takagi A, Noguchi J (2010) Structural dynamics of dendritic spines in memory and cognition. Trends Neurosci 33(3):121–129. https://doi.org/10.1016/j.tins.2010.01.001

13. Nishiyama J, Yasuda R (2015) Biochemical computation for spine structural plasticity. Neuron 87(1):63–75. https://doi.org/10.1016/j.neuron.2015.05.043

14. Hayashi-Takagi A, Yagishita S, Nakamura M, Shirai F, Wu YI, Loshbaugh AL, Kuhlman B, Hahn KM, Kasai H (2015) Labelling and optical erasure of synaptic memory traces in the motor cortex. Nature 525(7569):333–338. https://doi.org/10.1038/nature15257

15. Matsuzaki M, Honkura N, Ellis-Davies GC, Kasai H (2004) Structural basis of long-term potentiation in single dendritic spines. Nature 429(6993):761–766. https://doi.org/10.1038/nature02617

16. Nagerl UV, Eberhorn N, Cambridge SB, Bonhoeffer T (2004) Bidirectional activity-dependent morphological plasticity in hippocampal neurons. Neuron 44(5):759–767. https://doi.org/10.1016/j.neuron.2004.11.016

17. Zhou Q, Homma KJ, Poo MM (2004) Shrinkage of dendritic spines associated with long-term depression of hippocampal synapses. Neuron 44(5):749–757. https://doi.org/10.1016/j.neuron.2004.11.011

18. Trachtenberg JT, Chen BE, Knott GW, Feng G, Sanes JR, Welker E, Svoboda K (2002) Long-term in vivo imaging of experience-dependent synaptic plasticity in adult cortex. Nature 420(6917):788–794. https://doi.org/10.1038/nature01273

19. Harris KM (1999) Structure, development, and plasticity of dendritic spines. Curr Opin Neurobiol 9(3):343–348

20. Modla S, Czymmek KJ (2011) Correlative microscopy: a powerful tool for exploring neurological cells and tissues. Micron 42 (8):773–792. https://doi.org/10.1016/j.micron.2011.07.001

21. Knott GW, Holtmaat A, Trachtenberg JT, Svoboda K, Welker E (2009) A protocol for preparing GFP-labeled neurons previously imaged in vivo and in slice preparations for light and electron microscopic analysis. Nat Protoc 4(8):1145–1156. https://doi.org/10.1038/nprot.2009.114

22. Grabenbauer M, Geerts WJ, Fernadez-Rodriguez J, Hoenger A, Koster AJ, Nilsson T (2005) Correlative microscopy and electron tomography of GFP through photooxidation. Nat Methods 2(11):857–862. https://doi.org/10.1038/nmeth806

23. Maranto AR (1982) Neuronal mapping: a photooxidation reaction makes Lucifer yellow useful for electron microscopy. Science 217 (4563):953–955

24. Perkovic M, Kunz M, Endesfelder U, Bunse S, Wigge C, Yu Z, Hodirnau VV, Scheffer MP, Seybert A, Malkusch S, Schuman EM, Heilemann M, Frangakis AS (2014) Correlative light- and electron microscopy with chemical tags. J Struct Biol 186(2):205–213. https://doi.org/10.1016/j.jsb.2014.03.018

25. Martell JD, Deerinck TJ, Sancak Y, Poulos TL, Mootha VK, Sosinsky GE, Ellisman MH, Ting AY (2012) Engineered ascorbate peroxidase as a genetically encoded reporter for electron microscopy. Nat Biotechnol 30 (11):1143–1148. https://doi.org/10.1038/nbt.2375

26. Butko MT, Yang J, Geng Y, Kim HJ, Jeon NL, Shu X, Mackey MR, Ellisman MH, Tsien RY, Lin MZ (2012) Fluorescent and photooxidizing TimeSTAMP tags track protein fates in light and electron microscopy. Nat Neurosci 15(12):1742–1751. https://doi.org/10.1038/nn.3246

27. Shu X, Lev-Ram V, Deerinck TJ, Qi Y, Ramko EB, Davidson MW, Jin Y, Ellisman MH, Tsien RY (2011) A genetically encoded tag for correlated light and electron microscopy of intact cells, tissues, and organisms. PLoS Biol 9(4): e1001041. https://doi.org/10.1371/journal.pbio.1001041

28. Bishop D, Nikic I, Brinkoetter M, Knecht S, Potz S, Kerschensteiner M, Misgeld T (2011) Near-infrared branding efficiently correlates light and electron microscopy. Nat Methods 8 (7):568–570. https://doi.org/10.1038/nmeth.1622

29. Tanaka J, Matsuzaki M, Tarusawa E, Momiyama A, Molnar E, Kasai H, Shigemoto R (2005) Number and density of AMPA receptors in single synapses in immature cerebellum. J Neurosci 25(4):799–807. https://doi.org/10.1523/JNEUROSCI.4256-04.2005

30. Bosch M, Castro J, Saneyoshi T, Matsuno H, Sur M, Hayashi Y (2014) Structural and molecular remodeling of dendritic spine substructures during long-term potentiation. Neuron 82(2):444–459. https://doi.org/10.1016/j.neuron.2014.03.021

31. Maco B, Holtmaat A, Cantoni M, Kreshuk A, Straehle CN, Hamprecht FA, Knott GW (2013) Correlative in vivo 2 photon and focused ion beam scanning electron microscopy of cortical neurons. PLoS One 8(2): e57405. https://doi.org/10.1371/journal.pone.0057405

32. Schalek R, Kasthuri N, Hayworth K, Berger D, Tapia J, Morgan J, Turaga S, Fagerholm E, Seung H, Lichtman J (2011) Development of high-throughput, high-resolution 3D reconstruction of large-volume biological tissue using automated tape collection ultramicrotomy and scanning electron microscopy. Microsc Microanal 17(S2):966–967. https://doi.org/10.1017/s1431927611005708

33. Denk W, Horstmann H (2004) Serial blockface scanning electron microscopy to reconstruct three-dimensional tissue nanostructure. PLoS Biol 2(11):e329. https://doi.org/10.1371/journal.pbio.0020329

34. Knott G, Marchman H, Wall D, Lich B (2008) Serial section scanning electron microscopy of adult brain tissue using focused ion beam milling. J Neurosci 28(12):2959–2964. https://doi.org/10.1523/JNEUROSCI.3189-07.2008

35. Micheva KD, Smith SJ (2007) Array tomography: a new tool for imaging the molecular architecture and ultrastructure of neural circuits. Neuron 55(1):25–36. https://doi.org/10.1016/j.neuron.2007.06.014

36. Horstmann H, Korber C, Satzler K, Aydin D, Kuner T (2012) Serial section scanning electron microscopy (S3EM) on silicon wafers for ultra-structural volume imaging of cells and

tissues. PLoS One 7(4):e35172. https://doi.org/10.1371/journal.pone.0035172

37. Kuwajima M, Mendenhall JM, Harris KM (2013) Large-volume reconstruction of brain tissue from high-resolution serial section images acquired by SEM-based scanning transmission electron microscopy. Methods Mol Biol 950:253–273. https://doi.org/10.1007/978-1-62703-137-0_15

38. Titze B, Genoud C (2016) Volume scanning electron microscopy for imaging biological ultrastructure. Biol Cell 108(11):307–323. https://doi.org/10.1111/boc.201600024

39. Kamasawa N, Sun Y, Mikuni T, Guerrero-Given D, Yasuda R (2015) Correlative ultrastructural analysis of functionally modulated synapses using automatic tape-collecting ultramicrotome – SEM array tomography. Microsc Microanal 21(S3):1271–1272. https://doi.org/10.1017/s143192761500714x

40. Humpel C (2015) Organotypic brain slice cultures: a review. Neuroscience 305:86–98. https://doi.org/10.1016/j.neuroscience.2015.07.086

41. Collin C, Miyaguchi K, Segal M (1997) Dendritic spine density and LTP induction in cultured hippocampal slices. J Neurophysiol 77(3):1614–1623

42. Washbourne P, McAllister AK (2002) Techniques for gene transfer into neurons. Curr Opin Neurobiol 12(5):566–573

43. Murphy RC, Messer A (2001) Gene transfer methods for CNS organotypic cultures: a comparison of three nonviral methods. Mol Ther 3(1):113–121. https://doi.org/10.1006/mthe.2000.0235

44. Fortin DA, Davare MA, Srivastava T, Brady JD, Nygaard S, Derkach VA, Soderling TR (2010) Long-term potentiation-dependent spine enlargement requires synaptic Ca2+-permeable AMPA receptors recruited by CaM-kinase I. J Neurosci 30(35):11565–11575. https://doi.org/10.1523/JNEUROSCI.1746-10.2010

45. Lang C, Barco A, Zablow L, Kandel ER, Siegelbaum SA, Zakharenko SS (2004) Transient expansion of synaptically connected dendritic spines upon induction of hippocampal long-term potentiation. Proc Natl Acad Sci U S A 101(47):16665–16670. https://doi.org/10.1073/pnas.0407581101

46. Xu T, Yu X, Perlik AJ, Tobin WF, Zweig JA, Tennant K, Jones T, Zuo Y (2009) Rapid formation and selective stabilization of synapses for enduring motor memories. Nature 462(7275):915–919. https://doi.org/10.1038/nature08389

47. Wiedenmann J, Oswald F, Nienhaus GU (2009) Fluorescent proteins for live cell imaging: opportunities, limitations, and challenges. IUBMB Life 61(11):1029–1042. https://doi.org/10.1002/iub.256

48. Jaisser F (2000) Inducible gene expression and gene modification in transgenic mice. J Am Soc Nephrol 11(Suppl 16):S95–S100

49. Mikuni T, Nishiyama J, Sun Y, Kamasawa N, Yasuda R (2016) High-throughput, high-resolution mapping of protein localization in mammalian brain by in vivo genome editing. Cell 165(7):1803–1817. https://doi.org/10.1016/j.cell.2016.04.044

50. Suzuki K, Tsunekawa Y, Hernandez-Benitez R, Wu J, Zhu J, Kim EJ, Hatanaka F, Yamamoto M, Araoka T, Li Z, Kurita M, Hishida T, Li M, Aizawa E, Guo S, Chen S, Goebl A, Soligalla RD, Qu J, Jiang T, Fu X, Jafari M, Esteban CR, Berggren WT, Lajara J, Nunez-Delicado E, Guillen P, Campistol JM, Matsuzaki F, Liu GH, Magistretti P, Zhang K, Callaway EM, Zhang K, Belmonte JC (2016) In vivo genome editing via CRISPR/Cas9 mediated homology-independent targeted integration. Nature 540(7631):144–149. https://doi.org/10.1038/nature20565

51. Patterson MA, Szatmari EM, Yasuda R (2010) AMPA receptors are exocytosed in stimulated spines and adjacent dendrites in a Ras-ERK-dependent manner during long-term potentiation. Proc Natl Acad Sci U S A 107(36):15951–15956. https://doi.org/10.1073/pnas.0913875107

52. Woods G, Zito K (2008) Preparation of gene gun bullets and biolistic transfection of neurons in slice culture. J Vis Exp 12. https://doi.org/10.3791/675

53. Malinow R, Hayashi Y, Maletic-Savatic M, Zaman SH, Poncer JC, Shi SH, Esteban JA, Osten P, Seidenman K (2010) Introduction of green fluorescent protein (GFP) into hippocampal neurons through viral infection. Cold Spring Harb Protoc 4:pdb prot5406. https://doi.org/10.1101/pdb.prot5406

54. Feng G, Mellor RH, Bernstein M, Keller-Peck C, Nguyen QT, Wallace M, Nerbonne JM, Lichtman JW, Sanes JR (2000) Imaging neuronal subsets in transgenic mice expressing multiple spectral variants of GFP. Neuron 28(1):41–51

55. Tabata H, Nakajima K (2001) Efficient in utero gene transfer system to the developing mouse brain using electroporation: visualization of neuronal migration in the developing cortex. Neuroscience 103(4):865–872

56. Gogolla N, Galimberti I, DePaola V, Caroni P (2006) Preparation of organotypic hippocampal slice cultures for long-term live imaging. Nat Protoc 1(3):1165–1171. https://doi.org/10.1038/nprot.2006.168

57. Kwon HB, Sabatini BL (2011) Glutamate induces de novo growth of functional spines in developing cortex. Nature 474 (7349):100–104. https://doi.org/10.1038/nature09986

58. Belevich I, Joensuu M, Kumar D, Vihinen H, Jokitalo E (2016) Microscopy image browser: a platform for segmentation and analysis of multidimensional datasets. PLoS Biol 14(1): e1002340. https://doi.org/10.1371/journal.pbio.1002340

Chapter 8

Large-Scale Automated Serial Section Imaging with a Multibeam Scanning Electron Microscope

Anna Lena Eberle and Tomasz Garbowski

Abstract

Acquiring high-quality electron microscopic data begins with proper sample preparation reflecting the special requirements of the corresponding electron microscope. For the ZEISS MultiSEM, the special constraints of operating a multibeam scanning electron microscope on larger samples than traditionally possible require adapted sample preparation techniques for successful imaging. In this chapter, we introduce multibeam scanning electron microscopy for large-scale imaging experiments and describe sample preparation options as well as the corresponding sample preparation steps. We also summarize our experiences from application development and experimental work with the ZEISS MultiSEM.

Key words Multibeam, Scanning electron microscopy, Serial sectioning, Array tomography, Multi-SEM, Connectomics, Automated imaging, Large-scale imaging, 3D volume imaging

1 Introduction

There has been an increased interest in comprehensively mapping the connections within an organism's nervous system in the past years [1, 2], and a new research field coined "connectomics" has emerged since. On the macroscale, brain connectivity is often assessed by diffusion tensor imaging [3]; however, a detailed reconstruction of neural circuitry requires a spatial resolution only provided by electron microscopy [4]. As sample volumes for connectomics must be larger than previously common to EM for meaningful results, two main challenges had to be overcome: (1) While sample preparation for single-beam (SB) scanning electron microscopy of serial sections is well understood, and literature exhibits a wealth of protocols and operation guidelines (e.g., [5–7]), these protocols are not optimally suited for large sample sizes as necessary in connectomics. Only recently, EM compatible protocols have been established for larger sample volumes [8, 9]. (2) Imaging large volumes with high resolution and standard EMs results in prohibitively long experiment durations. As an

Irene Wacker et al. (eds.), *Volume Microscopy: Multiscale Imaging with Photons, Electrons, and Ions*, Neuromethods, vol. 155, https://doi.org/10.1007/978-1-0716-0691-9_8, © Springer Science+Business Media, LLC, part of Springer Nature 2020

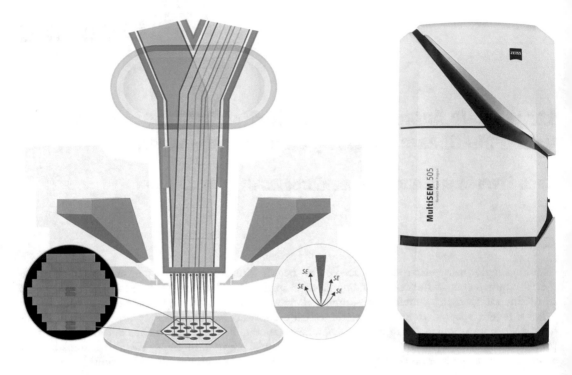

Fig. 1 Left: MultiSEM principle of operation. The primary electron beam array (green) is focused onto the sample while the resulting secondary electrons (red) are collected through the common objective lens. The beam arrays are separated by a beam splitter and the full image is formed by merging all image tiles. Right: Image of the ZEISS MultiSEM

example, a cubic millimeter of brain tissue prepared for serial section electron microscopy typically results in 20,000 ultrathin sections of a square millimeter area each. Image acquisition of the full series of sections with a state-of-the-art scanning electron microscope would take approximately 6 years, exceeding reasonable experiment duration in most cases. One solution for increasing imaging throughput is the use of multiple electron beams in parallel [10], as it has been realized in the multibeam scanning electron microscope by ZEISS (*see* Fig. 1). Its unique imaging principles result in a number of sample preparation requirements and operation procedures that we will explain in the following.

ZEISS MultiSEM has two electron optical columns. The illuminating electron optical column focuses a hexagonal pattern of electron beams onto the sample surface. All primary electron beams are focused near one focus plane orthogonal to the objective lens axis, with a working distance of 1.4 mm. The detection column projects the signal electrons onto the multidetector. We apply an electric potential to the stage and sample with respect to the objective lens ("stage biasing"). A magnetic beam splitter separates the illuminating electrons from the secondary electrons that form the signal (*see* Fig. 1).

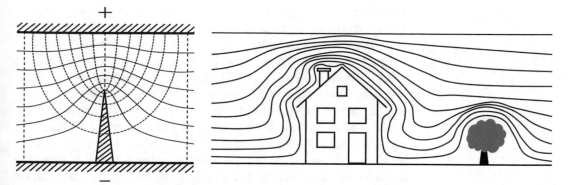

Fig. 2 Left: The shape of the sample surface influences the electrostatic potential (continuous lines) and the electrostatic field (dashed lines). In the example here, a protruding spike locally generates a very high electrostatic field, potentially causing a discharge. Right: The same principle is used by lightning rods to allow for a controlled discharge to ground

A sample on electrostatic potential can produce local field enhancements through sample topography (*see* Fig. 2). If the local field strength exceeds a critical value, electrical discharging may occur. This should be avoided as it may damage the sample and adjacent objects. Local field enhancement is avoided by proper high voltage design of the MultiSEM components in the vicinity of the sample, and by adequate sample preparation which we will describe next.

2   Challenges and Solutions for Sample Preparation

The electron trajectories in the secondary path, and thus the detected signals, are sensitive to sample properties such as conductivity, surface roughness, distance to edges, and sample tilt. Therefore, sample preparation must reflect the measures in this section to ensure high image quality and stable imaging performance.

2.1   *Conductivity*

Challenge: If a nonconducting sample is imaged in any SEM, charges deposited by the primary electron beam can build up an electrostatic field near the surface, affecting the trajectories of both the primary electron beams and the signal electrons in particular. In standard SB SEMs, this can be overcome by a number of methods, for example by introducing a small pressure of gas near the sample ("variable pressure mode"). Combining variable pressure mode with the MultiSEM stage bias is very challenging, and we do not recommend it.

Solution: To prevent charge buildup, the net sum of charges impinging onto and leaving the sample must be zero. One possible solution to this is to make the sample surface conductive and electrically connect the surface to the sample holder and the stage. This electrical connection can be made by double-sided sticky carbon tape, silver paint, or a combination of both. It will provide a proper mechanical fixation as well. The MultiSEM multi-purpose sample holders allow for mounting of several standard SEM sample carriers (e.g., SEM stubs of varying sizes, wafer chips, or cover glasses). Once the sample is connected adequately to the carrier, no additional fixation or electrical connection is needed. If the sample itself is intrinsically conductive enough (by, e.g., heavy metal contrast agents), connecting it to the holder is usually sufficient. In other cases, adding a conductive coating to the surface will be necessary. For example, an 8–10 nm thick carbon coating usually provides good imaging conditions.

2.2 Surface Roughness

Challenge: Secondary electrons are sensitive to sample surface topography. For example, if we expect an SE(2) [11] contrast mechanism, edge enhancement by strong sample topography can reduce or maybe even obscure this desired contrast. In addition, strong surface topography can affect the projection of the signal electrons in the detection path. Protruding parts of the sample can cause local field enhancements and a local electrical discharge which may damage the sample and parts nearby.

Solution: Ultrathin tissue sections cut with a diamond knife and mounted on a flat surface (e.g., a silicon wafer or a conductively coated cover glass) usually are sufficiently flat and exhibit good surface roughness. However, wrinkles and folds of the sections should be avoided, as these may impair the image quality. For bulk samples, a planar surface typically calls for mechanical polishing. Broad beam ion milling may further reduce surface roughness.

2.3 Sample Size/ Distance to Edge

Challenge: A height step at the edge of the sample can affect the detection path in the MultiSEM and thus the imaging in the same way as sample surface topography mentioned above. For example, an ultrathin section with up to ~100 nm section thickness is sufficiently thin to be imaged entirely without affecting the detection path, whereas a silicon wafer chip with ~700 μm height will lead to a distortion of the electron beams when imaging near the edge of the sample. By manually optimizing imaging parameters, acquiring useful images even close to a steep edge is still possible. As a rule of thumb, the imaged area should not be closer to an edge than twice the edge height to avoid frequent readjustments.

Solution: We recommend embedding the sample into a flat and conductive material (i.e., metal) as seamlessly as possible such that the MultiSEM automatic imaging parameter setting can be used to the maximum extent. For example, a piece of a silicon wafer can be mounted into a standard SEM stub by milling the shape of the sample into the stub surface to the same depth as the height of the sample. When inserted into the stub, the silicon wafer chip is then flush with the SEM stub surface. Alternatively, the sample can be mounted onto the sample holder with silver paint and the edges then masked with a fitting conductive shield, for example, made from thin metal sheets or silicon wafer pieces.

2.4 Sample Tilt

Challenge: As all the beams within the array are focused to a plane orthogonal to the objective lens axis, the sample surface must not be tilted with respect to the focal plane to ensure that all beams are in focus. The MultiSEM stage does not have a tilt option, and the working distance between sample and objective lens is 1.4 mm. If a sample is mounted with a strong tilt, it might collide with the objective lens while the stage moves over a larger distance.

Solution: The sample mount and the stage are aligned with low tolerances to the objective lens at the factory. Thus, if the sample surface is parallel to the surface of the sample holder after mounting, the sample is sufficiently parallel to the focal plane to ensure proper imaging conditions for all electron beams.

Next to the particular constraints due to the unique detection principle of the MultiSEM, there are more general requirements for sample preparation that apply to any SEM. Most electron microscopes require high vacuum conditions inside the object chamber, as gas molecules affect the electron beam which, for example, can lead to a loss of imaging resolution. Therefore, the sample needs to be compatible with a high vacuum environment and not contain any water or other strongly outgassing components. We recommended to degas every sample in a vacuum oven at approximately 25–30 °C and for at least 24 h prior to loading it into the Multi-SEM. These values might vary depending on the sample type and the materials used. For example, the adhesives used in various sticky tapes for SEM sample preparation can outgas over several days.

3 Methods

3.1 Preparation of Samples from Serial Ultrathin Tissue Sections

The first step for successful imaging in any electron microscope is the preparation of a suitable sample. Here, we will describe the preparation of the commonly used serial sections on wafers. A more detailed view on this topic can also be found in Chapters 4–7 within this book.

After extraction of the respective tissue under study (e.g., brain tissue), a proper fixation suited for electron microscopy (e.g., with glutaraldehyde) is applied to ensure ultrastructural preservation. Next, the tissue is stained to generate good signal contrast when interacting with the electron beam. Usually, electron dense elements, such as the heavy metals osmium, uranium, or lead, are introduced to selectively contrast certain tissue constituents such as lipids or proteins. This results in the selective labeling of membranes and cell organelles. Cutting sections that are only few ten nanometers thick requires embedding the sample into a suitable resin that can be cured to a very hard consistency. The sample is dehydrated with an increasing alcohol series, infiltrated with resin (e.g., epoxy), and the resin is polymerized under heat in an oven. The fully polymerized and hardened resin block is finally trimmed with a glass knife or a razor blade to remove excessive resin around the tissue, such that the resulting sections are as small as possible.

Several procedures for preparation of ultrathin tissue sections for electron microscopy are available and have been described in the literature. A very thorough and comprehensive overview of the various approaches to volume EM is given in [12]. In general, standard staining and fixation protocols established for transmission EM may work and give sufficient contrast for imaging in the MultiSEM. However, protocols that have been optimized for block-face imaging yield much better results. Some of these have been tested already, and we recommend them for use with the MultiSEM, for example, the different variations of the OTO protocol (OTO = osmium–thiocarbohydrazide–osmium) with or without the usage of reduced osmium and in varying sequences [7, 13, 14]. The BROPA protocol has been optimized for staining of very large tissue blocks or even whole mouse brains [15]. A very detailed description of a high contrast OTO staining protocol for neuronal tissue is given in [7]. A variation of the osmium protocol that yields excellent staining results is described in [9].

The fixed, stained, and resin-embedded tissue block is now ready for ultrathin sectioning to yield a series of consecutive slices that can be imaged with an electron microscope. Slice thicknesses between 30 and up to 100 nm call for using an ultramicrotome that repeatedly shaves off the top layer of the sample with a diamond knife. The knife is integrated into the front side of a small water boat, facing the sample, so that the tissue slice floats onto the water surface after being cut. From here, the sections can be transferred manually to a sample carrier, for example, a piece of silicon wafer or conductively coated glass coverslips. To increase efficiency, sections are routinely collected as ribbons consisting of numerous consecutive sections that stick together at the edges and that are handled as whole [16–20]. A more detailed description of such a device and its application can be found in Chapter 5 of this book.

Another approach is the Automated Tape Collecting Ultramicrotome (ATUMtome, RMC Boeckeler), which originally has been developed in the group of Jeff W. Lichtman, Harvard University [21, 22]. A more detailed description of this device and its application can be found in Chapter 7 in this book.

The following step-by-step guideline explains the preparation of a wafer with tape-carried ultrathin sections in detail and with special emphasis on working with the MultiSEM, considering the steps explained above.

1. Materials.

   Four-inch silicon wafers, p-doped and polished on one side (e.g., from ScienceServices, prod. # SC4CZp-525), may be used as sample carriers for serial sections. Use double-sided sticky carbon tape to adhere the tape with the ultrathin sections onto the wafer (e.g., from Ted Pella, prod. # 16084-8).

   We have made best experiences with the recommended materials. Whereas the type of wafer substrate is not as critical, especially carbon tape can be a source of undesired effects in the MultiSEM. There are, for instance, varieties of tape where only the top and the bottom surfaces are conductive, separated by an insulating layer in the middle. Such tapes can charge up and impair the imaging performance.

2. Covering the wafer with double-sided sticky carbon tape.

   Cut the carbon tape into strips of approximately 4″ (10 cm) length. Remove the white protective foil from the double-sided sticky carbon tape. Carefully apply the carbon tape strips side by side to the wafer, using only as much as necessary. Do not overlay the edges of the tapes. While applying the carbon tape to the wafer, avoid capturing air bubbles under the tape. The tape surface should be as flat as possible to ensure proper imaging conditions in the MultiSEM. In the end, remove the excessive carbon tape at the edge of the wafer with a scalpel or a razor blade. Always work cleanly and prevent dirt, dust or textile fibers from clothing from falling onto the sticky carbon tape.

3. Preparation of the Kapton tape.

   Cut the Kapton tape with the sections on it into strips that will fit onto the wafer. Keep at least 5 mm distance between the areas on the Kapton tape that shall be imaged (i.e., the tissue sections) and the nearest wafer edge. Make sure not to mix up the tape strips and to document the order of the serial sections. When mounting the Kapton tape strips onto the carbon tape, do not overlay the edges of the tape strips and avoid capturing air bubbles under the Kapton tape. The surface should be as flat as possible to ensure proper imaging conditions in the Multi-SEM. To remove air bubbles from underneath the Kapton tape, reuse the transparent protective foil from the double-sided sticky carbon tape clean from the step before. Cover the Kapton strips with this transparent foil (clean side facing downward), press and wipe the tape gently sideways with your glove-covered thumbs. This will push out excessive air bubbles from under the tape.

   The edges of the Kapton tape are often not sufficiently coated and therefore may be nonconductive. It is therefore helpful to "mask" these edges, either with silver paint or with narrow strips of carbon tape. Thin lines of silver paint applied besides the tape will alleviate the effects from the nonconductive tape edges. Carbon tape should be used sparsely, because most tapes are outgassing at least to some extent over a longer time in the vacuum chamber of the microscope, thereby leading to contamination. Also, the tape creates additional sample topography.

If the edge of the tape is too close to the imaging regions (i.e., the sections on Kapton tape), image distortion might be observed (*see* also imaging path as described in Subheading 2). We do not recommend using copper tape at all in the Multi-SEM because metal tapes may exhibit very sharp edges, especially after bending and twisting while handling. These edges can cause electrical discharges inside the MultiSEM chamber, potentially damaging the sample (*see* Fig. 2).

(a) Electrical contacting can be done using conductive ink (e.g., Circuit Scribe pen www.circuitscribe.com) or silver paint (e.g., PELCO conductive silver paint from Ted Pella # 16062) by drawing thin lines along the edges of the Kapton tape strips.

(b) Electrical contacting also can be done using carbon tape. Cut double-sided sticky carbon tape (e.g., from Ted Pella, prod # 16073) into approximately 1 mm wide strips and cover the edges of the Kapton tape with these narrow strips. Also cover the edge of the underlying carbon tape to the wafer substrate with these tape strips.

4. Sample degassing.

The solvent of the silver ink and the adhesive of the sticky carbon tape need to evaporate before the sample is transferred to the microscope chamber. After preparing a wafer, we recommend putting the wafer in a vacuum oven for at least 24 h at approximately 25–30 °C. For better results, the sample should be kept in the vacuum oven for several days if possible. This degassing procedure reduces chamber contamination and pumping time during sample transfer and helps to minimize chamber contamination. In addition, it should reduce air bubbles that may have been caught between the Kapton and carbon tape during mounting. It should be noted that keeping samples with a large amount of sticky carbon tape in the vacuum chamber of the microscope for a longer period of time (e.g., overnight or over the weekend) may lead to contamination of the chamber. This can happen even though the sample has been degassed thoroughly before. It can be reduced by covering up as much as possible of the carbon tape surface with Kapton tape strips. Additionally, we recommend to plasma clean the chamber on a regular basis as a preventive measure.

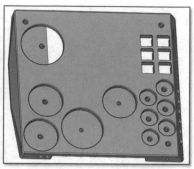

Fig. 3 Three different MultiSEM sample holders: the standard holder (left) and two multipurpose sample holders for biological (middle) and materials (right) samples

3.2 Mounting of the Sample onto the Sample Holder

After preparation, the sample needs to be mounted onto one of the MultiSEM sample holders. Three types are currently available: a plain sample holder, a life science multipurpose sample holder, and a materials science multipurpose sample holder (*see* Fig. 3).

The standard, plain sample holder is a blank plate. Samples can be attached to it with either silver paint or conductive double-sided carbon tape. This is the most flexible approach, but it increases contamination of both system and sample. It also increases preparation time due to the required fixation measures and therefore extended outgassing of the sample. Wafer samples prepared as described above are usually mounted onto these plain sample holders. The wafer is placed onto a plain MultiSEM sample holder, and the edge of the wafer is contacted to the sample holder with narrow double-sided sticky carbon tape strips or conductive silver paint.

The multipurpose sample holders have dedicated slots for taking SEM stubs of different sizes, 1 cm^2 sized wafer chips, ITO-coated cover glasses, or TEM grids, without the need to use silver paint or carbon tape. All MultiSEM sample holders are specifically designed for use with stage biasing.

3.3 Generating an Overview Image of the Sample

It is convenient to generate an overview image on a light microscope and use this image to navigate on the sample, which is supported by the ZEISS Shuttle & Find workflow. The MultiSEM sample holders are equipped with three standardized L-shaped fiducials (ZEISS Shuttle & Find L-markers). These markers can be registered both in the light microscope and the MultiSEM, such that the image coordinates can be transferred between both systems and correlated afterward. The more precise the mapping between these two coordinate systems is performed, the more precise desired locations can be targeted in the MultiSEM. Additionally, several preparatory steps for the actual imaging experiment can be performed on the light microscopic overview image, such as the automatic detection of ultrathin sections on a wafer or the

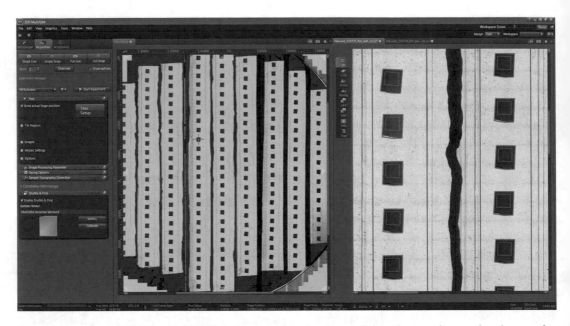

Fig. 4 Screenshot of ZEN for MultiSEM during experiment setup. A light microscopic overview image of a sample wafer with ultrathin sections cut with an ATUMtome and prepared as described before (left), regions of interest (in this case complete sections) labelled in green resulting from the automatic section detection (right) (Sample with courtesy of Jeff W. Lichtman, Harvard University)

definition of regions of interest (*see* Fig. 4). The magnification used at the light microscope depends on the size of the samples and the required accuracy of the navigation. Typically, a sample with serial sections lined up on a wafer as described previously is imaged with a 5× objective. A full 4″ wafer (*see* Fig. 4, left) produces a light microscopic image file of approximately 2 GB size with sufficient resolution to navigate and define regions of interest (*see* Fig. 4, right).

3.4 Imaging the Sample with ZEISS MultiSEM

The sample is now ready for imaging in the MultiSEM and can be transferred into the chamber through the airlock. After sample transfer, the stage can be moved to the imaging position. Now, the electron beams can be switched on. As before at the light microscope stage, the Shuttle & Find fiducials need to be registered in the MultiSEM. This allows for navigation and experiment setup on the basis of the previously acquired light microscopic overview image. The MultiSEM operation software ZEN guides the user through the steps of setting up the experiment recipe, including all required imaging parameters, such as pixel size, dwell time, and brightness and contrast, as well as the choice of an acquisition strategy, such as focusing and stigmation schemes. At first, the light microscopic overview image is imported to the MultiSEM operation software. To define regions of interest, an automatic section detection routine can be used on basis of this image

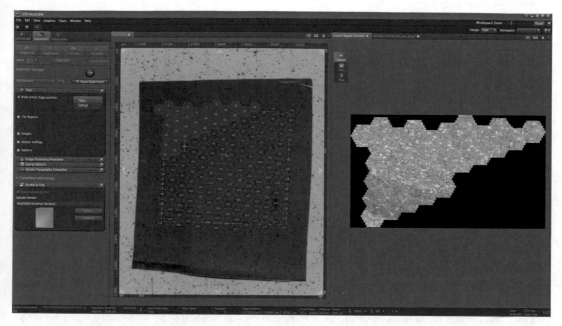

Fig. 5 Screenshot of ZEN for MultiSEM during a running experiment. Regions of interest are tessellated with hexagonal shapes to visualize the hexagonally arranged beam array, successfully imaged hexagons turn green (left). A preview image can be generated during imaging to allow for a quick assessment of the image quality (right) (Sample with courtesy of Jeff W. Lichtman, Harvard University)

(Fig. 4, right). Using one section as a template, the section detection algorithm finds all matching shapes within the marked areas (green polygons in Fig. 4, right). The closer the ribbon boundaries are to the sections, that is, the smaller the area that needs to be analyzed, the faster the algorithm will detect the sections. Within the section boundaries, the user can define one region of interest (ROI) or more, in case only a sub-fraction of the section shall be imaged. This needs to be done for one section only: the ROI will be applied to all sections. Additionally, they will be oriented with respect to the section margins, that is, if a section is rotated, the ROI will be rotated as well. The ROIs are then displayed as hexagonally tessellated areas (*see* Fig. 5). Next, the imaging parameters are optimized manually on a representative sample area. Setting up the experiment is finalized with the definition of automatic focusing and stigmation schemes that will maintain the image quality during acquisition. Finally, the experiment can be started. The acquired images are organized hierarchically in a tree data structure, starting top-level with an experiment description containing all ROIs as separate data sets, and finally going down to the level of individual image tiles generated by the individual electron beams. After the imaging of one entire ROI is finished, a number of corresponding higher-order metadata files are generated for the full region which, for example, facilitate the import of the raw image data to

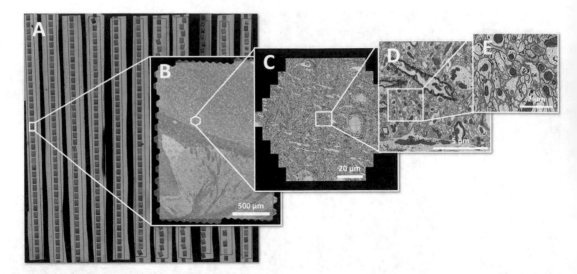

Fig. 6 Example of imaging results. (**a**) Light microscopic overview image of the full wafer, acquired with a ZEISS AxioImager Vario, (**b**) a full mouse brain section acquired with a 61-beam MultiSEM 505 at 4 nm pixel size and 100 ns pixel dwell time, (**c**) one hexagonal field of view consisting of 61 individual images, (**d**) image the central beam acquired and (**e**) further magnified cutout to assess expectable image quality (Figure adapted from [24], sample with courtesy of Jeff W. Lichtman, Harvard University)

commonly used open source software such as Fiji [23] for further reconstruction and analysis. Fig. 6 (adapted from [24]) shows an overview of imaging of a full wafer of sections at different zoom levels.

4   Outlook

ZEISS MultiSEM has been designed for continuous high-throughput imaging. As an example, complete imaging of a typical wafer containing 200 serial sections of 1 mm^2 each usually takes less than 2 days. Imaging a whole library of 100 wafers containing the sections from a full 1 mm^3 of brain tissue has been demonstrated in less than 6 months [25]. While there has been progress in processing and analyzing this data [26], storage of the data is still challenging. In the example above, 1 mm^3 of tissue sectioned with 50 nm thickness, and imaged at 4 nm pixel size, 2 PB of data will be generated. The storage cost therefore needs to be taken into account for such large-scale imaging experiments.

References

1. Sporns O, Tononi G, Kötter R (2005) The human connectome: a structural description of the human brain. PLoS Comput Biol 1:e42

2. Hagmann P, Kurant M, Gigandet X et al (2007) Mapping human whole-brain structural networks with diffusion MRI. PLoS ONE 2: e597

3. Mori S, Zhang J (2006) Principles of diffusion tensor imaging and its applications to basic neuroscience research. Neuron 51:527–539

4. Bargmann CI, Marder E (2013) From the connectome to brain function. Nat Methods 10:483–490

5. Echlin P (2009) Handbook of sample preparation for scanning electron microscopy and X-ray microanalysis. Springer US, Boston, MA

6. Fischer ER, Hansen BT, Nair V et al (2012) Scanning electron microscopy. In: Coico R, Kowalik T, Quarles J et al (eds) Current protocols in microbiology. John Wiley & Sons, Inc., Hoboken, NJ

7. Tapia JC, Kasthuri N, Hayworth KJ et al (2012) High-contrast en bloc staining of neuronal tissue for field emission scanning electron microscopy. Nat Protoc 7:193–206

8. Mikula S, Binding J, Denk W (2012) Staining and embedding the whole mouse brain for electron microscopy. Nat Methods 9:1198–1201

9. Hua Y, Laserstein P, Helmstaedter M (2015) Large-volume en-bloc staining for electron microscopy-based connectomics. Nat Commun 6:7923

10. Keller AL, Zeidler D, Kemen T (2014) High throughput data acquisition with a multi-beam SEM. Scan Microsc 2014:92360B. International Society for Optics and Photonics

11. Reimer L (1998) Scanning electron microscopy: physics of image formation and microanalysis. Springer, Berlin

12. Titze B, Genoud C (2016) Volume scanning electron microscopy for imaging biological ultrastructure: volume scanning electron microscopy. Biol Cell 108(11):307–323

13. Deerinck TJ, Bushong EA, Lev-Ram V et al (2010) Enhancing serial block-face scanning electron microscopy to enable high resolution 3-D nanohistology of cells and tissues. Microsc Microsanal 16:1138–1139

14. Wilke SA, Antonios JK, Bushong EA et al (2013) Deconstructing complexity: serial block-face electron microscopic analysis of the hippocampal mossy fiber synapse. J Neurosci 33:507–522

15. Mikula S, Denk W (2015) High-resolution whole-brain staining for electron microscopic circuit reconstruction. Nat Methods 12 (6):541–546

16. Blumer MJF, Gahleitner P, Narzt T et al (2002) Ribbons of semithin sections: an advanced method with a new type of diamond knife. J Neurosci Methods 120:11–16

17. Horstmann H, Körber C, Sätzler K et al (2012) Serial section scanning electron microscopy (S3EM) on silicon wafers for ultra-structural volume imaging of cells and tissues. PLoS ONE 7:e35172

18. Spomer W, Hofmann A, Wacker I et al (2015) Advanced substrate holder and multi-axis manipulation tool for ultramicrotomy. Microsc Microanal 21:1277–1278

19. Wacker I, Spomer W, Hofmann A et al (2016) Hierarchical imaging: a new concept for targeted imaging of large volumes from cells to tissues. BMC Cell Biol 17:38

20. Wacker IU, Veith L, Spomer W et al (2018) Multimodal hierarchical imaging of serial sections for finding specific cellular targets within large volumes. J Vis Exp. https://doi.org/10.3791/57059

21. Schalek R, Kasthuri N, Hayworth K et al (2011, 2011) Development of high-throughput, high-resolution 3D reconstruction of large- volume biological tissue using automated tape collection ultramicrotomy and scanning electron microscopy. Microsc Microanal:966–967

22. Hayworth KJ, Morgan JL, Schalek R et al (2014) Imaging ATUM ultrathin section libraries with WaferMapper: a multi-scale approach to EM reconstruction of neural circuits. Front Neural Circuits 8:1–18

23. Cardona A, Saalfeld S, Schindelin J et al (2012) TrakEM2 software for neural circuit reconstruction. PLoS ONE 7:e38011

24. Eberle AL, Garbowski T, Nickell S et al (2017, 2017) Pushing the speed boundaries of scanning electron microscopy. Imag Microsc:20–21

25. Haehn D, Hoffer J, Matejek B et al (2017) Scalable interactive visualization for connectomics. Informatics 4:29

26. DeWeerdt S (2019) How to map the brain. Nature 571:S6–S8

Chapter 9

Improving Serial Block Face SEM by Focal Charge Compensation

Ann-Katrin Unger, Ralph Neujahr, Chris Hawes, and Eric Hummel

Abstract

Serial block face SEM (SBFSEM) relies on particular fixation and embedding protocols introducing high amounts of different heavy metals into the sample to generate conductivity and sufficient contrast for imaging with back-scattered electrons. We describe two different preparation protocols to increase either overall metal content or to specifically enhance contrast of endomembrane systems. For samples containing large voids or lumina, such as kidney glomeruli or plant vacuoles this is not sufficient. They are charging, nevertheless, due to the absence in these lumina of molecules where heavy metals would bind in cytoplasm. We demonstrate how this structure-dependent charging can be alleviated by gas injection into the SEM chamber—highly localized to the vicinity of the sample surface. This is called focal charge compensation or focal CC. In addition, we explain in detail how to get resin blocks ready for imaging in an SEM using the GATAN 3view system.

Key words Scanning electron microscope, Serial block face SEM, 3D volume imaging

Abbreviations

CC	Charge compensation
CNTs	Carbon nanotubes
SEM	scanning electron microscope
EHT	Electrical high tension
FIB	Focused ion beam
TEM	Transmission electron microscopy

1 Introduction

Serial block face SEM imaging is a powerful tool to obtain high resolution 3D datasets of different sample types and tissues at nanometer resolution. The application was first developed for analyzing brain samples [1] but is now commonly used for a wide range of biological sample types and tissues [2]. A fixed sample is

Irene Wacker et al. (eds.), *Volume Microscopy: Multiscale Imaging with Photons, Electrons, and Ions*, Neuromethods, vol. 155, https://doi.org/10.1007/978-1-0716-0691-9_9, © Springer Science+Business Media, LLC, part of Springer Nature 2020

stained with heavy metal, dehydrated, and embedded in resin. Then 20–50 nm thick slices are removed from the block surface using an ultramicrotome positioned inside the SEM chamber. Repetitive cycles of cutting, imaging of the newly exposed block face using backscattered electrons, and raising of the sample block create a fully automated data acquisition workflow.

Quite a number of different applications such as focused ion beam scanning electron microscopy (FIBSEM, *see* also Chapters 11 and 12), array tomography (AT, *see* Chapters 4–8 of this volume), or transmission electron tomography (TEM tomography) exist to generate 3D volumes at nanometer resolution. While conventional focused ion beam systems deliver the smallest isotropic voxels but are limited in total volume size (cf Chapter 12), SBFSEM imaging enables the analysis of samples up to 1 mm^3 within a few days, albeit at larger voxels sizes. This results from the fact that an ion beam can remove thinner layers from a surface—usually in the range of several nanometers—than a diamond knife.

Since resins are electrical insulators heavy metals have to be introduced into the biological samples during sample preparation to combat charging of the finished block. The metals commonly used—osmium, uranium, and lead—have high atomic numbers Z which also helps to generate contrast when a detector for backscattered electrons is used for imaging. Some tissues with a high proportion of luminal space, such as kidney glomeruli, lung alveoli, or neuronal structures in the spinal cord tend to accumulate charges because their lumina are basically large areas with only resin which do not contain any biological material where the heavy metals would bind.

For such "difficult to image" samples Focal Charge Compensation (focal CC) has been developed in collaboration with the National Center for Microscopy and Imaging Research (NCMIR, UC San Diego, USA) as an extension of the Gatan 3View® SBFSEM system [3]. Focal charge compensation uses a gas injection system based on capillary needle. The needle is located close to the sample surface. Nitrogen is applied directly on the block surface and only here variable pressure conditions apply. The needle retracts automatically every time a slice is removed from the block face, meaning the workflow is uninterrupted and high image acquisition rates are maintained. Due to the fact that the variable pressure is only present within the focal plane, scattering effects of electrons which are prevailing when using global variable pressure conditions (e.g., in environmental SEM) are minimized. Using focal CC, image quality is highly improved without the need for long acquisition times or repeated imaging of the same position. Not only does this enable easy imaging of the most charge-prone samples, but it also significantly reduces beam exposure time. This in turn guards against sample damage, which is key to generate a reliable and reproducible 3D dataset. In this chapter we will give two examples for sample fixation and embedding protocols and describe how the finished sample blocks are further processed for SBFSEM. Loading of the mounted blocks and a few considerations about imaging parameters will close the chapter.

2 Materials

2.1 Fixation and Embedding

2.1.1 OTO Fixation

Cacodylate (Ted Pella Inc., Redding, CA): 0.3 M stock solution in H_2O, pH 7.4.

25% Glutaraldehyde (Sigma-Aldrich).

Paraformaldehyde (Electron Microscopy Sciences = EMS, Hartfield, PA).

Ringer's solution: 147 mmol/l Na^+, 4 mmol/l K^+, 2,3 mmol/l Ca^{2+}, 156 mmol/l Cl^- Primary fixative: 2.5% glutaraldehyde, 2% formaldehyde (fresh from paraformaldehyde) in 0.15 M cacodylate buffer pH 7.4 containing 2 mM calcium chloride.

4% OsO_4 in H_2O (Sigma-Aldrich).

Potassium ferrocyanide (Sigma-Aldrich).

Thiocarbohydrazide (TCH, Ted Pella Inc., Redding, CA).

L-aspartic acid (Sigma-Aldrich).

Lead nitrate (Sigma-Aldrich).

Hard Plus resin 812 kit (EMS catalogue # 14115).

Epon812 Embed Kit (Science Services, München, Germany).

Glass slides (EMS).

2.1.2 Endomembrane Infiltration

Zinc iodide (Sigma-Aldrich) in addition to the abovementioned chemicals.

2.2 Preparation of Sample Blocks

Razor blades.

Two-component glue aliquoted into two syringes (e.g., Uhu Plus Schnellfest).

Carbon-nanotubes Industrial Grade 5–20 μm (Nanolab).

Tweezers.

Toothpicks.

Safe lock reaction tubes (1 ml).

Stereomicroscope.

Isopropanol p.A.

Ethanol p.A.

Aluminum pins for 3View (EMS).

Weighing boats.

Hot plate (optional).

Sputter coater (Quorum, UK) with iridium or platinum target.

2.3 Scanning
Electron Microscopy

GeminiSEM 300 (Carl Zeiss Microscopy GmbH, Oberkochen, Germany).

Focal Charge Compensation needle (Focal CC).

Gatan 3View System (Gatan, Pleasanton, CA, USA) equipped with an oscillating Diamond knife (Diatome, Biel, Switzerland).

3 Methods

3.1 Fixation
and Embedding

3.1.1 OTO Protocol
for Brain Imaging, See *also*
Note 1

For SBFSEM sample preparation it is best to use a fixation and embedding protocol that includes high amount of heavy metal staining. This leads to less charging effects and beam damage during image acquisition and increases the contrast at low primary beam energies. Reducing beam damage also helps to minimize section thickness.

A protocol based on a twofold incubation with OsO_4 plus en bloc staining with uranyl acetate and Walton's lead [4, 5] is recommended to achieve best contrast (*see* **Note 1**):

1. Animals are anesthetized and perfused with normal Ringer's solution containing xylocaine (0.2 mg/ml) and heparin (20 units/ml) for 2 min at 35 °C followed by primary fixative (*see* Subheading 2.1.1) at 35 °C for 5 min.

2. Target tissues are removed and fixed for an additional 2–3 h on ice in primary fixative.

3. Some tissues such as brain should be cut into 80–100 μm thick vibratome sections in ice-cold 0.15 M cacodylate buffer containing 2 mM calcium chloride. Other tissues may be cut into small (<2 mm × 2 mm × 2 mm) pieces with a razor blade.

4. Tissues are washed 5 × 3 min in cold cacodylate buffer containing 2 mM calcium chloride.

5. Right before use, a solution containing 3% potassium ferrocyanide in 0.3 M cacodylate buffer with 4 mM calcium chloride is combined with an equal volume of 4% aqueous osmium tetroxide. The tissues are incubated in this solution for 1 h, on ice.

6. During the initial osmium incubation (**step 5** above) prepare TCH solution for the next step. This reagent needs to be fresh and available right at the end of **step 5**. Add 0.1 g TCH to 10 ml ddH_2O and place in a 60 °C oven for 1 h, agitate by swirling gently every 10 min to facilitate dissolving. Filter this solution through a 0.22 μm Millipore syringe filter right before use.

7. At the end of the first heavy metal incubation described in **step 5** (before adding the TCH) the tissues are washed with ddH$_2$O at room temperature 5 × 3 min (~15 min total).

8. Tissues are placed in the filtered TCH solution for 20 min, at room temperature.

9. Tissues are rinsed again 5 × 3 min in ddH$_2$O at room temperature and placed in 2% osmium tetroxide (NOT osmium ferrocyanide) in ddH$_2$O for 30 min, at room temperature.

10. The tissues are washed 5 × 3 min at room temperature in ddH$_2$O, then placed in 1% uranyl acetate (aqueous) and left in a refrigerator (~4 °C) overnight.

11. Next day, en bloc Walton's lead aspartate staining is performed: First, an aspartic acid stock solution is prepared by dissolving 0.998 g of L-aspartic acid in 250 ml of ddH$_2$O. Note: the aspartic acid will dissolve more quickly if the pH is raised to 3.8. This stock solution is stable for 1–2 months if refrigerated. To make the stain, dissolve 0.066 g of lead nitrate in 10 ml of aspartic acid stock and adjust pH to 5.5 with 1 N KOH. The lead aspartate solution is placed in a 60 °C oven for 30 minutes (no precipitate should form). The tissues are washed 5 × 3 min in ddH$_2$O at room temperature and then placed in the lead aspartate solution and returned to the oven for 30 min.

12. The tissues are washed 5 × 3 min in room temperature ddH$_2$O and dehydrated using ice-cold solutions of freshly prepared 20%, 50%, 70%, 90%, 100%, 100% ethanol (anhydrous), 5 minutes each, then placed in anhydrous ice-cold acetone and left at room temperature for 10 min.

13. Tissues are placed in room temperature acetone for 10 min. During this time, Epon 812 resin is prepared by weight as follows: 11.4 g part A, 10 g part B, 0.3 g part C, and 0.05–0.1 g part D, yielding a hard resin when polymerized. The resin is mixed thoroughly. Samples are placed into 25% Epon 812 / 75% acetone for 2 h, then into 50% Epon 812:50% acetone for 2 h and finally into 75% Epon 812:25% acetone for 2 h.

14. Tissues are placed in 100% Epon 812 overnight, then into fresh 100% Epon 812 for 2 h. Tissue sections are then mounted between liquid release agent-coated glass slides and tissue pieces are embedded in a thin layer of fresh resin in an aluminum weighing boat and placed in a 60 °C oven for 48 h (*see* **Notes 2** and **3**).

Flat embedding is highly recommended, as the sample has to be cut from the resin block for mounting on the 3View pin.

1. Maize (*Zea mays*) are germinated for 5 days in plates.

2. Primary root tips are fixed in 1% glutaraldehyde and 0.5% paraformaldehyde in 0.1 M sodium cacodylate buffer at pH 6.8 for 40 min and washed in buffer followed by distilled water [6].

3. The ZnI_2 solution is freshly prepared [7]: 1.5 g powdered zinc, 0.5 g resublimed iodine, and 10 ml distilled water. The solution is sonicated for 1 min, stirred for a further 5 min and filtered.

4. The ZIOs solution is made by adding equal volumes of 2% OsO_4 solution to the ZnI_2 solution. Tissue is allowed to impregnate for 2–6 h.

5. Dehydration is performed in 30 min cycles of 10%, 20%, 30%, 40%, 50%, 60%, 70%, 80%, and 90% acetone in water, followed by 3x100% acetone.

6. Embedding in Epon 812 (25%, 50%, 75%, 100% (3×).

3.2 Sample Preparation for 3View analysis

1. The maximum sample size for 3View analysis is 1 mm³. For this reason, the size of the block has to be reduced by trimming. To minimize charging effects and beam damage during SEM analysis, it is recommended to remove as much surplus resin as possible. In addition, the sample needs to have direct contact to the metal 3View pin to generate electrical conductivity. The trimming can be easily performed at a stereomicroscope using a razor blade. As for conventional ultramicrotomy it is best to trim a cube out of the sample block. To guarantee better cutting behavior and conductivity it is recommended to trim a pyramid-shaped block (Fig. 1a). Silver paint is used for surrounding the block for best possible conductivity.

2. Frequently even experienced users tend to lose the sample during the final trimming steps when the sample turns really small. To avoid flipping away the sample is placed on a piece of Scotch tape during the trimming process. After trimming it is washed with ethanol or isopropanol for 5–10 min in a reaction tube to remove the remaining glue of the scotch tape and finally air dried at room temperature.

3. The next step is to mount the sample on the 3View sample holder. The aluminum pin is placed into the preparation holder under the stereomicroscope and fixed by the screw (Fig. 1b). In a small weighing boat equal amounts of the two components of the glue are placed next to each other. To obtain a conductive glue roughly 100 mg carbon nanotubes are added and carefully mixed with a tooth pick to disperse them homogeneously. The texture of the resulting glue should not be too viscous but at the same time there need to be sufficient carbon nanotubes to facilitate conductivity.

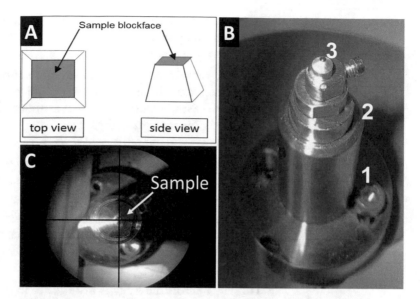

Fig. 1 Overview of sample preparation and mounting for SBFSEM. (**a**) Schematic overview of sample trimming in top and side view. (**b**) Setup of the sample mounting outside the SEM chamber 1 = Stereomicroscope adapter; 2 = SEM adapter; 3 = Aluminum pin with sample glued on top (**c**) Centered sample view inside the SEM chamber uslng the built-in stereomicroscope

4. A small droplet of the carbon nanotube–glue mixture is placed on the aluminum pin using a fresh toothpick, then the sample is placed on top using a fresh slightly wetted toothpick. The glue starts polymerizing 6 min after bringing both components together. It is therefore very important to work fast. The mounted assembly is dried for 25–30 min at room temperature. The drying process can be accelerated by placing the pin onto a heating plate at 60 °C.

5. After the sample is fixed on the aluminum pin it has to be fine trimmed. It is placed under a stereomicroscope into the sample holder adapter and fixed with a screw. Using a razor blade the sides of the block are trimmed at an angle of 35–45° making sure that leading and trailing edges are as parallel as possible. The trimming can be performed manually by simply using a razor blade or—more precisely—by using an ultramicrotome equipped with a diamond trimming knife or a glass knife (*see* also Chapter 5 of this volume for different trimming tools and geometries).

6. To further minimize charging during imaging the sample surface needs to be metal coated. A sputter coater adapter for the 3View pin is prepared by drilling a 2 mm hole into a standard SEM stub. The entire assembly of 3View pin plus adapter is coated with a few nanometers of metal. Iridium or Platinum are

best because they give very small grain sizes and do not disturb sectioning or damage the diamond knife of the in-chamber ultramicrotome.

7. The sample is removed from the sputter coater adapter and placed into the 3View sample holder under the stereomicroscope. Sample position is then centered as well as possible using an in objective grid (Fig.1c).

8. Sample preparation is now complete. The microscope front door is changed to the 3View system, controller and software are started, and the OnPoint BSE detector is inserted.

9. The diamond knife of the in-chamber microtome can be locked using a toothpick. This is just a safety mechanism to avoid diamond knife damage. After the diamond knife is in a safe position the 3View sample holder is placed into the ultramicrotome.

10. The microtome stage is brought to its lowest z-position before starting the knife/sample approach. This process is very similar to the adjustment on a conventional ultramicrotome. The goal is to bring the sample as close as possible to the diamond knife and to check if the cutting position of the sample and the cutting window of the diamond knife are set up correctly. For better visual control the stereomicroscope is mounted onto the 3View door.

11. To start the adjustment it is necessary to bring the diamond knife from the safe position to the cutting position. The switch to cutting position is performed using the software. For safety reasons it is recommended to block this first movement with the inserted toothpick and to manually perform the movement to the cutting position. A LED mounted on the 3View stage provides backlight illumination helping to judge the distance from the sample surface to the diamond knife edge. The retraction cycle of the 3View is coupled to the piezo function of the oscillating diamond knife so it is very important to turn on the stroke (equivalent to cutting mode with active piezo oscillation) in the software before starting the z-height adjustment of the sample. The sample is then manually moved up in Z-position under visual control. The light reflection between sample and knife becomes narrower with decreasing distance between knife and sample. This is the same mechanism as used for sample to knife adjustment on a stand-alone ultramicrotome.

12. When the distance between sample and knife is adjusted, the cutting window has to be checked. It is best when the knife starts cutting shortly before the sample is reached and stops cutting when it has passed the sample block face completely. To check this the cutting process is started in the software and carefully watched by eye.

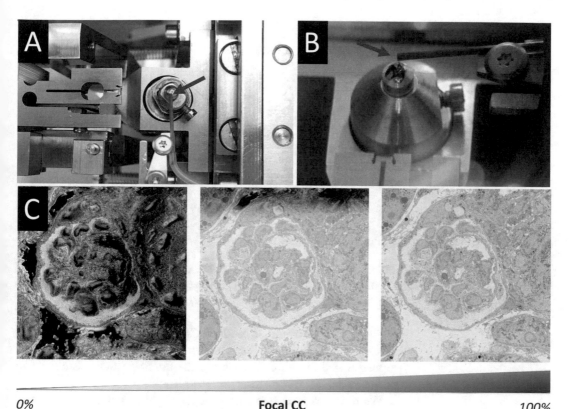

0%   **Focal CC**   *100%*

Fig. 2 Focal CC setup. (**a**) Top view illustrating how the FocalCC needle is attached to the Gatan 3View microtome stage. (**b**) Working position of the needle during block face imaging. The needle is perfectly adjusted to the sample block face. Red arrows in **a**, **b** point to the opening of the needle. (**c**) Imaging results obtained without and with focal charge compensation. Example shows highly charging kidney glomerulus (left without charge compensation; center with 50% charge compensation; right with 100% charge compensation). Sample courtesy of University of Freiburg, Germany

13. As final step the Focal CC (charge compensation) needle is brought close to the sample block face (Fig. 2b). Since the FocalCC works under high vacuum conditions the working distance is limited and the needle has to be placed <0.5 mm from the block surface. The cutting process can be started in the software now. When the first sections are generated, the system can be pumped.

14. The microtome is now under high vacuum.

3.3 Serial Block Face Imaging

Depending on the desired section thickness the interacting volume of the electron beam has to be adjusted accordingly (*see* **Note 4**). Charging effects and beam damage have also to be considered when balancing imaging parameters such as EHT, beam current, and dwell time with cutting thickness. We usually achieve best results with our machine at low acceleration voltages. Common parameters are 1.2 kV at a working distance (WD) of 5 mm, using

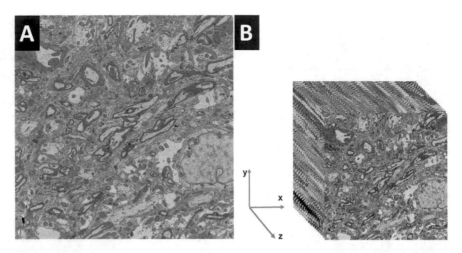

Fig. 3 Serial block face imaging of mouse brain. (**a**) 2D overview image of one slice of mouse brain imaged with Focal CC with a pixel size of 7 nm. (**b**) 3D image stack (20 slices) imaged with 7 nm pixel size and 30 nm slice thickness (i.e., voxels of 7 nm × 7 nm × 30 nm). Sample courtesy of Christel Genoud

the standard 30 μm aperture. Focal CC pressure is carefully lowered to 20–30% during stack acquisition. Depending on the sample properties and imaging parameters the cutting thickness can be in a range of 20–50 nm or more.

We have selected two different examples: First the most commonly used fixation and embedding technique for SBFSEM—the so-called osmium–thiocarbohydrazide–osmium (OTO) fixation which provides high contrast for all types of cellular structures. OTO has been used in different variations for brain imaging, a table comparing the different variants can be found in [5]. Figure 3 shows a brain sample treated with this method. Due to the high concentrations of heavy metal accumulating at the membranes optimal imaging quality without charging artifacts is observed.

The second technique, osmium impregnation was originally used to specifically stain the membranes of the endoplasmic reticulum (ER). In a first step, zinc binds preferably to these membranes and in the second step osmium interacts with zinc iodide leading to strongly stained ER membranes. The method was not only used for plant cells as shown in our example (Figs. 4 and 5), but may be also used for other types of cells such as brain and other tissues [7]. The osmium impregnation technique is used to selectively stain endomembranes such as ER and Golgi stacks as illustrated in Fig. 4. The comparison of Fig. 5a, b where no focalCC was used with Fig. 5c, d which were imaged with focalCC demonstrates that particularly tissues with a high proportion of "voids," such as the vacuolated plant cells shown here, profit from the application of focalCC.

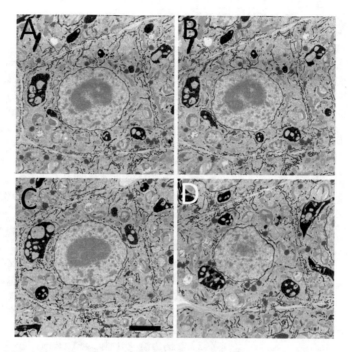

Fig. 4 Meristematic cells of *Vicia faba* (pea) root—zinc osmium impregnated. Representative slices (unprocessed data) of a meristematic cell shortly after division illustrate that the zinc iodide impregnation generates very high contrast for endomembranes due to their high osmium content. As vacuoles, which would potentially cause charging effects, are not present in meristematic cells it was not necessary to use Focal CC—the system could be operated under high vacuum conditions and still produce good images. Scale bar 500 nm

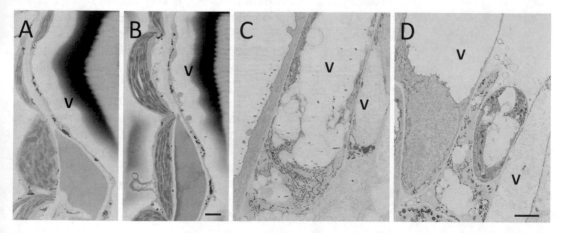

Fig. 5 Highly vacuolated plant cells imaged either without (**a**, **b**) or with (**c**, **d**) Focal CC. Charging artifacts are clearly visible as black patches in the areas of the vacuoles (V) in (**a**) and (**b**). The cells shown in (**c**) and (**d**) are free from these artifacts although they also contain large vacuoles. Scale bars 400 nm for (**a**, **b**) and 600 nm for (**c**, **d**)

4 Notes

1. This protocol was designed to enhance signal for backscatter electron imaging of epoxy-embedded mammalian tissue at low accelerating voltages (1–3 keV). However, it can easily be adapted for use with tissues from other species, tissue culture cells, plants and microbial cells by adjusting the buffer strength and the duration of relevant steps. This combinatorial heavy metal staining protocol employs a battery of contrasting steps after primary aldehyde fixation including: post fixation with ferrocyanide-reduced osmium tetroxide, thiocarbohydrazide–osmium liganding (OTO) and subsequent en bloc staining with uranyl acetate and lead aspartate. Calcium chloride is included in a number of steps to enhance membrane preservation and staining. This protocol was designed primarily to emphasize the contrast of membranes. Many other contrasting agents may be included to increase staining of other cellular and extracellular constituents. There are a number of variations of these protocols in the scientific literature (cf. table in [5]). Depending on the structures analyzed these protocols differ in the iterations of heavy metal staining.

2. A recent advance [8] introduced conductive material into the bulk of the embedding material by removing the sample from Epon resin before it cured and transferring it to a two-component epoxy glue with silver particles (Epo-Tek EE129-4). The ratio of the two components should be 1.25 parts A to 1.00 parts B.

3. During the embedding process it is very important to guarantee a good infiltration of the sample with the resin. During the cutting and imaging cycles heat is generated by the electron beam and can lead to damage of the resin, if the sample is not perfectly infiltrated or polymerized. An easy test to check the quality of the sample is to place it under constant scanning with the electron beam for 1–2 min and generate a few pictures. Outgassing artifacts appearing as vertical streaks in the image are clearly visible (Fig. 6). Generally, the embedding resin should be as hard as possible. Recommended resins are Epon 812 replacement and Spurr's resin which may be adjusted in hardness by varying the proportions of their different components. A recently published study identified Hard Plus resin as the most stable resin for 3D BioEM applications [9].

4. It is important to keep the acceleration voltage as low as possible for several reasons: As the penetration depth of the electron beam depends on the spot size (beam current), the acceleration voltage and the electron density of the sample material itself it is crucial to balance these parameters well. If the interaction

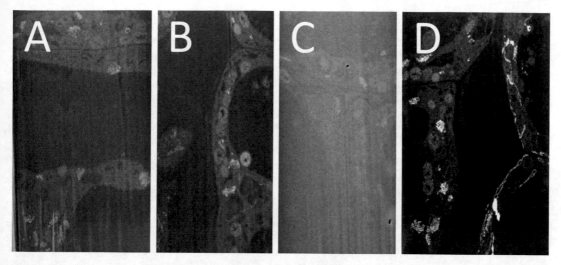

Fig. 6 Outgassing effects on the surface of a freshly prepared sample. (**a**) Sample block placed into the SEM directly after polymerization of the resin. The vertical streaking caused by outgassing is visible after 1 min of continuous beam exposure. (**b**) Freshly cut surface shows outgassing effects after 10 s. (**c**) Two minutes of continuous imaging increases this effect dramatically. (**d**) Block was kept in prevacuum overnight and then placed into the SEM: No outgassing effects are visible even after scanning continuously for 2 min. Note that in this series of images contrast is inverted compared to previously shown images

volume between beam and sample is too big, resin quality will degrade causing bad cutting results. A low electron dose will help to avoid damaging the resin during electron beam irradiation and will consequently preserve excellent cutting properties of the resin. Additionally, a large interaction volume will deteriorate image quality because the signal collected during scanning will then originate from deeper regions of the sample and adjacent scan spots may overlap. This will lead to less resolution in x and y for single images and adversely affect the Z-resolution of the entire image stack acquired. For this reason, it is important to keep the interaction volume of the electron beam as small as possible. The dependency of penetration depth into biological samples on accelerating voltage has recently been analyzed [5].

References

1. Denk W, Horstmann H (2004) Serial block-face scanning electron microscopy to reconstruct three-dimensional tissue nanostructure. PLoS Biol 2:e329. https://doi.org/10.1371/journal.pbio.0020329

2. Smith D, Starborg T (2019) Serial block face scanning electron microscopy in cell biology: applications and technology. Tissue Cell 57:111–122

3. Deerinck TJ, Shone TM, Bushong EA, Ramachandra R, Peltier ST, Ellisman MH (2018) High-performance serial block-face SEM of nonconductive biological samples enabled by focal gas injection-based charge compensation. J Microsc 270:142–149. https://doi.org/10.1111/jmi.12667

4. Tapia JC, Kasthuri N, Hayworth K, Schalek R, Lichtman JW, Smith SJ, Buchanan JA (2012)

High-contrast en bloc staining of neuronal tissue for field emission scanning electron microscopy. Nat Protoc 7:193–206. https://doi.org/10.1038/nprot.2011.439

5. Kubota Y, Sohn J, Hatada S, Schurr M, Straehle J, Gour A, Neujahr R, Miki T, Mikula J, Kawaguchi Y (2018) A carbon nanotube tape for serial-section electron microscopy of brain ultrastructure. Nat Commun 9:437. https://doi.org/10.1038/s41467-017-02768-7

6. Hawes C, Horne C (1983) Staining plant cells for thick sectioning: uranyl acetate, copper lead citrate impregnation. Biol Cell 48:207–210

7. Zhou D-S, Komuro A (1995) Ultrastructure of the zinc iodide-osmic acid stained cells in guinea pig small intestine. J Anat 187:481–485

8. Wanner AA, Genoud C, Masudi T, Siksou L, Friedrich RW (2016) Dense EM-based reconstruction of the interglomerular projectome in the zebrafish olfactory bulb. Nat Neurosci 19:816–825

9. Kizilyaprak C, Longo G, Daraspe J, Humbel BM (2015) Investigation of resins suitable for the preparation of biological sample for 3-D electron microscopy. J Struct Biol 189:135–146

Chapter 10

Using X-Ray Microscopy to Increase Targeting Accuracy in Serial Block-Face Scanning Electron Microscopy

Eric A. Bushong, Sébastien Phan, and Mark H. Ellisman

Abstract

In this chapter, we describe the use of X-ray microscopy (XRM) as a method for improving the accuracy and efficiency of volume electron microscopy (volume EM). By providing a means of nondestructively imaging EM specimens prior to performing volume EM, XRM allows the investigator to pinpoint specific regions of interest (ROIs) for imaging. In addition, given the excellent contrast and resolution that can be achieved with XRM when specimens are stained with protocols compatible with volume EM, it can also dramatically enhance the value of volume EM data, either by revealing how the EM data fits into a larger context and/or by improving the ability to perform correlated light microscopy (LM) and EM imaging. This chapter will focus on the combined use of XRM with diamond knife-based serial block-face scanning electron microscopy (SBEM). We also briefly describe software we have developed to ease tracking of ROIs across imaging modalities and allow direct targeting of ROIs in an SEM as guided by XRM volumes.

Key words X-Ray microtomography, MicroCT, Serial block-face scanning electron microscopy, Confocal microscopy, Correlated microscopy, CLEM

1 Introduction

The collection of serial block-face SEM datasets is generally a resource-intensive, time-consuming, and expensive endeavor. The acquisition of a single volume can require several days or weeks to complete. To improve the accuracy and efficiency with which data sets can be collected, several labs have turned to X-ray microscopy (XRM; also referred to as microcomputed tomography, or microCT) as a means of optimizing specimen preparation prior to volume EM imaging [1–8]. One motivating factor for employing XRM is the fact that the staining protocols used to yield biological specimens adequately electron-dense and conductive for SEM imaging also result in specimens that are completely opaque to

Electronic supplementary material The online version of this chapter (https://doi.org/10.1007/978-1-0716-0691-9_10) contains supplementary material, which is available to authorized users.

Irene Wacker et al. (eds.), *Volume Microscopy: Multiscale Imaging with Photons, Electrons, and Ions*, Neuromethods, vol. 155, https://doi.org/10.1007/978-1-0716-0691-9_10, © Springer Science+Business Media, LLC, part of Springer Nature 2020

visible light [1]. This creates a genuine challenge when attempting to track a region of interest for volume EM data collection as the investigator can only see the surface of the specimen in the SEM prior to collecting a volume. Finally, both serial block-face SEM and focused ion beam SEM techniques destroy the specimen during data acquisition, allowing for only a single opportunity to image a limited, targeted region of interest.

XRM offers the capability to nondestructively image dense specimens with high isotropic resolution. It used to be conducted using beam line sources at large facilities [9, 10], but nowadays is more accessible with several manufacturers offering lab-based XRM instruments. These instruments are capable of collecting volumes with submicron resolution, detecting elements within tissue as small as nucleoli and other subcellular organelles, given sufficient contrast (Fig. 1). XRM imaging is particularly powerful because it can reveal very subtle differences in density, allowing the researcher to virtually dissect the components of a specimen or even reveal specific staining patterns (Fig. 2). Osmium tetroxide has long been used as a stain for microCT imaging of biological specimens [11], so it is not surprising that the en bloc staining usually employed for volume EM imaging is ideal for generating excellent contrast in the XRM. In addition to allowing the researcher to create a three-dimensional (3D) map for subsequent volume EM imaging, microCT scans also provide a valuable means of prescreening specimens for delineating defects

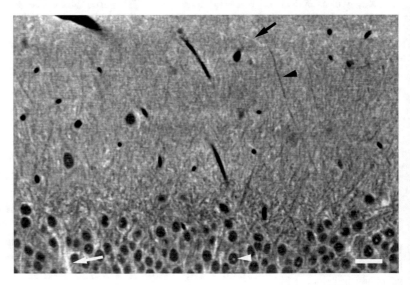

Fig. 1 A single computed slice from a microCT volume collected with a Zeiss Versa 510 XRM. The specimen was a 100 μm thick slice of mouse brain stained and embedded for SBEM imaging. Brighter pixels correspond to higher specimen density. Isotropic voxel size for this volume was 0.4159 μm. The specimen was imaged at 40 kVp using the 40× objective and an effective CCD array size of 1 k × 1 k (bin 2). It is possible to detect nucleoli (white arrowhead), dendrites (black arrowhead), glial fibers (white arrow), and myelinated axons (black arrow). Scale bar: 25 μm

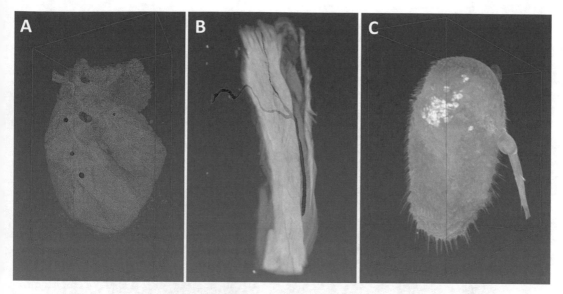

Fig. 2 Volume renderings of three representative specimens prepared for SBEM imaging and then scanned using an XRM. (**a**) Pancreatic islets (red) dispersed throughout a piece of mouse pancreas. (**b**) The location of a nerve and its terminals within a piece of intercostal muscle. (**c**) A *Drosophila* antenna with a subset of neurons specifically labeled using diaminobenzidine

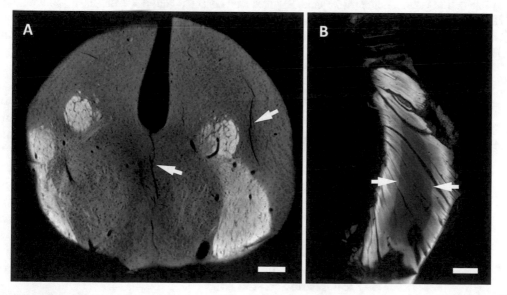

Fig. 3 MicroCT scans can reveal specimens of poor quality or areas within specimens to be avoided by volume EM. (**a**) Cracks are evident in this punch of brain tissue prepared for SBEM. (**b**) Inconsistent staining of a muscle specimen is revealed by microCT scan. Sufficient staining for SBEM imaging abruptly ends approximately 500 μm into the specimen. Stain penetration varies from tissue to tissue. Scale bars: (**a**) 200 μm, (**b**) 500 μm

and areas that are not suitable for further investigation, due to poor staining or structural damage (Fig. 3).

XRMs can image specimens as large as several centimeters in size, or even larger. However, as most XRM scanners use a CCD

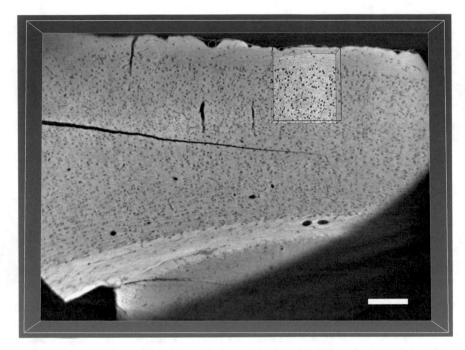

Fig. 4 Nested microCT: two microCT scans acquired at increasing resolution and subsequently registered using Amira. The larger volume (white bounding box) was collected with a 4× objective and voxel size is 2.179 μm. The interior tomography volume (black bounding box) was collected with a 40× objective and has 0.4133 μm voxels. The Zeiss XRM microscope stage allows for precise positioning of the specimen based on the lower resolution scan for acquisition of the interior tomogram. Scale bar: 200 μm

camera to collect two-dimensional (2D) projection images, the final field of view of a scan is limited by the number of pixels available on the CCD for a given resolution. Biological samples are generally small, less than a couple of millimeters in any dimension, to allow for penetration of the tissue with heavy metal stains and epoxy resin. But ultimately, the ROI for SBEM imaging will still be significantly smaller than the entire specimen. In order to collect high-resolution volumes of limited ROIs within a larger specimen, precise control of stage position can allow the researcher to collect a series of nested XRM volumes, starting with a low magnification scan to capture an entire specimen and subsequent interior tomography scans based on coordinates from the preceding low-resolution scan (Fig. 4).

The achievable resolution and contrast with an XRM depend on several factors, some related to the design of the microscope (e.g., X-ray source spot size, scintillator composition, and CCD design) and others to operator decisions when collecting data. A large palette of imaging parameters (e.g., beam energy, exposure time, sampling scheme, pixel size) is available depending on the needs. For instance, low-resolution microCT volumes are sufficient to reveal insights that are important when screening and prepping

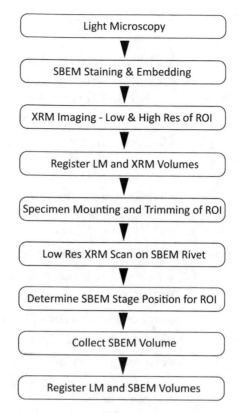

Fig. 5 Typical workflow for an experiment using XRM to target an ROI for SBEM imaging

specimens for volume EM imaging. On the other hand, high-resolution microCT volumes are useful when targeting cellular-level structures for volume EM imaging. In our typical specimen preparation workflow (Fig. 5), XRM can be used multiple times: (1) scanning the specimen at multiple resolutions before trimming and mounting for SBEM imaging to select and orient the desired ROI and (2) scanning the specimen after mounting on an SBEM mounting rivet to determine SEM stage coordinates for accurate targeting of ROI.

XRM imaging has been increasingly employed in volume EM approaches, and even conventional transmission EM work [1–8]. This chapter will describe a basic protocol to use XRM to target ROIs in brain slices initially selected based on LM for diamond-knife SBEM. The technique is precise enough to allow correlation of the confocal volume with the final EM volume. We will briefly describe the use of software (Navminator) that we have developed to take full advantage of the insights afforded by microCT volumes when preparing specimens for volume EM imaging.

2   Materials

2.1   Tissue Preparation and Confocal Imaging

VT1000 S vibrating blade microtome (Leica Biosystems).

Cacodylic acid, Sodium, Trihydrate (Ted Pella, Product No. 18851) (*see* **Note 1**).

DRAQ5 (ThermoFisher Scientific, Product No. 62251).

Olympus Fluoview microscope with 60× water objective (NA 1.20), 20× air objective (NA 0.75), and 10× air objective (NA 0.40).

Glass-bottomed petri dish, 35 mm (MatTek Corp., Product No. P35G-0-14-C).

2.2   SBEM Imaging and Specimen Preparation

Gemini 300 scanning electron microscope (Zeiss).

3View serial block-face imaging unit (Gatan)

SBEM specimen mounting rivets (Gatan).

Silver conductive epoxy (Ted Pella, Product No. 16043).

SEM sputter coater (Polaron Instruments, E5100).

Standard glass slides.

Liquid release agent (Electron Microscopy Sciences, Product No. 70880).

2.3   XRM and XRM Specimen Preparation

Versa 510 X-ray microscope (Zeiss).

Aluminum 3003 Seamless Round Tubing, 1/16″ OD, 0.0345″ ID, 0.014″ Wall, 12″ Length (Amazon).

Aluminum 3003 Seamless Round Tubing, 1/8″ OD, 0.097″ ID, 0.014″ Wall, 12″ Length (Amazon).

Five-minute epoxy.

2.4   Specimen Trimming and Ultramicrotomy

Ultracut UCT ultramicrotome (Leica Biosystems).

Glass strips for knives (Leica Biosystems).

Razor blades.

Low power stereoscope.

Cyanoacrylate glue.

Mounting cylinders (Ted Pella, Product Number 10580).

Aclar film, 7.8 mil (Ted Pella, Product Number 10501).

2.5   Image Analysis and Registration

Workstation (Intel 3GHz 8 Core, 512 GB RAM, NVIDIA Titan X DDR5 12 GB).

Amira 6 (ThermoFisher Scientific).

Anaconda (Python 2.7, https://www.anaconda.com/download/) for running Navminator.

Navminator   (https://confluence.crbs.ucsd.edu/display/ncmir/NCMIR+Software).

MB-Ruler (http://www.markus-bader.de/MB-Ruler/index.php).

3   Methods

3.1   General Considerations for XRM Imaging

A microCT scan consists of acquiring multiple projection images of a specimen rotating around an axis perpendicular to the X-ray beam. The collected 2D images are then used to generate a computed tomographic reconstruction, usually through a back-projection algorithm performed in reciprocal space. The intensity values in the final tomographic volume reflect the X-ray density within the specimen. There are numerous models of lab-based XRM scanners available that can differ in numerous aspects, but they all share the same basic components: (1) a microfocus X-ray tube, (2) a precise, stable motorized stage, which allows for specimen orientation and rotation, and (3) a detector, usually a scintillator screen coupled to a CCD array (Fig. 6). The scintillator

Fig. 6 The interior of a Zeiss Versa 510 XRM instrument. The X-ray tube is seen to the left, the stage and specimen holder in the center (white arrow), and the detector system on the right. The system is set up for a high-resolution scan of a specimen, and there is only about 10 mm separating the X-ray source and the detection scintillator

converts X-rays to visible light photons, which are collected by the CCD.

XRM detectors are usually flat panels. However, Zeiss XRMs employ an intermediate optical objective lens between the scintillator and CCD for additional magnification.

As with most forms of microscopy, there is no universal set of imaging parameters for collecting microCT volumes that works for any situation. Rather, the investigator must choose imaging conditions that will likely yield good results for the primary intended purpose, but bring drawbacks in other aspects. Overall, the final data quality is subject to the competing influence of numerous imaging parameters on properties such as resolution, contrast, and field of view. We will not attempt to provide step-by-step instructions for data acquisition on any particular instrument, but rather describe in this section the factors to be considered when collecting microCT volumes.

3.1.1 Beam Energy and Beam Filtering

Most lab-based XRM systems use X-ray tubes that will provide a range of X-ray energies spanning from approximately 20 kVp up to 150 kVp or more. Higher energy X-rays can penetrate larger, denser specimens and provide brighter beams, significantly reducing acquisition time. Lower energy X-rays on the other hand are more likely to be absorbed by X-ray dense regions of the specimen, resulting in greater contrast. Given the small dimensions of the specimens, the energies used for scanning volume EM specimens is generally limited to 40–80 kVp. Higher energies may lack in contrast, while lower energies would result in extremely long acquisition times and possibly suffer from specimen drift. Source filters can also be used to narrow the range of beam energies. Stronger filters may result in lower overall contrast but will reduce the presence of beam hardening artifacts in reconstructions (*see* **Note 2**).

3.1.2 Magnification and Detection

In XRMs, a magnified image of the specimen is usually generated through geometric magnification. Lab-based XRMs use point X-ray sources, which produce a cone (or sometimes, a fan) of illumination emanating from the X-ray tube. The X-ray beam interacts with the specimen and then spreads (and thus enlarges the image) until encountering the detector. The greater the distance between the specimen and the detector, the greater the geometric magnification. Decreasing the distance between the source and the specimen will also result in increased geometric magnification. The detectors of Zeiss XRMs are further equipped with multiple objective lenses for imaging at various magnifications. Each lens is fitted with a custom scintillator screen. The total image magnification is a combination of the geometrical magnification onto the scintillator and optical magnification of the image from the scintillator.

There are limits to the amount of magnification that can be achieved by increasing the distance of the detector from the specimen. First, increased distance also leads to an increase in blurring due to the finite spot size of the source. The confusion of X-rays originating from separate points will be exaggerated with increased geometric magnification. XRMs with smaller source spot sizes will therefore improve resolution. Second, the magnification (imaging geometry/objective) will affect acquisition time. Greater distance between the source and detector will result in a longer acquisition time for each image, potentially leading to increased specimen drift. Additionally, for the Zeiss models of XRMs, the higher the magnification power of the objective lens used, the lower the brightness and therefore the longer the exposure time needed.

3.1.3 Sampling Scheme

During acquisition, the specimen must be rotated during scanning to collect a set of 2D projection images. The sampling scheme can be modified depending on specimen geometry, desired resolution, and acquisition time goal. It is possible to collect images from approximately $-90°$ to $+90°$. This will yield sufficient resolution for many instances and limits the imaging time. However, due to the conical nature of the illumination beam for lab-based XRMs, a full $360°$ rotation of the specimen will deliver improved resolution.

The number of projection images that are collected will affect both scan time and resolution. More projection images (up to 3201 images, ~$0.1°$ tilt per image) can improve resolution, but given a constant exposure time per projection image, more images will obviously greatly extend acquisition time. Generally, for scans encompassing an entire specimen and lower resolution scans (binned CCD acquisitions, see below), 1801–2001 projection images will produce excellent results. Less projection images can be collected to decrease acquisition time. Interior tomography scans and higher resolution scans will benefit from increasing the number of projection images (up to 3201 or more).

3.1.4 Exposure Time

As discussed above, the exposure time choice is related to the beam energy, the objective (if used), and the distance between the source and objective. The density of the specimen also affects the imaging time, but with denser specimens one would typically use higher energy X-rays to achieve sufficient transmission rates with limited effect on image contrast. As a rule, a sufficiently long exposure time must be used to achieve adequate signal-to-noise in the projection images. However, exposure time should also be limited to minimize overall scan acquisition time and avoid specimen drift. In general, we use a minimal exposure time for each projection image while collecting as many projection images as possible in a reasonable total scan time.

Exposure time can also be reduced by using binning of the CCD during image acquisition at the expense of resolution. Two-fold binning of the detector results in a fourfold decrease in acquisition time while yielding the same pixel counts. Of course, this gain in acquisition time comes at the cost of a corresponding doubling of pixel size. If the specimen has a high aspect ratio (e.g., a vibratome slice or rectangular biopsy), then it is necessary to ensure that the exposure will be sufficient throughout the rotation of the specimen.

3.1.5 Specimen and Beam Stability

The stability of the specimen is of paramount importance for achieving quality microCT volumes. High-resolution scans (<1 μm voxels, 1 k × 1 k [binned by 2 on a Zeiss Versa]) can take 5–12 h to acquire with a lab-based XRM. During this time the specimen is continuously rotated and occasionally pulled completely out of the beam path for the acquisition of reference background images. The specimen therefore must be very well affixed to a stable support, preferably placing the ROI as close to the support as possible. To limit specimen drift, the specimen and its support should be given time to acclimate to the microscope temperature with the X-ray beam irradiating the sample. The source also needs to be stabilized before beginning any run. The lower the beam energy, the longer the source will need to stabilize. It may be necessary to allow the X-ray tube to warm up for an hour or more at low kVp.

3.2 Incorporating XRM Imaging in a Volume EM Workflow

XRM imaging of specimens can be performed solely as a survey step to assess specimen quality and identify ROIs, or it can serve as an integral step in a larger multimodal imaging workflow, bridging light microscopic and volume EM imaging of a specimen.

In the former scenario, the primary objective for microCT imaging is to determine an optimal subvolume within a specimen and an ideal orientation for mounting the specimen to access that subvolume by SBEM imaging. Once the specimen is mounted and prepared for SBEM imaging, an additional quick scan of the specimen as mounted on the SBEM pin provides an excellent map for determining the SEM stage coordinates for the SBEM run. Since the ROI will be hidden within the specimen itself during this step, the corners of the specimen block and small debris particles usually found on the block face can act as landmarks for calculating the position of the ROI. NCMIR has developed a software tool, called Navminator, that will coregister two images or volumes if provided with three or more corresponding landmarks. If the SEM stage coordinates of the landmarks are additionally provided, Navminator will provide the stage coordinates to target any arbitrary ROI within the coregistered microCT image.

In the latter scenario, where XRM imaging is being used to facilitate correlative LM and EM imaging, specimens will be stained and imaged by LM prior to prepping for EM imaging. The goal is to track the same ROI imaged by LM after the sample is rendered opaque by heavy metals. There are few intrinsic structures that can be seen across LM, XRM, and SBEM volumes. Some tissues will exhibit autofluorescent structures (e.g., lipofuscin), which are often rich in lipids and therefore osmiophilic. The resulting heavy osmium staining will make these structures both X-ray and electron dense. Vasculature can also often be seen in tissue by LM, given sufficient background fluorescence, and has been successfully used to register microCT datasets with other imaging modalities [3]. This approach relies on reducing the vasculature to skeletal structures, allowing for unambiguous identification of branch-points to use as fiducial landmarks.

For optimal alignment of volumes, the tissue should ideally contain punctate labeling that can be seen across imaging modalities. We have found that nuclear labeling with DNA intercalating dyes, such as DRAQ5, propidium iodide, DAPI, or TO-PRO-3, is a very good approach. It can quickly provide a very reliable, well-distributed punctate labeling pattern throughout a tissue specimen without the need for permeabilization with detergents. The coregistration of LM volumes to microCT volumes does not generally need to be exceptionally precise, since the goal is to guide the process of trimming down to and then approach an ROI within the specimen. Errors in registration on the order of a couple of microns will usually not affect the targeting of an ROI with SBEM. Simply using a handful of nuclei as landmarks is usually sufficient to map the ROI as defined by LM imaging within a microCT volume (Fig. 7). Our software tool (Navminator) can be used to quickly and easily register 2D and 3D images across imaging modalities. Navminator will keep a series of datasets registered with respect to each other and allow the user to map corresponding points between datasets. As described above, the SEM stage can also be registered with that ensemble, allowing the user to guide the microscope to any ROI that was pinpointed in a previously collected LM dataset.

After an ROI has been targeted and imaged by SBEM, the registration of confocal data with the final SBEM volume should ideally be as precise as possible. Again, in confocal volumes, a fluorescent DNA marker reveals numerous very discrete heterochromatin structures and nucleoli within each nucleus (with the latter being negatively stained by most DNA stains). These structures are also readily apparent in SBEM volumes. If there are at least a few nuclei within the final ROI, these landmarks can allow for very precise registration of confocal and SBEM volumes (Fig. 8, Supplemental Movie S1). For some tissues, it may be necessary to rely on vasculature or staining of other structures that can be detected in confocal and SBEM volumes. The poor axial resolution of confocal volumes is the greatest challenge in precisely aligning them with SBEM volumes. Even when confocal volumes are very well aligned

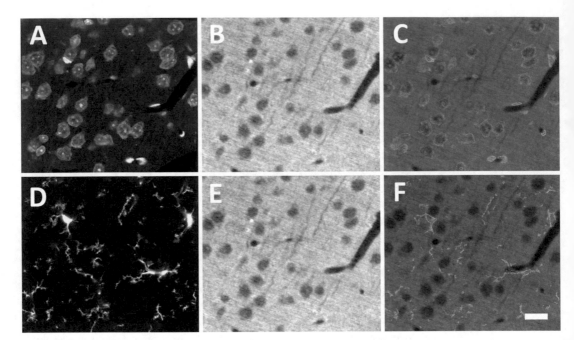

Fig. 7 Registration of LM data with XRM data. Single computed slice from a vibratome section of GFP-expressing brain, stained with DRAQ5, imaged by confocal, stained for SBEM, and then scanned with XRM. (**a**) DRAQ5 signal is used to register confocal and microCT volumes using nuclei as landmarks. (**b**) Computed slice from microCT volume. (**c**) Overlay of DRAQ5 registered with microCT slice. (**d–f**) GFP signal is now registered with microCT volume, revealing locations of GFP-labeled cells. Scale bar: 20 μm

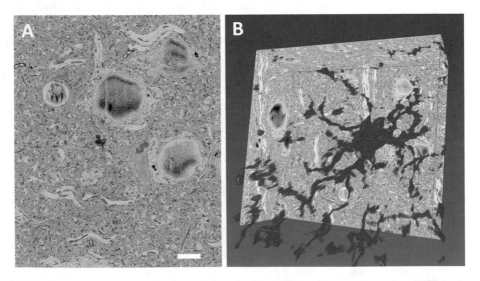

Fig. 8 GFP-labeled cell in brain slice is registered with SBEM volume, after using XRM to target the GFP-labeled cell within the SBEM specimen. (**a**) Single SBEM slice registered with GFP confocal volume and (**b**) volume rendering of GFP-labeled cell seen registered with full SBEM volume. Registration of confocal and SBEM volumes was accomplished through landmarks revealed by DRAQ5 labeling of heterochromatin and nucleoli. Scale bar: 5 μm

with SBEM volumes, the fluorescence signal will be found to spread beyond any labeled structures in the SBEM volume, especially in the axial direction.

3.3 Typical Protocol for Correlated LM–XRM–SBEM Imaging

3.3.1 Targeting ROI with LM

1. Perfuse an animal expressing GFP (or some other fluorescent protein) in target cell population with 4% paraformaldehyde and 0.5% glutaraldehyde in 0.15 M cacodylate buffer containing 2 mM $CaCl_2$. Remove brain and postfix in refrigerator in same fixative solution for 1 h.

2. Collect 100 μm thick sections of brain with a vibrating microtome (*see* **Note 3**).

3. Stain tissue with DRAQ5 (diluted 1:1000 in cacodylate buffer) for 1 h on ice. Wash the slices three times in buffer, 10 min per wash.

4. Image the ROI using a confocal microscope, collecting both GFP and DRAQ5 signals. A set of volumes should be collected at multiple magnifications without moving the specimen position. Place slice in a glass-bottomed petri dish filled with cacodylate buffer. Find and image a target cell with a 60× water objective. Subsequently, collect volumes of the same area with a 20× and 10× objective (*see* **Note 4**).

5. In Navminator, create a New Model to start your project.

6. Import using "Directory as Group" option to load one confocal stack (e.g., 60×), which is already stored as a collection of tiff files in a single directory.

7. Import using "Directory as Group" option to load second confocal stack (e.g., 20×). One volume will be visible in left viewer panel, and the second in right viewer panel. Click on a panel to make it active.

8. Create a Projection View of the relevant slices from both stacks. This will generate projection images made of multiple channels (minimum projection, maximum projection, average projection, standard deviation, and median) for each volume, which you can scroll through when choosing landmark features.

9. Identify at least three corresponding landmarks between confocal volumes (such as blood vessels, nuclei, or GFP-labeled cells) and tag these points as Registration Points in General Item Information panel. After three points, Navminator will begin the registration process using the coordinates between datasets.

10. Continue to add points as necessary, using the Registration—Compute function to occasionally check the accuracy of correspondence between volumes.

11. Once the registration is satisfactory, lock the landmark points.

12. In the 60× confocal volume, create a new point that corresponds to the center of the desired SBEM volume and label it "ROI."

3.3.2 Targeting ROI with XRM

1. Fix tissue in 2.5% glutaraldehyde in 0.15 M cacodylate buffer for 1 h in refrigerator. Wash 3× 10 min in 0.15 M cacodylate buffer.

2. Stain and embed the tissue for volume EM imaging [12]. For a step-by-step protocol *see* also Chapter 9 in this volume. Flat-embed the slice in epoxy between two glass slides precoated with liquid-release agent. Avoid applying too much pressure to slides when embedding to avoid cracks in tissue (*see* **Note 5**).

3. Under a low power stereoscope, use a razor blade to cut out the area imaged by light microscopy. To be safe, the specimen can be much larger than the final SBEM specimen at this point (several millimeters). Mount the specimen onto the end of a microCT specimen holder as seen in Fig. 9. Allow cyanoacrylate to completely set before proceeding.

4. Collect a low magnification microCT volume of the specimen, using the low magnification LM data as a guide. This volume

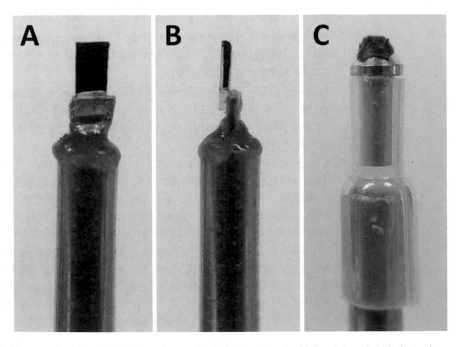

Fig. 9 Stably mounting small SBEM specimens for XRM imaging. (**a, b**) For flat embedded specimens, a short piece of 1/16″ OD aluminum tubing is flattened on one end and then epoxied into the end of a 1/8″ OD aluminum tubing. Specimen is glued to end of holder with cyanoacrylate glue. Alternatively, a chunk or biopsy of epoxy-embedded tissue can be glued directly to end of 1/8″ OD tubing. (**c**) For specimens already mounted on SBEM rivets, the end of a 1 mL plastic transfer pipette can be glued to the end of 1/8″ aluminum tube and the SBEM rivet snuggly secured in end of pipette tip for scanning

does not need to be high resolution, nor exhibit maximal contrast, and can be done with fewer than the optimal number of projection images. Therefore, it is possible to use a higher kVp (~80 kVp) and fewer projection images, usually reducing acquisition time to an hour or less. It is still wise to allow the specimen to sit in chamber at least 30 min before starting imaging and ensuring that X-ray tube is completely warmed up.

5. Following reconstruction of the low magnification microCT volume, load sections as groups into Navminator. As described above, using blood vessels and nuclei, register the microCT volume with the confocal data to locate the ROI that was imaged by confocal microscopy at 60× in the microCT volume (Fig. 10).

6. Collect a higher magnification microCT volume of subvolume imaged at 60× by confocal. Collect microCT volume with sufficient resolution to reliably detect nucleoli (between 0.5 and 1.0 µm voxel dimensions) (*see* **Note 6**).

3.3.3 Mounting ROI for SBEM Imaging

1. Examine the microCT volumes to determine the optimal mounting orientation of the specimen for approaching the ROI by SBEM. Determine which side of the specimen will be facing up (TOP) vs. down (BOTTOM) when mounted on the SBEM rivet (*see* **Note 7**).

2. Use the measurement tool in Amira or a tool such as MB-Ruler to measure the angles needed to approach the specimen. When mounted in the ultramicrotome, both the specimen and the knife can be tilted to adjust the final approach. The combination of these two angles will determine the final sectioning place in the SBEM. In additional to determining these two angles, measure the depth of the ROI within the specimen from the TOP side of the epoxy block (Fig. 11).

3. Use cyanoacrylic glue to attach a small piece of Aclar to the top of a mounting cylinder (*see* **Note 8**). Glue the specimen to the Aclar with cyanoacrylic glue, with the BOTTOM side facing up. Allow glue to set completely before proceeding.

4. Use a glass knife to trim into the face of specimen at the angle measured in Subheading 3.3.3, **step 2**, until sections begin to turn dark, meaning that tissue is being exposed.

5. Use razor blade to cut an approximately 1 mm × 1 mm square containing the ROI from the specimen. The specimen should easily detach from Aclar during this step.

6. Use silver conductive epoxy to mount the specimen to an SBEM specimen rivet. The specimen should now be mounted with the BOTTOM side facing down. Allow to sit at room temperature for a couple of hours and then place in a 60 °C oven overnight.

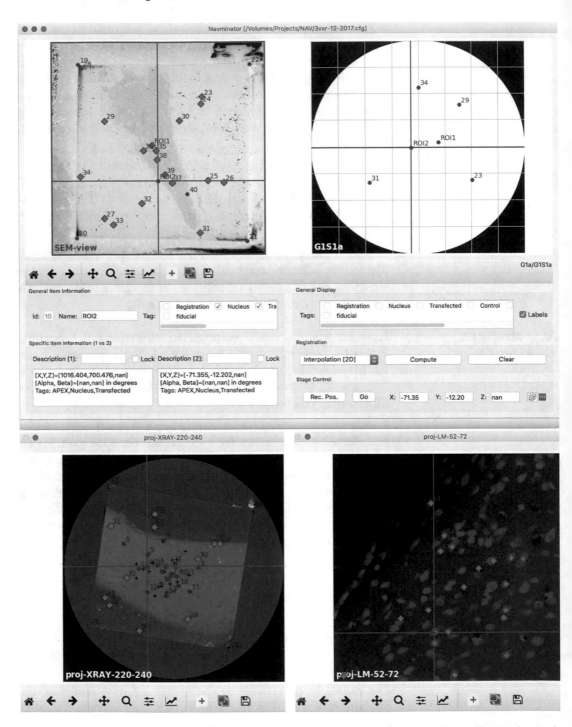

Fig. 10 The Navminator interface. Multiple views of a specimen can be opened simultaneously, each displaying either a single imaging product (e.g., a confocal volume), or the product of one imaging modality mapped onto another imaging modality. The upper two panels show the SBEM block-face (left) and stage view (right). The lower two panels show the microCT scan of the specimen mounted on the SBEM rivet (left) and the 60× confocal scan of DRAQ5 (right). All these volumes are coregistered with the landmarks (yellow diamonds). A few of the landmarks visible on the block face are mapped onto stage view, allowing for computation of the stage coordinates of the ROI

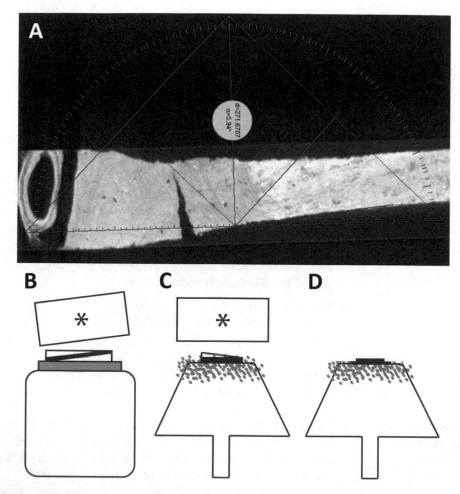

Fig. 11 Process for using microCT volume to guide mounting of specimen on SBEM rivet. (**a**) Once the ROI has been located, the orientation of the tissue relative to the BOTTOM surface of the epoxy block is measured (here with MB-Ruler). (**b**) The specimen is glued to a piece of Aclar (gray), in turn glued to a mounting cylinder. The knife (asterisk) and mounting cylinder are tilted at the angles determined from the microCT volume and epoxy is removed from BOTTOM surface of specimen. (**c**) The specimen has been flipped over and mounted to SBEM rivet using silver epoxy (gray speckles). (**d**) The specimen can now be approached with the knife directly and epoxy removed from top and sides of specimen

7. Use a glass knife to trim silver epoxy from the four sides of specimen (*see* **Note 9**). With the glass knife now approaching the TOP face, with no tilt applied to the specimen or knife, section into the specimen to a depth 5–10 μm above the ROI, as determined in Subheading 3.3.3, **step 2** (*see* **Note 10**).

8. Coat the specimen with a thin layer (~2–3 nm) of gold-palladium in SEM specimen coater (*see* **Note 11**).

9. Collect a low resolution microCT scan of specimen on SBEM specimen rivet, mounted as seen in Fig. 9c. Use a large enough FOV to include entire block face so that corners of block can be used as landmarks for registration with SEM image in next step.

10. Register the block face from microCT volume collected in Subheading 3.3.3, **step 9** with Navminator.

11. Mount specimen in SBEM and approach with the diamond knife, but do not start cutting the block. Image the block face at low magnification. Use Navminator to register the SEM image of the block face with the microCT volume collected in Subheading 3.3.3, **step 9**, using landmarks such as the corners of the block and debris particles created during the trimming process (*see* **Note 12**).

12. Place Stage View on right panel of Navminator and the SEM block face image on the left panel. For each landmark visible on the block face, record the Gatan stage position in Navminator.

13. Shift-click on Compute in Navminator to transfer all previous points (including the ROI point) to the Stage View. When the ROI point is now selected, Navminator will report the predicted stage coordinates for the point.

14. Move stage to the desired position, finish the final approach with the diamond knife, and begin SBEM volume acquisition.

15. Bin the SBEM volume in XY dimensions to reduce the size of the volume to no more than a few gigabytes. The binned version does not need be particularly high resolution, as long as the subnuclear structures are resolvable.

3.3.4 Using Amira to Register Confocal and SBEM Volumes

1. Load both the binned SBEM volume, the DRAQ5 and fluorescent protein channels of the $60\times$ confocal volume into Amira. Ensure that the voxel dimensions for the three volumes are correct and in the same units in the Crop editor.

2. Connect an Orthoslice module to the SBEM and DRAQ5 volumes. Select a slice for each module that approximately matches in both volumes. It may be helpful to connect a Slice module to the DRAQ5 volume and use the Rotation tool to select a slice orientation that more closely matches the cutting orientation of the SBEM volume.

3. Use the Crop Editor of the DRAQ5 and GFP volumes to flip X, Y, or Z directions as necessary to match the same orientation as the SBEM volume.

4. Use the Transform Editor of the DRAQ5 volume to rotate and pan the volume relative to the SBEM volume, bringing the selected slices into rough alignment manually.

5. Connect a Copy Transformations module to the GFP volume, set the DRAQ5 volume as the reference, and apply.

6. Use the Crop Editor for the DRAQ5 and GFP volumes to crop a subvolume slightly larger than the SBEM volume.

7. Connect a Resample Transformed Image module to both the DRAQ5 and GFP volumes, set the SBEM volume as the reference for each, select the Lanczos method, and apply.

8. Create a Landmarks-2-sets module and attach two Landmark View modules, with the first Landmark View associated to Point Set 1 and the second to Point Set 2.

9. Set the Amira viewer to have two panels and view the SBEM volume and Landmark View—Point Set 1 in the top panel, and the DRAQ5 and Landmark View—Point Set 2 in the bottom panel.

10. With the Landmark-2-sets module in Add mode, choose as many matching landmarks as possible by first clicking on the SBEM Orthoslice and then clicking on the corresponding point in the DRAQ5 panel.

11. Attach a Landmark Image Warp module to the Landmarks-2-sets module and set the DRAQ5 volume as its image data. Set the module to use a direction of $2 \rightarrow 1$, set the method to Bookstein, and apply.

12. Attach a new Orthoslice module to the SBEM volume and attach a Colorwash module to it. Connect the Colorwash module to the warped DRAQ5 volume and confirm the accuracy of the fit between the confocal and SBEM volumes.

13. Repeat Subheading 3.3.3, **step 11** and **12** with the GFP volume.

14. Versions of the SBEM volume with less binning applied can now be loaded and will coregister with the confocal volumes.

4   Notes

1. Cacodylate buffer is made as 0.3 M stock solution by dissolving 64.21 g cacodylic acid in 900 mL of double distilled water in a safety hood. Stir until dissolved and then add 0.2 M HCl to bring pH to 7.4. Bring final volume to 1 L. The solution is very toxic and must be used and disposed with care.

2. There are several reconstruction artifacts that can occur in microCT volumes, including metal artifacts, motion artifacts, ring artifacts, and beam hardening artifacts. Some of these artifacts can be avoided through careful specimen preparation, selection of imaging parameters, and post-processing of volumes. There is no space in this chapter to review the details of all these artifacts and the reader is encouraged to refer to more in-depth reviews of microCT imaging [10, 13].

3. It may be advantageous to use a small tissue punch (~ 2 mm diameter) to remove a region of interest from the tissue slices before proceeding. Small pieces of tissue are less likely to break

during EM processing. If LM correlation is not being performed, it is possible to use small chunks of tissue (<1 mm in any dimension) and proceed directly to EM staining and embedding.

4. If the final goal is to merge confocal data with SBEM data, then it is preferable to use Nyquist sampling for the 60× volume in *XY* and *Z*. Lower *XY* and *Z* resolution is acceptable for lower magnification confocal volumes, and even a single optical slice can suffice for orientation of a high-resolution volume with the tissue slice/punch. Slice can be held down during imaging using a small harp or a small piece of coverslip weighed down with a small chunk (few mm cube) of cured epoxy glued to the coverslip.

5. If working with small pieces of tissue or biopsy, the specimens can be embedded in a small aluminum weigh dish or in a silicon embedding mold.

6. This scan can be done with just enough resolution to allow for mapping of GFP fluorescence and targeting of SBEM volume collection, or it can be a very high-quality scan if microCT volume will be providing additional contextual information for the final SBEM volume that will be provided by the additional resolution.

7. The specimen should be mounted such that the ROI is closer to the block face than the SBEM rivet. This will reduce the chance of the diamond knife striking the rivet and allows enough space for the silver epoxy to secure the specimen to the rivet at the same time.

8. If the specimen needs to be approached at a high knife angle to achieve the desired final cutting place, then the edges of the top of the mounting cylinder can be trimmed down to yield a smaller area before gluing on the Aclar. This will allow the knife to approach the specimen without striking the mounting cylinder.

9. It is best to trim down the silver epoxy on each side of the specimen far enough so that the diamond knife will not need to cut through silver epoxy during the SBEM run. Ideally at least 10–20 μm of tissue will remain embedded in the silver epoxy after trimming the sides of the block. This will also significantly improve the ability to acquire a microCT volume of the specimen on the SBEM rivet. Attempting to image the tissue through silver epoxy generates a great deal of X-ray imaging artifacts and reduces contrast within the tissue.

10. It is possible at this point to account for the angle of the SBEM knife by placing an old specimen already imaged by the SBEM into the ultramicrotome, aligning the knife with the block face,

and then replacing the old specimen with the new specimen. This can help reduce the time necessary for approaching with the diamond knife in the SBEM.

11. Coating with metal is helpful with specimens that have significant amounts of empty resin at the block face, to eliminate charging when trying to image the block face during registration steps.

12. Registering the block-face image with the microCT data is particularly useful when there is little or no tissue exposed at the beginning of the SBEM volume. If tissue is already exposed when starting the SBEM run, then observing the position of the ROI point in Navminator relative to the microCT data and comparing with the SEM image of the block face may be sufficient for choosing the desired stage position.

Acknowledgments

The National Center for Microscopy and Image Research is supported by a grant to Mark Ellisman from the National Institutes of General Medical Sciences (P41 GM103412). We would like to acknowledge Chih-Ying Su (R01 DC015519), Angela Tsang, Katerina Akassoglou (R35 NS097976), and Victoria Rafalsky for providing some of the specimens used as examples in figures.

References

1. Bushong EA, Johnson DD Jr, Kim KY, Terada M, Hatori M, Peltier ST, Panda S, Merkle A, Ellisman MH (2015) X-ray microscopy as an approach to increasing accuracy and efficiency of serial block-face imaging for correlated light and electron microscopy of biological specimens. Microsc Microanal 21:231–238

2. Handschuh S, Baeumler N, Schwaha T, Ruthensteiner B (2013) A correlative approach for combining microCT, light and transmission electron microscopy in a single 3D scenario. Front Zool 10:44

3. Karreman MA, Ruthensteiner B, Mercier L, Schieber NL, Solecki G, Winkler F, Goetz JG, Schwab Y (2017) Find your way with X-Ray: using microCT to correlate in vivo imaging with 3D electron microscopy. Methods Cell Biol 140:277–301

4. Morales AG, Stempinski ES, Xiao X, Patel A, Panna A, Olivier KN, McShane PJ, Robinson C, George AJ, Donahue DR, Chen P, Wen H (2016) Micro-CT scouting for transmission electron microscopy of human tissue specimens. J Microsc 263:113–117

5. Ng J, Browning A, Lechner L, Terada M, Howard G, Jefferis GS (2016) Genetically targeted 3D visualisation of Drosophila neurons under electron microscopy and X-ray microscopy using miniSOG. Sci Rep 6:38863

6. Parlanti P, Cappello V, Brun F, Tromba G, Rigolio R, Tonazzini I, Cecchini M, Piazza V, Gemmi M (2017) Size and specimen-dependent strategy for x-ray micro-ct and tem correlative analysis of nervous system samples. Sci Rep 7:2858

7. Sengle G, Tufa SF, Sakai LY, Zulliger MA, Keene DR (2013) A correlative method for imaging identical regions of samples by micro-CT, light microscopy, and electron microscopy: imaging adipose tissue in a model system. J Histochem Cytochem 61:263–271

8. Zheng Z, Lauritzen JS, Perlman E, Robinson CG, Nichols M, Milkie D, Torrens O, Price J, Fisher CB, Sharifi N, Calle-Schuler SA, Kmecova L, Ali IJ, Karsh B, Trautman ET,

Bogovic J, Hanslovsky P, Jefferis GSXE, Kazhdan M, Khairy K, Saalfeld S, Fetter RD, Bock DD (2017) A complete electron microscopy volume of the brain of adult *Drosophila melanogaster*. bioRxiv

9. Mizutani R, Takeuchi A, Uesugi K, Takekoshi S, Osamura RY, Suzuki Y (2008) X-ray microtomographic imaging of three-dimensional structure of soft tissues. Tissue Eng Part C Methods 14:359–363

10. Stock SR (2009) Microcomputed tomography : methodology and applications. CRC Press, Boca Raton, FL

11. Metscher BD (2009) MicroCT for comparative morphology: simple staining methods allow high-contrast 3D imaging of diverse non-mineralized animal tissues. BMC Physiol 9:11

12. Williams ME, Wilke SA, Daggett A et al (2011) Cadherin-9 regulates synapse-specific differentiation in the developing hippocampus. Neuron 71:640–655

13. Stauber M, Muller R (2008) Micro-computed tomography: a method for the non-destructive evaluation of the three-dimensional structure of biological specimens. Methods Mol Biol 455:273–292

Chapter 11

FIBSEM Analysis of Interfaces Between Hard Technical Devices and Soft Neuronal Tissue

Antje Biesemeier, Birgit Schröppel, Wilfried Nisch, and Claus J. Burkhardt

Abstract

State of the art electrophysiological experiments use technical devices where an electrode–cell pair can be used to stimulate neurons and to record neuronal answers to these stimuli vice versa. 3D reconstruction of such biological technical interfaces is helpful to gain morphological insights to these interfaces. Questions like whether and how cellular structures do interact with the technical surface are important to interpret the functionality of the whole system and need high resolution imaging of the interface. Routine electron microscopical preparation methods like grinding, polishing or ultramicrotomy known from the material sciences on the one hand and biological approaches on the other hand are not valid as they can easily destroy the biological-technical interface in question due to their different composition and hardness. Only their combination with 3D FIB-SEM tomography allows for site-specific nanoanalytics of the complex device containing both soft organic matter and hard material and possessing a biological technical interface of interest considerably hidden inside the sample. Correlative light and electron microscopical investigation is easily obtained using specific sample holders that can be used in all instruments needed. Voxel sizes of less than 10 nm and fields of view of more than 40 μm are feasible, and can be adapted depending on the research question. Here we show two examples of how FIB SEM tomography can be used to investigate such cell–electrode pairs. However, the technique can be widened and used for almost any topic where soft and hard materials have to be investigated together.

Key words 3D-FIB-SEM tomography, Biological–technical interface, Plastic embedding, Mechanical preparation methods, Serial sectioning, Correlative microscopy, Microelectrode array (MEA), Retina chip

Abbreviations

AuPD	Gold/palladium
EDX/EDS	Energy dispersive X-ray spectroscopy
EM	Electron microscopy
EsB	Energy selective backscatter detector
FIB-SEM	Focused ion beam-scanning electron microscopy
GA	Glutaraldehyde
Ga^+	Gallium ion

Irene Wacker et al. (eds.), *Volume Microscopy: Multiscale Imaging with Photons, Electrons, and Ions*, Neuromethods, vol. 155,
https://doi.org/10.1007/978-1-0716-0691-9_11, © Springer Science+Business Media, LLC, part of Springer Nature 2020

LM   Light microscopy
MEA   Microelectrode array
OsO4   Osmium tetroxide
PFA   Paraformaldehyde
ROI   Region of interest
SE2   Everhart-Thornley secondary electron detector

1   Introduction

For decades, materials science and biology were two fields of electron microscopical research that hardly met each other. Due to the immense improvements in the fields of medical diagnostics and therapy or biotechnology, more often questions arose that targeted the interface between cells or biological tissues and technical surfaces as for example prostheses or biosensors. In basic research, microchip-based implants for hearing or vision are already being implanted and microelectrode arrays (MEA), small microstructured devices consisting of a culture dish, the microelectrode array, and contact pads on a glass substrate, are used in dedicated electrophysiological setups.

Close contact between the cells and the surface of the bioelectric devices is crucial for signal transduction in both directions (electrical stimulation and recording). Thus, for successful application of active microimplants, coupling and connection of the biological system to the technical surface of the microimplant is of crucial relevance. Beside any chemical modifications added to the technical interface and any material induced biological responses of the tissue to the micro implant surface, gaining morphological information of the biological technical interface with high spatial resolution is an important issue.

1.1 FIB-SEM on Biological and Technical Interfaces

To correlate functionality demonstrated, for example, by electrophysiological recordings with the morphology of the device–tissue interface at high spatial resolution is not a trivial task due to the different properties of the materials present.

The focus of this book chapter lies in the description of a workflow that allows for the combined investigation of in vitro electrophysiological features of neuronal cells (from two different devices, *see* below Subheading 1.2), with high-resolution cell-specific imaging of the underlying cellular structures of those cell–electrode pairs that show a beneficial performance using "FIB-SEM" tomography. This includes the following:

1. An easy selection of the region of interest previously defined by the electrical and electrophysiological measurement.

2. Site-specific preparation of cross sections by removing material (ion milling) with a focused ion beam (FIB).

3. Low-voltage scanning electron microscopy (SEM) imaging of the FIB prepared cross sections using dedicated detectors.

However, challenges for sample preparation result from to the very different properties of the materials constituting the device–tissue interface. Sectioning (or even serial sectioning for 3D analysis; explained in Chapters 4–7 of this volume) of the sample with an ultramicrotome, as it is usually performed for biological specimens is not possible due to the different hardness of the technical and biological parts of the interface (*see* Subheading 4).

Instead, a work flow is used that combines embedding the sample in epoxy resin (a standard approach in biological EM to dehydrate and harden the soft biological tissue) with mechanical preparation like grinding and polishing followed by FIB-SEM tomography, techniques usually applied for material sciences samples. The Ga beam of the FIB is able to handle the different hardness of the sample without the typical cutting artifacts. The region of interest, for example, a specific electrode–cell pair, can be followed visually in all steps. Thereby, the interface between hard materials of technical surfaces (ceramics, metals, glass, …) and adjacent layers of plastic embedded single cells or tissues is cut and imaged with nanoscale resolution.

For 3D FIB-SEM tomography a block face or a cross section is prepared. Smallest layers of material (down to few nanometers) are milled from this micro block face by using the focused Gallium ion beam (Ga^+). The resulting block face is imaged using low keV SEM and milled again, leading to a stack of well aligned, parallel serial 2D images that are reconstructed to a 3D model during postprocessing [1]. Thereby, an image stack may be recorded with comparable resolution in x, y, and z direction with voxel sizes of 10–40 nm. Highly sophisticated larger volume FIB analysis on neuronal tissue with volumes of $250 \times 250 \times 250 \ \mu m^3$ and a retained voxel size of only 8 nm are described in Chapter 12 of this volume.

While FIB-SEM nowadays is used on a regular basis for purely technical [2] or biological [3] questions by others and also in this laboratory [4–11], the investigation of biological technical hybrid is relatively novel and therefore described here in detail using two examples of our work.

Fig. 1 Setup for measuring signals from the ganglion cell layer of a "retina–chip" interface, retina marked by yellow circle (Image courtesy of MJ Lee and G Zeck (NMI))

1.2 Exemplary Studies

1.2.1 Example 1: 3D FIB-SEM Analysis of an In Vitro Chip–Retina Interface

The "retina chip" RI alpha AMS® (Retina Implant AG, Reutlingen, Germany) was made to assist patients suffering from retinitis pigmentosa, an illness leading to a progressive loss of retinal photoreceptors. As the disease does not affect the inner neuronal network of the retina or the optic nerve, the chip acts in light perception, amplification and electrical stimulation of the downstream neurons. The chip is composed of a device with 40×40 pixels, each pixel consisting of a light sensitive photo diode, an amplifier and an electrode for electrical stimulation. As soon as light is hitting the photo diode, the signal is amplified and transmitted to the ganglion cells via an electrical stimulus [12]. During development of the retina chip, several studies were performed to study the electrical stimulus needed to evoke a reaction from the ganglion cells. Electrophysiological experiments were performed with different setups, for example, retina on a chip with the chip facing the outer retina for stimulation, but also "upside down"with the ganglion cell layer facing the device to record the ganglion cell potentials evoked by electrical stimulation upstream [13]. Such a setup (cf Fig. 1) was also used to perform the first FIB-SEM tomogram of a retina–chip interface (Subheading 3.4.3).

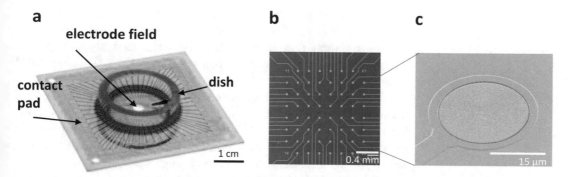

Fig. 2 The microelectrode array (MEA). (**a**) Image of a microelectrode array (MEA): a small microstructured device, consisting of a microelectrode array in the middle of the device and contact pads in the outer area on glass substrate with a frame-like structure on top that can be used as a culture dish. (**b**) Detailed view of a standard device with electrodes made of titanium nitride (TiN): 60 electrodes, with an electrode diameter of 30 μm and rough electrode surface. It is used for extracellular electrophysiological measurements, for example, for recording of action potentials of neuronal cell cultures or for stimulation experiments combined with recording of the answer of the respective neuronal network or tissue. (**c**) Magnified view of one single electrode (diameter 30 μm) [19]

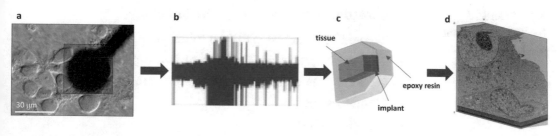

Fig. 3 Exemplary workflow for the analysis of a cell–electrode interface: From living cells, (**a**) here on one electrode (boxed black area) of an MEA, action potentials are recorded (**b**). After fixation and embedding in resin (**c**) the interface has to be exposed prior to FIB-milling by mechanical preparation methods such as grinding (*see* below, Fig. 4). FIBSEM nanotomography finally produces a 3D dataset, here of two cells sitting on the TiN electrode (**d**)

1.2.2 Example 2: 3D FIB-SEM Analysis of Neuronal Cell Cultures on Microelectrode Arrays (MEA) for Electrophysiological Purposes

In the second example we are analyzing neuronal cells cultured directly on an MEA device (Figs. 2 and 3). As primary cell cultures of dorsal root ganglion cells (DRG) are used, a small population of additional cell types (Schwann cells and glia cells) is also present in the culture.

In contrast to patch clamp setups, MEA devices allow recording and stimulation of electrophysiological experiments in cell culture systems (e.g., neuronal cells or cardiomyocytes) over a long period of time. The signals are extremely small in electrophysiological MEA measurements; therefore, the physical properties of the microelectrodes are extremely important and there is still ongoing development of new microelectrode materials to allow for even better recording and stimulation [14].

In the example presented here, after electrophysiological measurements, the MEA–cell set was fixed, stained, and embedded. Promising areas (single cell–electrode pairs that showed proper action potentials) were investigated with 3D FIB-SEM tomography.

This allows the direct correlation of the electrophysiological measurements obtained in vitro and imaged by light microcopy with the three dimensional ultrastructure of the relevant cells in the dish after finishing the experiment and embedding of the whole set up for 3D FIB-SEM tomography (Fig. 3).

This approach proved feasible and showed that electrical signals from DRG neurons could be recorded by the MEA, although they were not in direct contact with the electrode, but usually found on top of other cells, probably of glial origin. Typical orthoslices of the setup, showing the tomograms from different viewing angles (xz, xy, and yz direction), are presented in Subheading 3.4.3.

2   Materials

2.1   Chemicals for Chemical Fixation, Staining and Embedding, See also Note 1

- 2.5% glutaraldehyde (GA) in 0.1 M buffer.
- 4% paraformaldehyde (PFA) in phosphate buffered saline.
- 2% osmium tetroxide (OsO_4) in 0.1 M buffer.
- 0.1 M, 0.2 M cacodylate buffer or phosphate buffer.
- Uranyl acetate.
- Propylene oxide.
- Ethanol.
- EPON epoxy resin.

2.2   Tools for Mechanical Prepreparation

- *Diamond wire saw (Well; used for Example 2)*: with a diamond wire with 40 μm grain size: for removal of the culture dish from the embedded sample and for reduction of the sample size from the MEA side.
- *Grinding and Polishing with the Labopol (Struers; used for Example 2, Fig. 4a)*: for grinding: Diamond pad, 20 μm grain size; For polishing: Diamond suspension with 9 and 3 μm grain size on adequate polishing cloth; For removal of EPON from the cell side until a final thickness of about 30–50 μm.
- *Grinding and Polishing with the mba (Abele GmbH; used for Example 1; Fig. 4b)* with a diamond grinding disc of 20 μm and abrasive paper (corundum 500 granularity); Abrasive paper (silicon carbide 1000/2500 granularity), Clay for polishing: Thin ground process can be conducted with the Abele system using a diamond grinding disc and dedicated abrasive papers of proper granularity.

Fig. 4 Grinding and polishing devices: Different machines can be used to grind or polish the sample, two typical devices are depicted here. (**a**) The Labopol 5 (Strues) as it was used to prepare sample 2. (**b**) The mba (Abele GmbH) used for sample 1. Different grinding disks, abrasive papers, clay or suspensions containing diamond dust of specific grain sizes, or even clay can be used to ground the sample surface

2.3 Materials for Light Microscopy

Standard light microscopes allowing transmitted and incident illumination with adequate optics (e.g., Zernike-phase contrast, DIC) can be used. Suitable long distance objectives allow samples with highly varying height. Sample holders that can fit both in the LM and SEM for proper correlative investigation are recommended.

2.4 Materials for FIB-SEM Analysis

2.4.1 Sample Preparation for FIB-SEM

- *Crystal bond (e.g., by EMS)*: for sample mounting: Heat aluminum stub to 120 °C, apply Crystal Bond and mount sample.

- *Sputter Coating (Balzers SCD 040)* with 15 nm thick layer of Au (80%)Pd(20%).

- *Conductive silver lacquer (Plano, EM grade)*: for contacting the sample with the stub; ensures good conductivity for FIB-SEM analysis.

2.4.2 FIB-SEM Analysis

For FIB-SEM analysis, the Zeiss Auriga 40 was used. The Canion FIB column (Orsay Physics) allows for a resolution smaller 7 nm at 30 keV and probe currents of 1 pA to 20 nA. The Gemini I column of the SEM has a resolution of 1.0 nm at 15 keV and 1.9 nm at 1 keV with probe currents of 4 pA to 20 nA. It has three different detectors installed: SE2 detector (Everhart-Thornley), in-lens detector, and EsB (energy selective backscatter) detector.

2.4.3 Software for 3D Reconstruction

Data post processing was done using Fiji [15]. Slices within one stack were automatically aligned using the StackReg plugin [16] in translation mode (correcting only for shift in *x*- and *y*-direction). 3D representations of the tomograms were done using the 3D Viewer plugin [17] or VolumeJ [18].

3   Methods

In the projects presented here, tissues (example 1) and cell cultures (example 2), previously investigated by electrophysiology are fixed and embedded for electron microscopy directly after finishing the live cell experiments (Subheading 3.1). The resulting plastic blocks containing the biological-technical interface are prepared by mechanical systems (Subheading 3.2). The region of interest selected by electrophysiological measurements and correlative light microscopy (Subheading 3.3) is then analysed by FIB-SEM (Subheading 3.4). Thereby, only those neurons showing advantageous recordings and their respective electrode are investigated.

3.1 Chemical Fixation, Staining, and Embedding

Biological systems usually contain a lot of water, which has to be removed and substituted by hard resin in order to be handled for electron microscopy and FIB analysis in the vacuum. It also has to be made conductive to allow scanning the surface by the electron beam without charging the sample.

Therefore, classical chemical fixation with aldehydes (paraformaldehyde, glutaraldehyde), staining of cellular structures with heavy metals, dehydration in ethanol and embedding in epoxy resin was performed here. This yields a hard plastic block containing both the cells on the one side and the technical material on the other side.

Staining with heavy metals (osmium tetroxide, uranyl acetate) is needed for most biological material, as cellular structures have low mass contrast in the electron microscope. Using osmium and uranium salts, cell membranes and organelles appear electron dense and can easily be resolved. Osmium also fixes lipid structures.

3.1.1 Step by Step Protocol: Fixation and Embedding

Cell cultures or tissue explants containing a microchip or other technical material can be prepared using conventional biological preparation methods:

1. The whole system—cells and technical device within the culture dish, as used for electrophysiological experiments—is washed with buffer to remove culture medium.

2. Then, the dish is filled with 4% PFA solution and the cells prefixed for 10 minutes at 36 °C, followed by a 2% glutaraldehyde step for 2 h (cell monolayer)—12 h (tissue) in the cold. After washing with buffer.

3. The samples are postfixed with 1% osmium tetroxide for 1 h at room temperature, washed several times with water.

4. Dehydrated in a graded series of ethanol (30%, 50%, 70%; each 15 min).

5. Blockstained with saturated uranyl acetate in 70% ethanol (overnight at 4 °C).

6. washed in 70% ethanol and further dehydrated in 80%, 90%, 95% ethanol for each 15 min.

7. and 2× 100% ethanol (each 20 min).

8. Wash in propylene oxide for 20 min (*see* Subheading 4).

9. The dish is filled with a mixture of epoxy resin (here EPON) with the last dehydration solvent (here propylene oxide) 1:1 for 1 h, 2:1 for each 1 h.

10. Finally, the sample is moved to embedding molds, filled with fresh resin and polymerized in the oven according to standard procedures (60 °C, 48 h).

3.2 Mechanical Prepreparation (See also *Note 2*)

In routine embedded samples, the biological–technical interface is usually located some hundreds of microns or even millimeters inside the sample. The thicknesses of chip and biological tissue themselves additionally complicate the access to the interface. By utilizing mechanical preparation methods commonly applied to materials science samples such as conventional sawing, grinding and polishing, a well-defined thin section (or thin ground processed sample) can be obtained providing a sample with the chip–retina interface being accessible to FIB–SEM analysis over a wide range of the interface and not only in a limited cross section. The interface to be characterized should be not more than 50 μm below the sample surface; otherwise a cross-sectional preparation with FIB would not be possible (Fig. 5).

One possibility is to prepare a classical thin ground processed sample (as done for example 1), so thin that the silicon chip itself becomes transparent again allowing also light microscopical evaluation. The other possibility is to thin the sample parallel to the interface of interest by grinding and polishing. This is usually done coming from the epoxy side of the resin bloc. Therefore, also very large objects with both chip and tissue having diameters in the millimeter range, can be handled and prepared to a final size for imaging of about 30–50 μm.

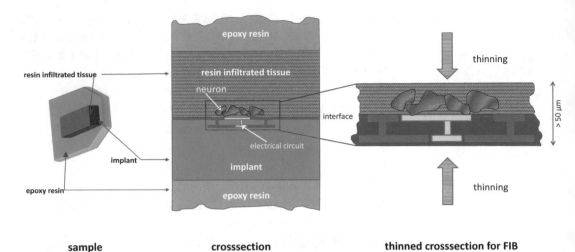

Fig. 5 The challenge of FIB-SEM tomography: FIB-SEM analyses are limited to a near-surface layer of approx. 3–50 µm. Therefore, the sample, usually several mm thick, has to be thinned either from one side or from both sides, that is, the tissue side and the side facing the device, respectively. Different approaches are possible and described below. Note, in green the tissue infiltrated by resin is depicted. In contrast, the implant (blue) usually cannot be infiltrated by resin and thus has a different, usually higher hardness. The interface is therefore prone to rupture if being cut. Only FIB-SEM is able to handle the different hardness without artifacts. After the various mechanical processing steps the sample is finally thinned to such an extent that the interface may be visualized using a light microscope (*see* Fig. 6; Figure adapted from [20])

In the following, a typical step by step protocol is presented. It was used for example 1. An alternative protocol, used for example 2, is provided in Subheading 4.

3.2.1 Step by Step Protocol: Mechanical Prepreparation

Example 1:

1. First, superfluous EPON is removed from the side with the technical device using a polishing machine and abrasive paper with corundum (granularity 500) to remove the plastic without removing material from the silicon based chip. This is done till the bottom side of the device is exposed completely.

2. The EPON on the tissue side of the sample is reduced to a remaining thickness of 200 µm by grinding on a diamond grinding disc with 20 µm grain size, polished using abrasive paper (Silicon Carbide) with a granularity of 1000 respectively 2500 and finished using a polishing suspension (aluminum oxide).

3. Then, the device is thinned from the bottom side of the silicon based chip parallel to the biological-technical interface using a diamond grinding disc with 20 µm grain size. Polishing is achieved by abrasive paper (Silicon Carbide) with granularity of 1000 respectively 2500 and finished using a polishing

thin section in LM   **SEM surface imaging**   **FIB-SEM cross section**

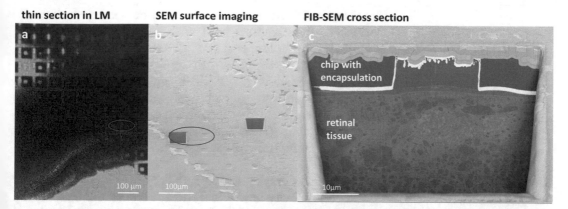

Fig. 6 Correlated LM-SEM for targeting the ROI on the retina–chip interface (example 1): In (**a**) the final thin ground processed sample can be observed. In the upper right corner, the sample is still too thick on the device side, in the central part of (**a**), the device is thin enough and translucent allowing for a first glimpse on the retina sitting below. The oval points to two electrodes, one of them being cut in **b**. (**b**) Corresponding SEM image showing two trapezoid ditches where FIB-SEM tomography was performed. (**c**) Detailed view of one of the ditches milled showing the chip-retina interface (*see* also Fig. 1)

suspension (aluminum oxide). Spacers common to such grinding and polishing machines suitable for preparation of thin ground processed samples allow fast thinning till a dedicated thickness of few microns is reached.

After preparation, the sample will be thin and transparent enough to allow light microscopic imaging. This permits correlation of overview images, obtained by LM with the SE image that is presented in SEM live mode for determination of the region of interest for FIB-SEM analysis (Fig. 6).

3.3 Light Microscopy (See also Note 3)

Light microscopy is used to get an overview of the region of interest. All electrodes of a chip (or other technical features) are visible and the ones showing particular good signals in recording or stimulation can be addressed. By correlating the light microscope images and the images obtained by the SEM this region of interest can be relocated in the FIB-SEM instrument for milling the micro block-face that will later be used for 3D data acquisition.

1. To confirm that the position of the relevant structures (cell attached to device) is not repositioned after termination of the electrophysiological experiment and during the preparation and embedding procedure, it is important to image this region of interest first during the in vitro measurements and after successful embedding. Any repositioning has to be taken into account when selecting the ROI for the FIB analysis.

2. Correlation of the ROI selected by LM to the situation in the FIB-SEM machine is crucial for the intended analysis. If

possible, dedicated sample holders can be used that allow for correlative investigations both with LM and SEM. If such a multifunctional holder is applied together with dedicated software tools (usually supplied with the FIB-SEM) the ROI can be selected on the basis of the LM image, and the stage is then automatically moved to the corresponding position on the sample. Alternatively, this correlation can be done by the operator by applying specific markers on the sample surface and manually moving the stage into position.

3.4 FIB-SEM Analysis (See also Note 4)

3.4.1 Sample Preparation for FIB-SEM Analysis

For FIB-SEM tomography, a high stability of the sample together with good conductivity in the FIB-SEM is crucial. Therefore, the sample is mounted on the sample holder using crystal bond and sputter coated with gold–palladium (80%:20%). This approach yields proper conductivity of nonconductive sample surfaces such as most biological tissues and plastic embedded samples. To ensure good conductivity from the sample surface to the sample holder additional conductive bonding can be obtained by using conductive silver lacquer.

Step by step protocol: sample preparation for FIB-SEM analysis:

1. Mount the sample on a suitable stub using crystal bond: Apply the crystal bond to the stub sitting on a heating plate, heated to 120 °C, and mount the sample.

2. Sputter the sample with a layer of gold-palladium about 15 nm thick.

3. If needed, apply conductive silver lacquer.

3.4.2 FIB-SEM Analysis

For 3D FIB-SEM tomography [1], a Crossbeam® instrument (Zeiss Auriga 40) equipped with a gallium FIB and an SEM with dedicated low-voltage capabilities, is used. Therefore, the sample surface is tilted perpendicular to the Ga^+ beam and moved into the coincidence point of ion and electron beam to allow simultaneous milling and imaging. After an initial preparation of the cross section at the ROI, the gallium FIB produces a series of cross sections containing the biological technical interface at the region of interest previously selected. Each of these cross sections is imaged by the low keV SEM using one of the detectors available for image acquisition. For samples containing heavy metal stained biological tissue, the energy selective backscattered (EsB) detector is preferable yielding images with exceptionally high contrast due to the material contrast of the staining within the tissue and the material contrast of the technical device. Additionally, the images acquired by an EsB detector show almost no FIB induced artifacts such as curtaining because of the missing topographical information in backscattered electron images (*see* also Fig. 9 in Subheading 4). In this way, layers

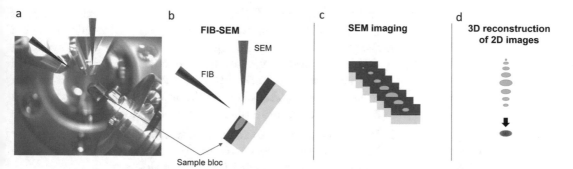

Fig. 7 Workflow of FIB-SEM analysis. (**a, b**) Consecutive performance of several FIB milling and SEM imaging steps is possible when the sample is sitting at the coincidence point of the Ga + beam (red) and the electron beam (green). (**c**) The 2D images are stored and (**d**) then reconstructed to a 3D model of the material, in a post processing step. Typical voxel size 10 × 10 × 10 nm

of down to few nanometers may be removed sequentially by the ion beam and the new block-face may be imaged and added to a 3D image stack (Fig. 7). Depending on the FIB-SEM system, SEM imaging is possible even while the ion beam is polishing the sample and hence speeding up the acquisition time. Using typical voxel sizes in the range of 10 × 10 × 10 nm, morphological information with high spatial resolution is obtained in the volume of interest.

Step by step protocol: FIB-SEM analysis:

1. Sample is loaded into the FIB-SEM system, tilted to 54° (sample surface perpendicular to the Ga⁺ beam in a Zeiss Crossbeam System) and the ROI is moved into the coincidence point for the electron and ion beam by correlating the light and electron images.

2. For rough surfaces, an additional protective layer is deposited by electron or ion beam induced deposition (e.g., platinum or carbon) over the ROI to reduce curtaining artifacts while doing FIB-SEM tomography; for sufficiently smooth sample surfaces, this step can be omitted.

3. A broad FIB trench is milled using 10 nA FIB-current at an acceleration voltage of 30 kV followed by appropriate polishing steps (e.g., 2 nA at 30 kV and 500 pA at 30 kV) to provide a smooth cross section to start a 3D FIB-SEM tomography; this broad FIB trench has to be considerably larger than the size of the ROI as milling artifacts tend to appear at the edges of the trench.

4. For FIB-SEM tomography of the present samples, a FIB current of 2 nA was chosen to achieve an adequate milling time per slice and hence to reduce cutting artifacts. SEM imaging was accomplished by using a current of approximately 1 nA at an acceleration voltage of 2 kV (example 1) or 1.8 kV (example 2)

and by using the EsB detector with a detector grid voltage of 1.5 kV. SEM scanning parameters were adapted to enable imaging with adequate signal-to-noise ratio by combining dose factor and image averaging possibilities (e.g., scan speed 3–5 and line average 30–50 on a Zeiss Crossbeam system). In total, milling and imaging should be accomplished in a way that it does not take more than 2 min to acquire one image.

5. If autofocus and autostigmation are not available on the FIB-SEM system, focus and astigmatism have to be controlled regularly during FIB-SEM data acquisition.

3.4.3 Data Processing and 3D Reconstruction

The resulting stack of 2D images is utilized for 3D reconstruction. Data post processing was done using Fiji [15]. Therefore, the images of the 3D FIB-SEM analysis were merged to appropriate stacks containing the correct voxel size information. Subsequently, slices within one stack were automatically aligned using the Stack-Reg plugin [16] in translation mode (correcting only for shift in x- and y-direction) and the images of the resulting stacks were

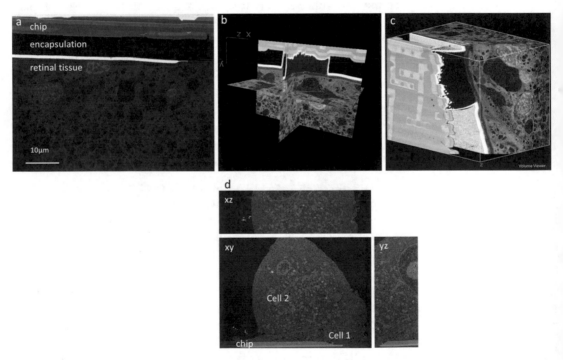

Fig. 8 Exemplary FIB-SEM tomogram orthoslices of the retina implant (**a–c**; example 1; voxel size 60 nm^3; adapted from Schroeppel et al., 2011) and the interface between MEA (TiN) and cell (**d–f**; example 2; voxel size 30 nm^3). (**a**) Cross section of the retina–chip interface. The white line between the organic encapsulation layer and the tissue is the stimulation electrode. **b** + **c**) Different presentations of the final 3D stack. The threshold is adjusted in a way, that the encapsulation material appears transparent so one can have a better view on the stimulating electrode. (**d**) orthoslices of the TiN electrode–neuronal cell interface. Note the high quality of the resolution of these images, especially in z direction

restricted to the overlapping area. 3D representations of the tomograms were done using the 3D Viewer plugin [17] or VolumeJ [18]. The resulting 3D reconstruction helps to analyze morphological aspects of the interface (Fig. 8).

4 Notes

1. Notes about chemical fixation, contrast enhancement, and embedding

   Note that all reagents for biological fixation and embedding are hazardous substances (most of them are toxic and highly volatile) and have to be handled, stored and disposed according to local regulations. If available prefer EM grade chemicals or if not chemicals with p.a. grade.

   In general, EM fixatives should be made fresh or stored at -20 °C to avoid unwanted reorganization of the aldehyde groups that can lead to artifacts only visible with ultrastructural resolution.

   In general, several EM protocols exist and each laboratory uses its own set of protocols depending on the material to be embedded and the research question. Therefore, the protocols provided here are just one way to do it. For example, choosing the right buffer can already be a question of "taste": Here, the protocols for cacodylate buffer (contains arsenic (!), long stability in the fridge, ideal buffering capacity) and phosphate buffer (nontoxic for humans, but maybe not suitable for investigation of mitochondria in plant cells) (e.g., www.microscopy.berkeley.edu/Resources/instruction/buffers.html; Accessed 28 Nov 2018) are provided. Alternative resin species are Araldite, Durcupan, and LR white.

   (a) Chemicals

   - *Glutaraldehyde (GA, 25%, EM grade)*: prepare a 2.5% solution in 0.1 M buffer, adjust pH to 7.4 using HCl; Immediately freeze aliquots at -20 °C in screw cap flasks; gently thaw in the fridge directly before use.

   - *Paraformaldehyde (PFA)*: Prepare a 4% solution in phosphate buffered saline according to standard procedures (e.g., www.rndsystems.com/resources/protocols/protocol-making-4-formaldehyde-solution-pbs;Accessed 28 Jan 2018), freeze aliquots and thaw directly before use.

   - *Osmium tetroxide (OsO₄)*: stock solution 4% solution in water, stored in the fridge; Prepare a 1:1 dilution in 0.2 M cacodylate buffer directly before use gaining 2% OsO_4 in 0.1 M buffer.

- *Cacodylate buffer (dimethylarsinic acid sodium salt tri-hydrate powder)*: Prepare a 0.1 M dilution with pH 7.4 in ddH$_2$O; store at 4 °C.

- *Phosphate buffer*: for 1 L of a 0.1 M buffer (pH 7.4): 20.21 g dibasic sodium phosphate; 3.39 g monobasic sodium phosphate in 800 mL of distilled water, adjust pH (HCl, NaOH), finally add water until 1 L.

- *Uranyl acetate*: Use a saturated dilution of uranyl acetate in 70% ethanol for block staining; prepare directly before use.

- *Propylene oxide:! Note, propylene oxide* can damage plastic parts like cell culture plates; to avoid the use of propylene oxide a very dry ethanol (100%, kept on a molecular sieve) can be used after the 99% ethanol step.

- *Ethanol (99.9% p.a.)*: Keep the ethanol stock dry using a molecular sieve (0.3 nm beads)): Prepare a series of 30–50–70–90–95% in ddH$_2$O and store at room temperature.

- *EPON epoxy resin*: Carefully add all chemicals in the following order in a beaker set on scales (weight for 0.5 L in brackets): 53.7% Glycid ether (177.89 g), 6.7% DDSA (22.19 g), 38.2% MNA (126.55 g); finally, add 1.5% DMP-30 (5.0 g); cover the beaker with Parafilm and stir in the dark until all chemicals have blended. Aliquot and store at −20 °C. Avoid repeated thawing and freezing.

(b) Can cryo methods be used instead of plastic embedding?

No, cryo preparation methods are still not suitable: besides the sophisticated work flow and prerequisites, the size of the implants, together with their high heat capacity do not allow rapid enough freezing needed for vitrification of the sample. Therefore, routine plastic embedding procedures typically used for biological tissue embedding are recommended and used here.

2. Notes about mechanical prepreparation

(a) Typical artifacts in preparing interfaces of technical devices and tissue

Cutting artifacts:

As discussed in the introduction, biological material (even after embedding in resins) and technical devices have extremely different hardness; therefore, the interface is especially prone to disintegration when preparing samples from the specimen block. Shearing or breaking off of sample pieces can occur when grinding and polishing a cross section of such an interface or when using an

Mechanical preparation   **FIB-SEM**

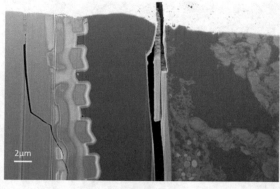
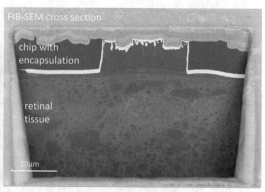

Bad   GOOD

Disintegration of the edges between soft and hard
matter and within the device material after grinding   No artefacts

Fig. 9 SEM images of a typical example of an artifact after mechanical pre-preparation perpendicular to the interface (left) and another sample with mechanical pre-preparation parallel to the interface and FIB-SEM milling (right). The material shown is retinal tissue on a silicon chip

ultramicrotome (Fig. 9). This can be avoided when using the FIB Gallium beam. For the ion beam, different hardness yield only different etch rates, but not any shearing at the interface between hard and soft materials. Therefore, FIB-SEM is the only valid possibility for tomography of such interfaces.

(b) Step-by-step protocol for example 2
- Using a diamond wire saw (40 μm grain size), the sample is reduced in size such that the culture dish of the MEA and the surrounding material are removed and only the region of interest containing the MEA–cell interface remained positioned centrally.

- Then, the sample is mounted to an appropriate holder and grinded on a diamond grinding disc with 20 μm grain size and manually thinned to a final thickness of 30–40 μm. The thickness of the EPON is regularly controlled under a light microscope.

- Finally, the EPON surface is polished on dedicated polishing cloths with 9 μm respectively 3 μm diamond suspension to smooth the surface and hence to prevent unwanted curtaining (cf also **Note 4** last paragraph) during FIB-SEM analysis.

3. Notes about light microscopy

For light microscopy any modality can be chosen, for example, light or fluorescence in vivo imaging of cells using

a) FIB-SEM cross section of a transwell cell culture system imaged with SE2 **under different acceleration voltages.**

3kV   15kV   30kV

b) FIB-SEM cross section of a transwell cell culture system **imaged with different detectors** at 3 kV acceleration voltage.

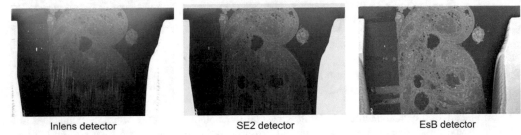

Inlens detector   SE2 detector   EsB detector

Fig. 10 SEM acquisition conditions have high impact on image quality. Choosing low acceleration voltages (**a**) combined with the right detector for each question (**b**) is crucial for investigation of stained and plastic embedded biological material. Curtaining artifacts can be minimized when imaging with the right settings

phase contrast or in vivo fluorescent markers and later investigation of the fixed surface by light microscopy. The glutaraldehyde and some of the resins can give a high background in fluorescence imaging, which is therefore not always possible after embedding. Note also that if the electrode material is not translucent, only incident light microscopy might be possible which demands a different setup with specific objectives and illumination.

4. Notes about FIB-SEM tomography

Basic requirements for FIB-SEM tomography are low-voltage mode of state-of-the-art SEMs and dedicated detectors for material contrast:

By using an SEM in low-voltage mode, only the information of the cross section surface itself is detected. As an example for the influence of the electron beam acceleration voltage, a FIB cross section of a transwell cell culture was imaged with three different acceleration voltages (3, 15, and 30 keV, *see* Fig. 10a). As the information depth increases with increasing acceleration voltage, the images become more blurry and loose detailed information from the cross section itself. It is crucial that the information depth of the SEM imaging is less than z-slice thickness.

Also, proper choice of the detector is important (Fig. 10b). The energy selective backscattered (EsB) detector is preferable as the images acquired show high material contrast and hence high contrast for heavy metal stained biological samples. In addition, the EsB detector shows almost no topographical information as backscattered electrons are used. An example of a worst-case scenario of a cross section preparation is given in Fig. 10: the sample block face shows remarkable curtaining at the bottom of the cross section as preparation artifact. This curtaining is clearly visible in the in-lens detector image and slightly visible in the SE2 detector image. Within the EsB detector image, the curtaining vanishes completely as the image is free of topographical information.

References

1. Holzer L, Muench B, Wegmann M et al (2006) FIB-nanotomography of particulate systems—part I: particle shape and topology of interfaces. J Am Ceram Soc 89:2577–2585

2. Zankel A, Wagner J, Poelt P (2014) Serial sectioning methods for 3D investigations in materials science. Micron 62:66–78

3. Titze B, Genoud C (2016) Volume scanning electron microscopy for imaging biological ultrastructure. Biol Cell 108:307–323

4. Steinmann U, Borkowski J, Wolburg H et al (2013) Transmigration of polymorphnuclear neutrophils and monocytes through the human blood-cerebrospinal fluid barrier after bacterial infection in vitro. J Neuroinflammation 10:832

5. Burkhardt C, Nisch W (2005) Electron microscopy on fib prepared interfaces of biological and technical materials: first results. Pract Metallogr 42:161–171

6. Schmid G, Zeitvogel F, Hao L et al (2014) 3-D analysis of bacterial cell-(iron) mineral aggregates formed during Fe (II) oxidation by the nitrate-reducing Acidovorax sp. strain BoFeN1 using complementary microscopy tomography approaches. Geobiology 12:340–361

7. Desbois G, Urai JL, Hemes S et al (2016) Multi-scale analysis of porosity in diagenetically altered reservoir sandstone from the Permian Rotliegend (Germany). J Pet Sci Eng 140:128–148

8. Hemes S, Desbois G, Urai JL et al (2015) Multi-scale characterization of porosity in Boom Clay (HADES-level, Mol, Belgium) using a combination of X-ray μ-CT, 2D BIB-SEM and FIB-SEM tomography. Microporous Mesoporous Mater 208:1–20

9. Schraermeyer U, Schultheiss S, Oltrup T et al (2017) Unravelling the mystery of the Stiles Crawford effect. Invest Ophthalmol Vis Sci 58:5603–5603

10. Dahm T, Adams O, Boettcher S et al (2018) Strain-dependent effects of clinical echovirus 30 outbreak isolates at the blood-CSF barrier. J Neuroinflamm 15:50

11. Neuhaus J, Schröppel B, Dass M et al (2018) 3D-electron microscopic characterization of interstitial cells in the human bladder upper lamina propria. Neurourol Urodyn 37:89–98

12. Zrenner E, Bartz-Schmidt KU, Benav H et al (2010) Subretinal electronic chips allow blind patients to read letters and combine them to words. Proc R Soc Lond B Biol Sci 278:1489–1497

13. Stett A, Mai A, Herrmann T (2007) Retinal charge sensitivity and spatial discrimination obtainable by subretinal implants: key lessons learned from isolated chicken retina. J Neural Eng 4:S7

14. Fejtl M, Stett A, Nisch W et al (2006) On micro-electrode array revival: its development, sophistication of recording, and stimulation. In: Advances in network electrophysiology. Springer, New York, NY, pp 24–37

15. Schindelin J, Arganda-Carreras I, Frise E et al (2012) Fiji: an open-source platform for biological-image analysis. Nat Methods 9:676

16. Thevenaz P, Ruttimann UE, Unser M (1998) A pyramid approach to subpixel registration based on intensity. IEEE Trans Image Process 7:27–41

17. Schmid B, Schindelin J, Cardona A et al (2010) A high-level 3D visualization API for Java and ImageJ. BMC Bioinformatics 11:274

18. Abràmoff MD, Viergever MA (2002) Computation and visualization of three-dimensional soft tissue motion in the orbit. IEEE Trans Med Imaging 21:296–304

19. Schroeppel B, Roehler S, Samba R, Stamm B, Burkhardt CJ (2015) FIB-SEM tomography of neuronal cells on microelectrode arrays (MEA). In: Microscopy conference LS5.068 3D, Goettingen

20. Schroeppel B, Nisch W, Stett A, Burkhardt C (2013) Analytische Elektronenmikroskopie an Implantatoberflächen und biologisch-technischen Grenzflächen. WOMAg 5:1–5

Chapter 12

Transforming FIB-SEM Systems for Large-Volume Connectomics and Cell Biology

C. Shan Xu, Song Pang, Kenneth J. Hayworth, and Harald F. Hess

Abstract

Isotropic high-resolution imaging of large volumes provides unprecedented opportunities to advance connectomics and cell biology research. Conventional focused ion beam scanning electron microscopy (FIB-SEM) offers unique benefits such as high resolution (<10 nm in x, y, and z), robust image alignment, and minimal artifacts for superior tracing of neurites. However, its prevailing deficiencies in imaging speed and duration cap the maximum possible image volume. We have developed technologies to overcome these limitations, thereby expanding the image volume of FIB-SEM by more than four orders of magnitude from 10^3 μm^3 to 3×10^7 μm^3 while maintaining an isotropic resolution of $8 \times 8 \times 8$ nm^3 voxels. These expanded volumes are now large enough to support connectomic studies, in which the superior z resolution enables automated tracing of fine neurites and reduces the time-consuming human proofreading effort. Moreover, by trading off imaging speed, the system can readily be operated at even higher resolutions achieving voxel sizes of $4 \times 4 \times 4$ nm^3, thereby generating ground truth of the smallest organelles for machine learning in connectomics and providing important insights into cell biology. Primarily limited by time, the maximum volume can be greatly extended.

In this chapter, we provide a detailed description of the enhanced FIB-SEM technology, which has transformed the conventional FIB-SEM from a laboratory tool that is unreliable for more than a few days to a robust imaging platform with long-term reliability: capable of years of continuous imaging without defects in the final image stack. An in-depth description of the systematic approach to optimize operating parameters based on resolution requirements and electron dose boundary conditions is also explicitly disclosed. We further explore how this technology unleashes the full potential of FIB-SEM systems, revolutionizing volume electron microscopy (EM) imaging for biology by gaining access to large sample volumes with single-digit nanoscale isotropic resolution.

Key words Focused ion beam scanning electron microscopy (FIB-SEM), Volume electron microscopy, 3D imaging, Large volume, 3D structure, Isotropic resolution, Connectomics, Cell biology, *Drosophila*, *Mouse brain*, *Mammalian* cell

1 Introduction

Connectomics aims to decipher the functions of brains by mapping their neural circuits. The findings will not only guide the next generation of development in deep learning and artificial intelligence but also transform our understanding of the brain, in both

Irene Wacker et al. (eds.), *Volume Microscopy: Multiscale Imaging with Photons, Electrons, and Ions*, Neuromethods, vol. 155, https://doi.org/10.1007/978-1-0716-0691-9_12, © Springer Science+Business Media, LLC, part of Springer Nature 2020

healthy and diseased states. It can enhance research to find cures for brain disorders and perhaps even pave the way to ultimately comprehend the human mind.

Connectomics extracts the connectivity of neurons in the brain as a basis for understanding its function. The brain features that must be imaged with high resolution vary greatly in size, from the nm scale of the synapses to the mm length scale of the neurons that form even the smallest circuits. In three dimensions, these dimensional scales, taken to the third power, translate to 3D image stacks that can easily reach $10,000^3$ or Tera voxels. Furthermore, the data must maintain a high degree of accuracy and continuity, because even a few missing image planes could compromise the ability to reconstruct a neuron and trace its contribution to a neuronal network, invalidating months of data acquisition. These technical requirements have shaped the adaptation and development of FIB-SEM as a 3D volume imaging technique. The enhanced FIB-SEM technology developed at Howard Hughes Medical Institute's Janelia Research Campus [1, 2] has overcome the limitations of conventional volume EM methods, emerging as a novel approach for connectomic research.

FIB-SEM is introduced in more detail in Chapter 11. It is a technique that has been used in materials science and the semiconductor industry for multiple decades. It has more recently been applied in biological imaging since 2006 [3]. FIB-SEM uses scanning electron microscopy to raster scan the surface of a planar sample with a fine electron beam, a few nanometers in diameter, and monitors that surface by the back scattered electrons along with secondary electrons. Biological tissues are typically stained with heavy metals, such as osmium, which binds preferentially to the cell membranes and lipids thus enhancing the electron scattering signal at such locations. After imaging, the focused ion beam (FIB), typically comprising 30 keV gallium ions, strafes across the imaged surface and ablates a few nanometers from the top of the sample to expose a new slightly deeper surface for subsequent imaging. Cycles of etching and imaging gradually erode away the sample while enabling the collection of a stack of consecutive 2D images, usually requiring tens of seconds to a few minutes per cycle.

Compared with serial thin section imaging [4–6], block face-based approaches [1, 3, 7, 8] provide greater consistency and stability of the image acquisition, thereby resulting in much better self-aligned image stacks. This is particularly important for connectomic studies: a well-registered and defect-free image stack is the foundation for successful automated segmentation. Serial sections incorporate folds and other imperfections in the sections thus posing significant challenges in registration and segmentation. Both diamond knife and FIB are options for the precise removal of material necessary for block face-based 3D imaging. FIB-based removal, unlike diamond knife, removes tissue at the atomic level without any mechanical moving components. It therefore offers the potential for nanometer control of the z-axis resolution. FIB milling is also less sensitive to damage by electron radiation from the SEM beam, so it can tolerate a higher electron dose to provide better signal-to-noise ratio (SNR) imaging (details of which are discussed in Subheadings 2.1, 2.2, and 2.3). The major disadvantages of conventional FIB-SEM include the slow imaging acquisition rate and lack of long-term stability. Additionally, the process of material ablation depletes the FIB gallium source in 3–4 days, so that just a few tens of microns in z thickness can be imaged before a pause is required to replenish the gallium source. After a pause, imprecise beam position could then result in excessive material loss while reengaging the beam. Other factors such as room temperature fluctuations can disturb the fine control of the increment in z-axis milling. These limitations constrain conventional FIB-SEM to small volumes.

To meet the large volume demands of connectomics, a number of enhancements to conventional FIB-SEM are discussed below. These enhancements have substantially improved long-term reliability, and hence enabled uniform defect-free z increments that can support imaging of hundreds of microns in sample thickness. This represents a new regime in sample size and resolution for 3D volume imaging, with minimal trade-off between large volume and fine resolution. To probe biological questions with the most appropriate technology, it is necessary to characterize various platforms through a unified metric—minimum isotropic resolution, defined by the worst case in the x, y, or z axis. Because each imaging technology yields different resolutions in three axes, inferior resolution in any dimension can limit useful resolution in the other two, and impair the quality of subsequent image processing and analysis. Figure 1 provides an overview of the volume EM operating space. Specifically, it highlights the sample volume and minimum isotropic resolution that can be accessed only by long-term FIB-SEM imaging, and compares this to that by other volume EM imaging modalities. For example, diamond-knife cut serial section TEM with tomography, using ~500-nm-thick sections, can give better spatial resolution at the cost of imaging volume, shown at the lower left of Fig. 1. Conversely, even larger volumes with lower spatial resolution can be collected by diamond-knife cut serial section TEM or diamond-knife cut serial block-face SEM (SBFSEM) (Chapter 9), both enclosed for reference at the top of Fig. 1. The space at the lower right of Fig. 1 invites a technology for connectomic studies: the fine neurites can be traced at any random orientations without degradation of resolution, while neurons expanded over long distance can be fully captured by the large volume. Our enhanced FIB-SEM system delivers a larger volume with fine spatial resolution, addressing the need for connectomic studies. The red diagonal dotted lines indicate contours of constant imaging time:

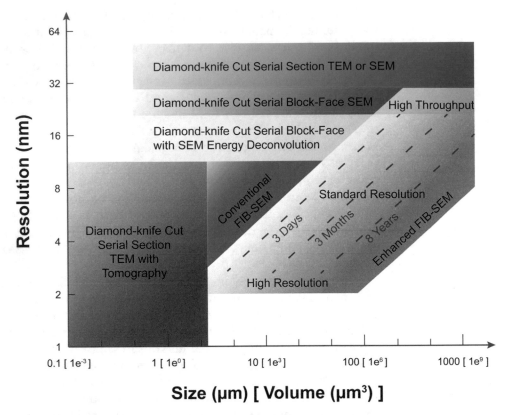

Fig. 1 A comparison of different 3D volume EM imaging modalities in the application space defined by minimum isotropic resolution and total volume. The operating regime of enhanced FIB-SEM is divided into three zones: Standard Resolution, High Resolution, and High Throughput. The three red dotted lines indicate the general trade-off between resolution and total volume during FIB-SEM operations of 3 days, 3 months, and 8 years, respectively, using a single FIB-SEM system. These contours are sensitive to staining quality and contrast. The boundaries of the different imaging technologies outline the regimes where they have a preferential advantage, though in practice there is considerable overlap and only a fuzzy boundary

For a given time allotted for data acquisition, one may choose either finer resolution at the expense of sample volume or vice-versa. To explore this FIB-SEM application space, we break the regime into three domains, labeled simply: high resolution, standard resolution, and high throughput (reduced resolution). Application examples and opportunities of each will be discussed later in this chapter.

The enhanced FIB-SEM system accelerates image acquisition while greatly improving reliability, advancing the operating period from days to years, and generating continuously imaged volumes larger than $10^7 \, \mu m^3$ [1]. These volumes are large enough for many connectomic studies, in which the excellent isotropic resolution enables automated tracing of small neurites and reduces the time-consuming human proofreading effort that is particularly crucial for dense reconstruction studies. The two most important technological advances are improvements in (1) imaging speed and

(2) system reliability, including error detection at all known failure modes and seamless recovery from those interruptions.

More than 10× improvement in imaging speed of the back-scattered electron signal could be achieved without contrast degradation through a positive sample biasing strategy. In a Zeiss Gemini SEM column, this new configuration transforms the traditional in-column (InLens) detector into an effective backscattered electron detector [1]. Compared with a traditional energy-selective backscattered (EsB) detector, the InLens detection via the biased scheme captures a larger fraction of the backscattered electrons, hence achieving a significant gain in imaging speed. Even higher throughput without sample bias is possible only when the steady state FIB-SEM imaging generates tolerable artifacts (e.g., artifacts like mild streaks which can be filtered out in the Fourier domain).

A major milestone in system reliability is accomplished through the following approaches: (1) using multiple layers of error and disturbance protection to prevent catastrophic failures; (2) comprehensive closed-loop control of the ion beam to maintain stability, allowing for a seamless restart of the imaging cycle after interruptions; (3) repositioning the FIB column to be 90° from the SEM column to enable a shorter working distance and thus enhancing signal detection and image quality. Furthermore, features such as a zero overhead in-line image auto-optimization (focus, stigmation, and beam alignment) routinely ensure optimal images that are consistent throughout the entire volume. In depth descriptions of these improvements can be found in the Technology and Methods section of Xu et al. [1] and US Patent of Xu et al. [2]. Together, armed with innovative hardware architectures, system control designs and software algorithms, our enhanced system surpasses deficiencies in platform reliability against all known failure modes thereby overcoming the small volume limitations of conventional FIB-SEM techniques.

This transformative technology has empowered researchers to explore large volume connectomics and cell biology with an optimal balance of resolution, volume, and throughput. The operating regimes can be grouped into three general categories. Details of seven exemplary cases including resolution requirement, FIB-SEM parameters, electron dose boundary, throughput estimate, and corresponding image reference are summarized in Table 1.

- Standard resolution (6–8 nm voxel) with and without parallel processing (by means of hot-knife partitioning), for large volume connectomics with traceability of small processes down to 15 nm; and for overview study of cellular structures.

- High resolution (<5 nm voxel) for observing the finest details of synaptic ultrastructure in connectomic studies; and for fine structures of cell biology.

Table 1
Summary of FIB-SEM imaging conditions for Standard Resolution, High Resolution, and High Throughput Modes

	Case 1	Case 2	Case 3	Case 4	Case 5	Case 6	Case 7
Imaging mode	High Resolution	Standard Resolution	Standard Resolution	Standard Resolution	High Throughput	High Throughput	High Throughput
Application example	Mammalian single cell	Mammalian single cell	*Drosophila* CNS	*Drosophila* CNS	*Drosophila* CNS	*Drosophila* CNS	Mouse cortex
Embedding material	Epon with Durcupan cap	Epon with Durcupan cap	Epon with Durcupan cap	Durcupan	Durcupan	Durcupan	Durcupan
Hot-knife partitioning used	No	No	Yes	No	No	No	No
Voxel resolution (nm³)	4 × 4 × 4	8 × 8 × 8	8 × 8 × 8	8 × 8 × 8	12 × 12 × 12	16 × 16 × 16	16 × 16 × 16
SEM frame size (μm²)	30 × 30	100 × 10	300 × 25	300 × 80	200 × 200	300 × 300	500 × 500
SEM probe current (nA)	0.25	2	3–4	3–4	4	4–6	8
SEM scanning rate (MHz)	0.2	0.25	2–3	2–3	4	4	8–12
Electron dose (e⁻/voxel)	7800	50,000	6200–12,400	6200–12,400	6200	6200–9300	4100–6200
Electron radiation energy density (keV/nm³)	97.7	117.2	14.6–29.3	14.6–29.3	4.3	1.8–2.7	1.2–1.8
FIB milling probe (nA)	15	15	15	15–30	15–30	30	30
Throughput (μm³/month)	0.03×10^6	0.2×10^6	$(1–1.5) \times 10^6$	$(1.8–2.4) \times 10^6$	12×10^6	24×10^6	$(33–50) \times 10^6$
Sample image	Fig. 5	Fig. 4	Fig. 9a	Fig. 9a	Fig. 9b	Fig. 9c	Fig. 9d

Application examples and corresponding sample images are listed for reference. SEM electron landing energies used for electron energy dose estimate are 0.8 kV for High Resolution Mode and 1.2 kV for Standard and High Throughput Modes. SEM scanning rates are based on samples prepared by standard staining protocol. Faster scanning rates are expected for samples prepared by improved staining protocol (Throughput estimates include FIB milling and other operational overheads)

- High throughput at reduced resolution (>10 nm voxel), for large volume connectomics with traceability of processes larger than 20 nm.

2   Methods

In this section, we present sample datasets from the *Drosophila* Central Nervous System (CNS), mammalian neural tissue, cultured mammalian cells, and the green alga *Chlamydomonas reinhardtii* to illustrate the power of this novel high-resolution and high throughput technique to address questions in both connectomics and cell biology. We explicitly report exemplary protocols in the three operating regimes with the goal of serving as a reference for readers to explore and optimize parameters for their own applications.

2.1   Standard Resolution Mode

The initial motivation of our FIB-SEM platform development was mainly to satisfy the requirements of Janelia's *Drosophila* connectomic research [9–12], and to allow connectome mapping of small pieces of mouse and zebrafish nervous systems. Connectome studies comes with clearly defined resolution requirements—the finest neurites must be traceable by humans and should be reliably segmented by automated algorithms [13]. For example, the very finest neural processes in *Drosophila* can be as little as to ~15 nm [14], although such dimensions are only seen in short twigs attached to long-distance neurites of stouter caliber [15]. In the case of mouse cortex, the finest long-distance axons can shrink to ~50 nm, while dendritic spine necks can shrink to ~40 nm [16]. These fundamental biological dimensions determine the minimum isotropic resolution requirements for tracing neural circuits in each case.

To optimize the FIB-SEM operating conditions for each studied connectome, besides the resolution requirement, it is crucial to consider the effects of electron dose and electron radiation energy density. Electron dose, the number of electrons per voxel, is a determining factor for both SNR and imaging throughput. Electron radiation energy density, the product of electron dose and electron beam energy, represents the amount of irradiated energy per unit volume. It has a direct impact on milling rate or z removal consistency, which will become relevant later in the discussion. For reliable automated segmentation of *Drosophila* datasets, using 8 nm isotropic voxels, an electron dose of 6200–12,400 e$^-$/voxel (3–4 nA electron beam, 2–3 MHz imaging rate) is required to achieve a sufficient SNR. We use a 1.2 kV electron beam to achieve the desired z resolution. Together this means that the electron radiation energy density is 14.6–29.3 keV/nm^3 during steady state imaging. In comparison, the mammalian brain has larger neurites than that of *Drosophila*. For test datasets of well stained mouse cortex, 16 nm isotropic voxels and an electron dose as low as

4100 e^-/voxel (1.2 keV/nm^3) appear sufficient for reliable automated segmentation. However more rigorous tests must be performed to determine the optimal voxel resolution and electron dose for mammalian connectomics using FIB-SEM. Applications using a voxel resolution larger than 10 nm will be further explored in Subheading 2.3.

Utilizing these Standard Resolution Mode parameters, we have acquired large (by typical FIB-SEM standards, ~1 × 10^6 μm^3) connectomic datasets spanning parts of the *Drosophila* optic lobe [17, 18], mushroom body [19], and antennal lobe [20]. However, with attempts to extend the imaged volume's depth in the direction of the FIB beam beyond approximately 80 μm, thick-thin milling wave artifacts arose on the trailing edge of the block thus impairing the *z* resolution requirement. Interestingly, this milling wave phenomenon is dependent upon the electron radiation energy density. That is, lowering the total electron radiation energy density (keV/nm^3) lessens the magnitude of milling wave artifacts which, in turn, allows the imaged volume to be made considerably longer in the direction of the FIB beam. Conceptually, one could multiply the SEM scanning rate on the heavily stained high contrast samples, thus expanding the imaged volume significantly. This observation suggests that the wave artifacts result, at least in part, from electron beam-induced modification of the plastic resin on the block surface.

Moreover, such electron beam-induced artifacts seem to be inherent to all block face-based imaging techniques, manifesting as thick-thin alternations in SBFSEM. Electron beam-induced modification and its effect on SBFSEM diamond-knife sectioning has been studied [21]. The reported electron dose of SBFSEM to achieve consistent 25 nm sectioning is limited to 7.3 e^-/nm^2, which equates to an electron radiation energy density of 0.73 keV/nm^3 using a 2.5 kV beam. Significantly higher electron doses cause SBFSEM sectioning to alternate between thick and thin slices, as if the surface layer had become a hardened crust. Recalling the above discussion, the value of electron radiation energy density of 0.73 keV/nm^3 is actually 40 times lower than the value used in our *Drosophila* connectomics FIB-SEM datasets, implying that significant electron radiation-induced surface modification occurs during our standard FIB-SEM runs, while our FIB milling is less sensitive to electron radiation damage. These values are also generally consistent with the literature on radiation-induced chemical modification of polymers suggesting that a significant percentage of chemical bonds are modified at energy levels above 1 keV/nm^3 [22].

The sample volume for connectome studies is typically set to encompass a particular circuit of interest. We were interested in imaging an entire central complex and mushroom body of *Drosophila*, which required an imaging volume of ~250 × 250 × 250 μm^3. Comparable research on mammalian brains would require considerably larger volumes [16]. Such studies are clearly well beyond the 80 μm limit of Standard Resolution Mode FIB-SEM discussed here. To overcome this limitation, we developed an ultra-structurally smooth thick partitioning approach whereby heavy metal-stained, plastic-embedded samples could be subdivided into 20 μm thick slabs [23]. These thick slabs are subsequently reembedded and mounted so that their minimum dimension is oriented in the direction of the FIB beam, thus avoiding any milling wave artifacts. Each thick slab is FIB-SEM imaged separately and the resulting volume datasets are stitched together computationally.

To be effective, the cut surfaces of the slabs must be smooth at the ultrastructural level and have only minimal material loss. Specifically, for connectomic research, all long-distance processes must remain traceable across sequential slabs. In our hands, traditional approaches using vibratome slicing or room-temperature microtomy failed to meet these requirements. Instead, we modified an existing hot-knife microtomy procedure [24] to use a heated, oil-lubricated diamond knife [23]. These modifications allowed us to section both *Drosophila* and mammalian brain tissue at up to 25 μm thickness with an estimated material loss between consecutive slabs of ~30 nm—sufficiently minimal to allow us to trace long-distance neurites in both fly and mammal [23].

In our largest study to date, we used this hot-knife approach to section an entire male *Drosophila* ventral nerve cord (VNC) into 25 slabs, each at 25 μm thick. This volume, $220 \times 200 \times 600$ μm^3 ($\sim 2.6 \times 10^7$ μm^3) in total, was imaged in parallel across six FIB-SEM machines in about 6 months. In addition, a female *Drosophila* "hemi-brain" that spans the entire central complex, a unilateral mushroom body and optical lobe, was sectioned in a sagittal plane into 20-μm-thick consecutive slabs (Fig. 2). Thirteen such slabs were imaged in two FIB-SEM machines [11]. The fully segmented "hemi-brain," $250 \times 250 \times 250$ μm^3 ($\sim 1.6 \times 10^7$ μm^3) in volume, containing $\sim 25 \times 10^3$ neurons with $\sim 60 \times 10^6$ synaptic connections, is considered to be the largest connectome in the world in terms of the number of neurons and synapses being traced [12].

The *Drosophila* connectome project demonstrates a significant advantage of the hot-knife approach—it allows many FIB-SEM machines to operate in parallel on a single imaging task. However, the limitations of this technique—that it is incompatible with heavy metal stained samples and Durcupan resin [23], inevitably limits its adoption. As of today, hundreds of samples with smaller required volumes have been FIB-SEM imaged without hot-knife sectioning, while Durcupan, the preferred resin for FIB-SEM imaging, has been readily used for infiltration and embedding. More importantly, most recent improvements via the progressive lowering of temperature and low temperature staining (PLT-LTS) heavy metal enhancement protocol [25] enables even faster imaging rate without any degradation in quality (Fig. 3). Lower radiation resulting

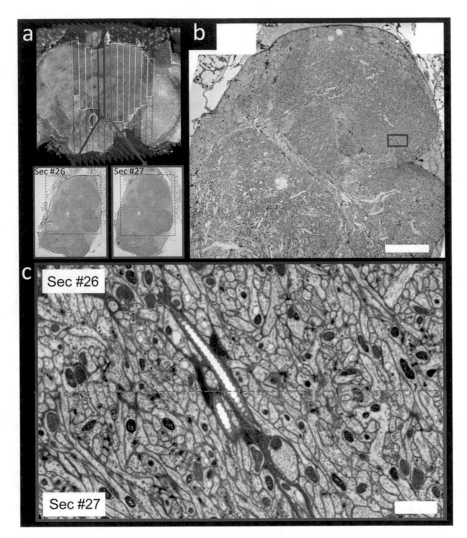

Fig. 2 FIB-SEM imaging of an entire *Drosophila* central complex and a complete unilateral mushroom body. (**a**) X-ray micro-CT image of a *Drosophila* brain showing the locations of the 13 consecutive 20 μm thick hot-knife sections that were FIB-SEM imaged for this study. Yellow highlighting is used to designate imaged volume. Light micrographs of two of these sections (labeled #26 and #27 in the overall series) are shown as well. Dashed boxes in the light micrographs designate regions that were FIB-SEM imaged. (**b**) Cross section through the FIB-SEM volume of Sec #26. Scale bar, 40 μm. (**c**) Example zooming in on the boundary between hot-knife sections #26 and #27 whose FIB-SEM images have been computationally 'volume stitched'. Yellow dashed line designates stitch line. The location of this stitched region is designated by the red rectangle in (**b**). Scale bar, 2 μm. Sample was prepared by Zhiyuan Lu (Dalhousie University)

from faster imaging significantly extends the FIB milling depth to hundreds of microns, thereby enabling much larger volume to be collected at standard resolution without the complexity of hot-knife partitioning. As an example, we have imaged an entire *Drosophila* L1 larval CNS (embedded in Durcupan) with $8 \times 8 \times 8$ nm^3 voxel resolution at 10 MHz, achieving synaptic

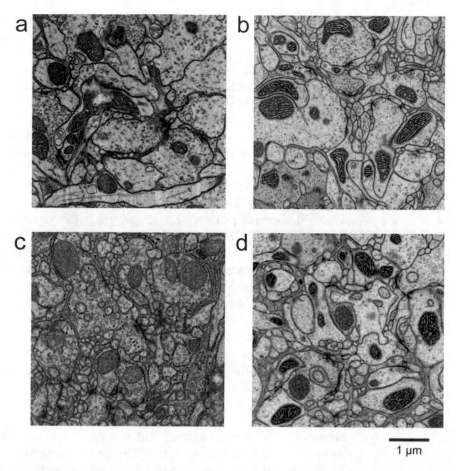

1 μm

Fig. 3 Faster imaging rate is achievable using improved staining *Drosophila* brain samples without image quality degradation. (**a**) standard staining sample $4 \times 4 \times 4$ nm^3 at 200 kHz. (**b**) improved staining sample $4 \times 4 \times 4$ nm^3 at 2 MHz. (**c**) standard staining sample $8 \times 8 \times 8$ nm^3 at 3 MHz. (**d**) improved staining sample $8 \times 8 \times 8$ nm^3 at 10 MHz. Samples were prepared by Zhiyuan Lu (Dalhousie University). PLT-LTS progressive heavy metal enhancement staining protocol [25] was used in improved staining samples

resolution without any gaps, all in less than 3 weeks. This dataset will be used to generate a full CNS connectome using automated methods. A typical *Drosophila* L1 larva CNS has a volume of ~5×10^6 μm^3 in volume, and the faster SEM scanning rates enabled by the improved staining contrast allows us to extend the FIB milling depth to 200 μm with a sufficient margin. Primarily limited by time, the sample block in the z direction can be further expanded to its geometric limit. Empowered by much enlarged imaging volumes and improved system throughput, we are embarking on a journey to tackle many biological questions. Promisingly, the 3D ultrastructure of tissue from the adult mouse hippocampus revealed by the enhanced FIB-SEM, has validated nonconcentricity in myelinated axons [26], and discovered membrane-bound lipid-dense structures in neurons [27].

In conjunction with connectomic studies, the improved imaging speed of our FIB-SEM system also allows rapid sampling of neuronal cultures or an entire mammalian cultured cell at 8 nm isotropic voxel in a week or less [28, 29]. Note that the minimum dimension of cell samples grown on a cover glass is usually less than 20 μm so that the imaging procedure can be simplified without hot-knife partitioning. Furthermore, a much higher electron dose of 50,000 e$^-$/voxel can be used to boost SNR without uneven milling, the detailed condition of which is listed in Table 1 Case 2. This straightforward isotropic imaging mode opens up a new application space for cell biology, providing a better alternative to the serial section TEM or SEM cut on a diamond knife. The FIB-SEM datasets allow direct visualization of an entire cell in three dimensions, thus one no longer relies on sampling from 2D EM sections to infer its 3D organization. The comprehensive 3D overview of the structure and distribution of intracellular organelles permits examination of any arbitrary slices hence offering new insights, and could be mined for statistics. An exemplary dataset of a U2OS cell imaged by FIB-SEM at $8 \times 8 \times 8$ nm^3 voxels in 5 days is illustrated in Fig. 4, where the nucleus, mitochondria, endoplasmic reticulum, and Golgi are visible in one slice plane cropped out of the 3D data volume. Furthermore, whole-cell imaging at a Standard Resolution Mode enables correlative light and electron microscopy (CLEM) applications, revealing a comprehensive picture of intracellular architecture, where subcellular components can be protein labeled and unknown EM morphologies can be classified without ambiguity [29]. From the enlarged view in Fig. 4b, one can see that the standard resolution is sufficient to resolve, for example, the cristae inside mitochondria. However, such resolution is challenged to distinguish, for example, between actin filaments and microtubules. In order to render finer details, we have therefore developed new capabilities for applications at a resolution finer than 6 nm.

2.2 High Resolution Mode

The requirement for a High Resolution Mode is considerably more time consuming per unit volume than that of the Standard Resolution Mode described above, while further improvements in resolution should enable important scientific advances by rendering finer details of cell biology and ground truth of connectomics. Even higher isotropic resolution is possible at the expense of imaging speed (or total volume within a given time), the significant challenges stem from the requirement to improve resolution in x, y, and z directions simultaneously.

To image the block face of a sample with SEM, a sharply focused electron beam is raster scanned across the surface. Lateral resolution in the xy direction is dependent primarily on the blur of the incoming beam, and to a lesser extent on the lateral scattering that generates secondary or backscattered electrons

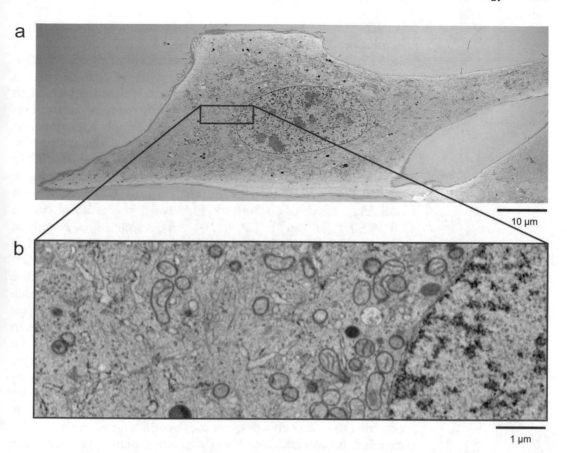

Fig. 4 One slice plane from an isotropic image stack of a U2OS cell at $8 \times 8 \times 8$ nm^3 voxel resolution. (**a**) The overview of the entire cell shows various intracellular organelles and their distributions. (**b**) 10× zoom of the red box in (**a**) provides a closer view of mitochondria, Golgi, endoplasmic reticulum, and actin filaments near the nucleus. Scale bar, 10 μm in (**a**) and 1 μm in (**b**). Sample was prepared by Kathy Schaefer, David Hoffman, Gleb Shtengel, and Amalia H. Pasolli (Howard Hughes Medical Institute, Janelia Research Campus)

[30]. Additionally, the major contributions to the beam blur include spherical aberrations and Coulomb repulsion from the electron lenses. The easiest way to mitigate this blur for finer *xy* resolution is simply to shrink the beam aperture, with concurrent loss of imaging current and thereby imaging speed. The *z* resolution, on the other hand, is dependent upon the incoming electron landing energy that determines its probing depth [1]. For a typical osmium-stained resin-embedded biological sample, a landing energy of 800–1200 eV offers a *z*-axis resolution of 5–8 nm at contrasts of 20–40%, respectively. Contrast between the heavy metal stain and the background signal of embedding resin deteriorates rapidly if energies below 800 eV are used. Even higher *z* resolution can be achieved by lowering electron landing energy to reduce the point spread function size along *z*-axis at the cost of reduced contrast.

Here we present typical operating conditions for the High Resolution Mode. Lateral resolutions of 1.5–3 nm (using the 25–75% edge transition definition) should be possible with commercial SEMs operating at beam currents of 0.2–0.3 nA. A corresponding sampling interval of 2–4 nm is then used to match such high spatial detail. Similar to the SNR requirement in Standard Resolution Mode, an electron dose of approximately 6000–8000 e^{-}/voxel is needed to achieve reasonable contrast in the High Resolution Mode, which inevitably relies on the heavy metal staining level. With an average staining contrast, using a 0.8 kV 0.25-nA electron beam, such a dose corresponds to a scanning rate of ~200 kHz, which translates to a $30 \times 30 \times 30$ µm^3 (~2.7×10^4 µm^3) volume in a month at $4 \times 4 \times 4$ nm^3 voxels. Note that the electron radiation energy density of this imaging condition at ~97.7 keV/nm^3 is more than triple compared with that of the Standard Resolution Mode (Case 1 vs. Cases 3 and 4 in Table 1). As a result, the imaged volume depth in the direction of FIB beam without milling artifacts drops to 20–30 µm, but this is manageable since the minimum dimension of cultured cells (typically less than 20 µm) can be aligned to the FIB milling direction. Ultimately, this mode opens up a unique application space that compliments standard EM tomography which can achieve higher spatial resolution at the cost of smaller and less thick samples. Although the tomographic approach can be extended to thicker effective volumes by stitching multiple samples together, it takes considerable effort with diminishing returns for stitching a larger number of sections, compared with the ease of the FIB-SEM approach.

Figure 5 shows a typical image of a portion of a HeLa cell that was grown on a cover glass, then high pressure frozen with standard freeze substitution, osmium tetroxide staining, and Durcupan resin embedding. After being trimmed to a pedestal of ~$80 \times 80 \times 80$ µm^3 in size, it was FIB-SEM imaged over a 50×8 µm^2 region (8 µm dimension of imaging aligned with the direction of the FIB beam) with $4 \times 4 \times 4$ nm^3 voxels at 200 kHz, using an electron landing energy of 1 keV. The Figure exemplifies the resolution and quality that can be achieved with this technique. The nuclear envelope in both a perpendicular and a tangential slice plane shows the double membrane and nuclear pores. Chromatin is also visible as the dark granular structure inside the nucleus. On the surface of the tangential plane, nuclear membrane polyribosome chains are visible. The resolution is sufficient to see the hollow center of the 20 nm diameter microtubules that lay on the surface of the nuclear membrane. They are easily identified as parallel lines when bisected by the image plane along their axis. The same is true of the microtubule structure of the centrosome. Golgi, endoplasmic reticulum, mitochondria, and so on are all identifiable in the cytoplasm. Clearly, these structures are much better resolved than

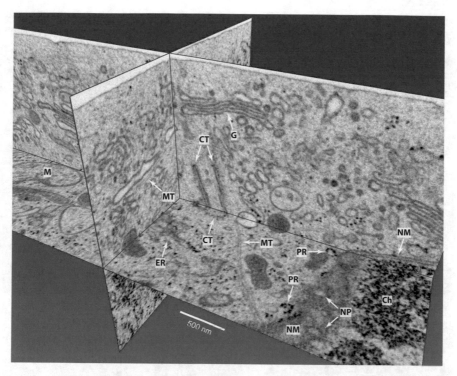

Fig. 5 FIB-SEM image of a HeLa cell using High Resolution Mode at 1 keV, 0.25 nA, 200 kHz, and $4 \times 4 \times 4$ nm^3 voxels. Many intracellular cellular organelles are clearly resolved. They are labeled as: chromatin (Ch), centrosome (CT), endoplasmic reticulum (ER), Golgi (G), mitochondrion (M), microtubule (MT), nuclear membrane (NM), nuclear pore (NP), and polyribosome (PR). Sample was prepared by Aubrey Weigel, Gleb Shtengel, and Amalia H. Pasolli (Howard Hughes Medical Institute, Janelia Research Campus)

those using a standard resolution of $8 \times 8 \times 8$ nm^3 shown in Fig. 4. Additional examples, such as neural tissues and single cells of higher resolution FIB-SEM datasets are presented in the Refs. 1, 19, 31. Figure 6 offers a side-by-side comparison of the High Resolution Mode over the Standard Resolution Mode. Figure 7 demonstrates the power of high-resolution datasets in revealing and classifying 3D cellular structures, and such details could be objectively quantified and extracted for statistical purposes. Likewise, these high-resolution images can also help to detail typical synaptic morphology and aid in deciphering the extremely fine processes of neural connectivity (Fig. 8) in connectomic studies.

With the recently improved staining protocol [25], it is encouraging that without imaging degradation we could demonstrate an additional $10\times$ improvement in the SEM scan rate: from 200 kHz (Fig. 3a) to 2 MHz (Fig. 3b) at $4 \times 4 \times 4$ nm^3 using *Drosophila* brain samples. At such a rate, we have imaged sub-compartments of the *Drosophila* brain rather rapidly. For example, a fan-shaped body middle column, $45 \times 55 \times 45$ μm^3 ($\sim 1 \times 10^5$ μm^3), was imaged in 2 weeks, while the mushroom body α lobe, $50 \times 50 \times 120$ μm^3

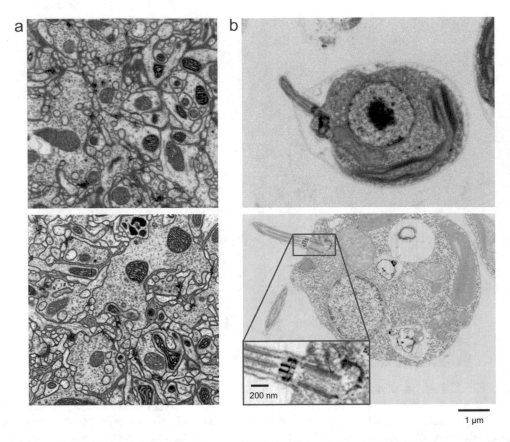

Fig. 6 Improved FIB-SEM resolution reveals more detailed cellular structures in biological samples. Typical images of (**a**) *Drosophila* central complex and (**b**) *Chlamydomonas reinhardtii*, using standard $8 \times 8 \times 8$ nm^3 voxel imaging condition are shown in the top panels. The bottom panels show the corresponding high-resolution images at $4 \times 4 \times 4$ nm^3 voxels. Scale bar, 1 μm. Inset scale bar, 200 nm. Reproduced from [1]

($\sim 3 \times 10^5$ μm^3), could be potentially imaged within a month, both at $4 \times 4 \times 4$ nm^3 resolution. Evidently, a successful strong staining protocol can significantly benefit high-resolution image acquisition because the faster imaging rate substantially improves the throughput.

In summary, high-resolution datasets, while limited in volume, can provide additional important scientific details, and serve as an accurate gold standard to aid in the interpretation of much larger data volumes obtained through Standard Resolution and High Throughput Modes. Furthermore, such datasets can potentially serve as ground truth for machine learning.

2.3 High Throughput Mode

With an initial focus on the *Drosophila* connectome, we chose 8 nm isotropic voxel size as our standard operating baseline to enable tracing *Drosophila*'s very fine neurites. To meet the demands of diverse collaborations, we have expanded 3D FIB-SEM operation space to High Throughput Mode, in which a larger volume can be

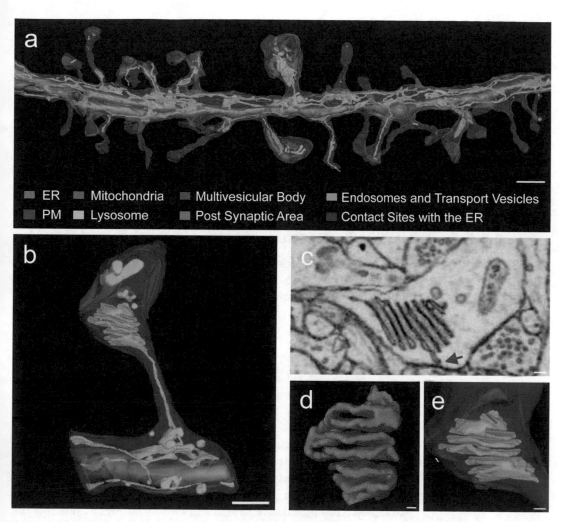

Fig. 7 Visualization of 3D structures of a dendritic segment using high resolution 4 × 4 × 4 nm³ resolution FIB-SEM data. (**a**) 3D model showing all membranous organelles present in a dendrite: endoplasmic reticulum (ER), plasma membrane (PM), mitochondria, lysosome, multivesicular body, postsynaptic area, endosomes and transport vesicles, and contract sides of PM with ER are highlighted in red. (**b**) Zoomed in view of one dendritic spine. (**c**) Single FIB-SEM image showing a cross-section of the spine apparatus, which comprises seven cisternae, one of which makes a contact with the PM in the plane of the image (red arrow). (**d**) 3D reconstruction of the spine apparatus shown in (**c**). (**e**) Two contacts of spine apparatus with the PM (red). Scale bar, 800 nm in (**a**), 400 nm in (**b**), and 80 nm in (**c–e**). Reproduced from [31]

FIB-SEM imaged at a 10× or more improvement in volume rate with sufficient resolution to yield biologically significant details for select questions.

FIB-SEM ability to acquire an isotropic 3D dataset enables visualization of uniform resliced planes at any random angle and permits extension to resolutions of any voxel size. One can choose larger (>10 nm) voxels, in which the imaging rate is substantially accelerated by the cubic power reduction in the number of voxels.

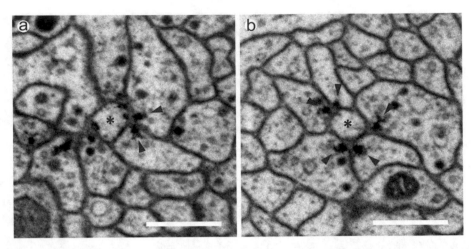

Fig. 8 Images from the higher resolution dataset allow catalog of various newly observed synaptic motifs in *Drosophila* mushroom body with greater confidence. (**a**) A triangular motif of two adjacent Kenyon cells synapse onto a mushroom body output neuron. (**b**) A rosette motif of five Kenyon cells surround a mushroom body output neuron. Kenyon cells and mushroom body output neurons are labeled by red arrowheads and red asterisk, respectively. Scale bars, 500 nm. Reproduced from [19]

Additionally, higher SEM probe currents can be used for faster scan rates without sacrificing SNR. In theory, either larger voxels or faster imaging with lower electron dose and SNR can translate into higher volume imaging rates. To determine the boundary conditions, neuron traceability as a function of voxel size and SNR needs to be carefully evaluated. We generated a dataset of the *Drosophila* brain with various voxel resolutions from $8 \times 8 \times 8$ nm^3 to $16 \times 16 \times 16$ nm^3 and electron doses from ~3000 to ~12,000 e$^-$/voxel. Examples and corresponding images of selected conditions are shown in Table 1 and Fig. 9, respectively. A comparison is drawn among Standard Resolution Mode with (Case 3) and without (Case 4) hot-knife processing, and High Throughput Mode without hot-knife (Cases 5 and 6). To best balance between traceability and throughput, we then imaged multiple *Drosophila* larval CNS (L1 and L3) samples (embedded in Durcupan) at $12 \times 12 \times 12$ nm^3 voxel resolution for tracing the skeletons of all neurons. The typically volumes of L1 or L3 are about 5×10^6 μm^3 or 1×10^7 μm^3, respectively. For samples prepared by our standard staining protocol, at an SEM scanning rate of 3 MHz, it took roughly 20 days for L1, and 40 days for L3. For the recent improved-staining samples, at an SEM scanning rate of 10 MHz, we finished several L1 samples in 10 days each, and expect an L3 sample in 20 days.

Even though *Drosophila* brain tissues are usually more challenging to image because of their lower staining contrast and smaller processes compared with mammalian neural tissues, it is encouraging that not only neurites as small as 50 nm in diameter are

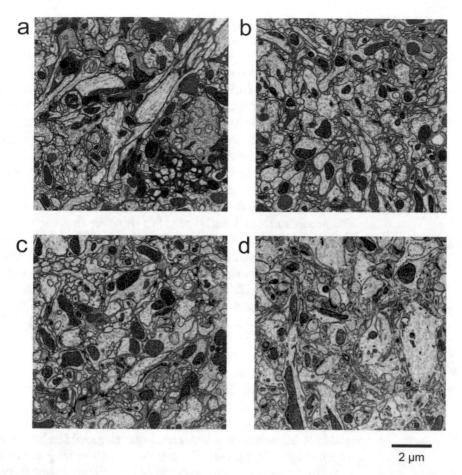

2 µm

Fig. 9 Sample images of connectomic studies using the enhanced FIB-SEM system. SEM probe landing energy was fixed at 1.2 kV. (**a**) *Drosophila* brain with $8 \times 8 \times 8$ nm^3 voxels scanned by a 4-nA SEM probe at 3 MHz. (**b**) *Drosophila* brain with $12 \times 12 \times 12$ nm^3 voxels scanned by a 4-nA SEM probe at 4 MHz. (**c**) *Drosophila* brain with $16 \times 16 \times 16$ nm^3 voxels scanned by a 4-nA SEM probe at 4 MHz. (**d**) Mouse cortex with $16 \times 16 \times 16$ nm^3 voxels scanned by an 8-nA SEM probe at 12 MHz. Scale bar, 2 µm. *Drosophila* brain samples were prepared by Zhiyuan Lu (Dalhousie University) and mouse cortex sample was prepared by Graham Knott (École Polytechnique Fédérale de Lausanne)

well distinguished and traceable, but T-bar synapses are also clearly visible using High Throughput Mode in all test conditions. The readily extractable neuron shapes can also aid in cell type identification in comparison with those obtained from optical images. Moreover, analysis of automated segmentation and human proofreading suggests that neuron tracing is more sensitive to isotropic voxel size than SNR. Further increases in voxel size and SEM scanning rate have diminishing returns for volume throughput once the volume rate of SEM imaging exceeds that of FIB milling. As illustrated by Case 7 shown in Table 1, the SEM volume imaging rate at $16 \times 16 \times 16$ nm^3 voxel resolution is close to the maximum FIB removal rate of ~50 µm^3/s using a 30-nA milling probe. Therefore, while all conditions yield traceable results, we recommend not to exceed a $16 \times 16 \times 16$ nm^3 voxel size for connectomic studies prepared for automated segmentation. Ultimately, it is a delicate balance between traceable resolution and throughput.

Once the resolution boundary conditions of High Throughput Mode have been determined, it is imperative to characterize the corresponding electron dose and radiation energy density. As shown in Table 1, the electron doses used in the three operating modes (for all cases except Case 2) remain rather consistent at ~6000 e$^-$/voxel due to the SNR requirement. In order to accommodate large volumes, the electron radiation energy density decreases monotonically from 97.7 to 1.2 keV/nm^3 as the voxel size increases from 4 to 16 nm. We discover that the reduction of electron radiation damage to specimen widens the margin of uniform milling significantly, which in turn allows larger dimension samples to be directly imaged without the complication and overhead from hot-knife partitioning and post data stitching. Consequently, the material loss at each hot-knife interface (~30 nm) can be completely avoided. For example, to image the entire *Drosophila* VNC, a volume of $220 \times 200 \times 600$ μm^3 (~2.6×10^7 μm^3), requires roughly two FIB-SEM-years at $8 \times 8 \times 8$ nm^3 voxel resolution in Standard Resolution Mode with the hot-knife procedure; in contrast, it can be accomplished efficiently without hot-knife using High Throughput Mode (Cases 5 and 6 in Table 1), primarily for tracing the lower order branches of all neurons. Moreover, the faster scanning rate enabled by the recently improved staining protocol extends the FIB milling depth beyond 200 μm; therefore, such a volume can be accomplished in two FIB-SEM-months at $12 \times 12 \times 12$ nm^3, or less than five FIB-SEM-months at $8 \times 8 \times 8$ nm^3 voxel resolution with minimal milling artifacts.

More importantly, while High Throughput Mode overlaps with SBFSEM in the resolution-volume space (Fig. 1), FIB-SEM provides higher SNR images than SBFSEM, because FIB milling is less sensitive to electron radiation damage compared with the diamond-knife cutting. To obtain consistent cutting, SBFSEM caps the electron dose and radiation energy density at 2000 e$^-$/voxel and 0.73 keV/nm^3, respectively [21]. The 3× or more electron dose in FIB-SEM directly translates into higher SNR images, thereby enabling higher accuracy in the subsequent analysis and interpretation.

Altogether, the High Throughput Mode offers biologists a viable option to obtain an overview of large volume neural circuitry in a reasonable amount of time. This regime should be particularly attractive to researchers who focus on structures larger than 16 nm and who are interested in mining statistical information from multiple samples. The speed improvement of High Throughput Mode is rather promising: the same volume of tissue can be imaged at least

$10\times$ faster than that of Standard Resolution Mode. Coupled with the staining improvement, the High Throughput Mode opens up exciting new avenues for connectomic studies. As can be seen in Fig. 9d, thanks to a much improved staining contrast of mammalian brain tissues [32], we are able to raise the SEM imaging rate to 12 MHz using an 8-nA electron beam without image degradation (Table 1, Case 7). Under such conditions, the projected imaging time of a 1 mm^3 volume is only 2–3 FIB-SEM-years, a significant improvement over nearly 1 FIB-SEM-century using the Standard Resolution Mode!

3   Summary

Three-dimensional imaging offers tremendous value to elucidate biological structures and decipher their function. The enhanced FIB-SEM technology has addressed the limitations of existing imaging modalities thus effectively expanding the operating space of volume EM, delivering fine isotropic resolution, and high throughput with long-term reliability to image sufficiently large volumes encompassing the entire region of interest. The expanded volumes open a vast new regime in scientific learning, where nano-scale resolution coupled with meso and even macro scale volumes is critical. The largest connectome in the world has been generated using this enhanced FIB-SEM platform, where the superior z resolution empowers automated tracing of neurites and reduces the time-consuming human proofreading effort. Increased resolution further improves the interpretation of otherwise ambiguous details. Nearly all organelles can be resolved and classified with whole-cell imaging at 4 nm voxel resolution. We have routinely imaged entire mammalian cells at this resolution to study the close contacts among various organelles. Furthermore, new CLEM applications enabled at the whole-cell level, can readily probe cell biology questions that are otherwise intractable.

At the forefront of volume EM imaging innovations, enhanced FIB-SEM technology pushes the envelope of image acquisition capability and system reliability, offering a novel package suited for large volume connectomics and cell biology.

Acknowledgments

We would like to thank David Peale and Patrick Lee for consulting support in system modification. We also thank Zhiyuan Lu, Gleb Shtengel, David Hoffman, Amalia H. Pasolli, Kathy Schaefer, Aubrey Weigel, Nadine Randel, Michael J. Winding, and Graham Knott for EM sample preparation. We gratefully acknowledge Patrick Naulleau, Ian A. Meinertzhagen, and Steve Plaza for reviewing the manuscript and providing timely feedback. Our gratitude extends to Janelia FlyEM connectome program, in particular Gerry Rubin and Steve Plaza for their leadership. We were solely funded by the Howard Hughes Medical Institute.

References

1. Xu CS, Hayworth KJ, Lu Z et al (2017) Enhanced FIB-SEM systems for large-volume 3D imaging. eLife 6:e25916. https://doi.org/10.7554/eLife.25916

2. Xu, CS, Hayworth KJ, Hess HF (2020) Enhanced FIB-SEM systems for large-volume 3D imaging. US Patent 10,600,615, 24 Mar 2020

3. Heymann JA, Hayles M, Gestmann I et al (2006) Site-specific 3D imaging of cells and tissues with a dual beam microscope. J Struct Biol 155:63–73. https://doi.org/10.1016/j.jsb.2006.03.006

4. Harris KM, Perry E, Bourne J et al (2006) Uniform serial sectioning for transmission electron microscopy. J Neurosci 26:12101–12103. https://doi.org/10.1523/JNEUROSCI.3994-06.2006

5. Bock DD, Lee WC, Kerlin AM et al (2011) Network anatomy and in vivo physiology of visual cortical neurons. Nature 2011 (471):177–182. https://doi.org/10.1038/nature09802

6. Hayworth KJ, Kasthuri N, Schalek R et al (2006) Automating the collection of ultrathin serial sections for large volume TEM reconstructions. Microsc Microanal 12:86–87. https://doi.org/10.1017/S1431927606066268

7. Denk W, Horstmann H (2004) Serial blockface scanning electron microscopy to reconstruct three-dimensional tissue nanostructure. PLoS Biol 2:e329. https://doi.org/10.1371/journal.pbio.0020329

8. Knott G, Marchman H, Wall D et al (2008) Serial section scanning electron microscopy of adult brain tissue using focused ion beam milling. J Neurosci 28:2959–2964. https://doi.org/10.1523/JNEUROSCI.3189-07.2008

9. Scheffer LK, Meinertzhagen IA (2019) The fly brain atlas. Annu Rev Cell Dev Biol 35:737–653. https://doi.org/10.1146/annurev-cellbio-100818-125444

10. Takemura SY, Bharioke A, Lu Z et al (2013) A visual motion detection circuit suggested by *Drosophila* connectomics. Nature 500:175–181. https://doi.org/10.1038/nature12450

11. Xu CS, Januszewski M, Lu Z et al (2020) A connectome of the adult *Drosophila* central brain. bioRxiv:2020.01.21.911859. https://doi.org/10.1101/2020.01.21.911859

12. Scheffer LK, Xu CS, Januszewski M et al (2020) A connectome and analysis of the adult *Drosophila* central brain. bioRxiv:2020.04.07.030213. https://doi.org/10.1101/2020.04.07.030213

13. Januszewski M, Kornfeld J, Li PH et al (2018) High-precision automated reconstruction of neurons with flood-filling networks. Nat Methods 15:605–610. https://doi.org/10.1038/s41592-018-0049-4

14. Meinertzhagen IA (2016) Connectome studies on *Drosophila*: a short perspective on a tiny brain. J Neurogenet 30:62–68. https://doi.org/10.3109/01677063.2016.1166224

15. Schneider-Mizell CM, Gerhard S, Longair M et al (2016) Quantitative neuroanatomy for connectomics in *Drosophila*. eLife 5:e12059. https://doi.org/10.7554/eLife.12059

16. Helmstaedter M (2013) Cellular-resolution connectomics: challenges of dense neural circuit reconstruction. Nat Methods 10 (6):501–507. https://doi.org/10.1038/nmeth.2476

17. Takemura SY, Xu CS, Lu Z et al (2015) Synaptic circuits and their variations within different columns in the visual system of *Drosophila*. PNAS 112:13711–13716. https://doi.org/10.1073/pnas.1509820112

18. Shinomiya K, Huang G, Lu Z et al. (2019) Comparisons between the ON- and OFF-edge motion pathways in the *Drosophila* brain. eLife 8:e40025. doi: https://doi.org/10.7554/eLife.40025

19. Takemura S, Aso Y, Hige T et al (2017) A connectome of a learning and memory center in the adult *Drosophila* brain. eLife 6:e26975. https://doi.org/10.7554/eLife.26975

20. Horne JA, Langille C, McLin S et al. (2018) A resource for the *Drosophila* antennal lobe provided by the connectome of glomerulus VA1v. eLife 7:e37500. doi: https://doi.org/10.7554/eLife.37550

21. Titze B (2013) Techniques to prevent sample surface charging and reduce beam damage

effects for SBEM imaging. Dissertation, Heidelberg University, pp 1–112

22. Calcagno L, Compagnini G, Foti G (1992) Structural modification of polymer films by ion irradiation. Nucl Instrum Methods Phys Res, Sect B 65(1–4):413–422. https://doi.org/10.1016/0168-583X(92)95077-5

23. Hayworth KJ, Xu CS, Lu Z et al (2015) Ultrastructurally smooth thick partitioning and volume stitching for large-scale connectomics. Nat Methods 12:319–322. https://doi.org/10.1038/nmeth.3292

24. McGee-Russell SM, De Bruijn WC, Gosztonyi G (1990) Hot knife microtomy for large area sectioning and combined light and electron microscopy in neuroanatomy and neuropathology. J Neurocytol 19(5):655–661. https://doi.org/10.1007/BF01188034

25. Lu Z, Xu CS, Hayworth KJ et al (2019) *En bloc* preparation of *Drosophila* brains enables high-throughput FIB-SEM connectomics. bioRxiv:855130. https://doi.org/10.1101/855130

26. Gao R, Asano SM, Upadhyayula S et al (2019) Cortical column and whole-brain imaging with molecular contrast and nanoscale resolution. Science 363(6424):eaau8302. https://doi.org/10.1126/science.aau8302

27. Ioannou S, Jackson J, Sheu S et al (2019) Neuron-astrocyte metabolic coupling protects against activity-induced fatty acid toxicity. Cell 177(6):1522–1535. https://doi.org/10.1016/j.cell.2019.04.001

28. Nixon-Abell J, Obara CJ, Weigel AV et al (2016) Increased spatiotemporal resolution reveals highly dynamic dense tubular matrices in the peripheral. Science 354(6311):433–446. https://doi.org/10.1126/science.aaf3928

29. Hoffman DP, Shtengel G, Xu CS et al (2019) Correlative three-dimensional super-resolution and block face electron microscopy of whole vitreously frozen cells. Science 367 (6475): eaaz5357. https://doi.org/10.1126/science.aaz5357 10.1101/773986

30. Hennig P, Denk W (2007) Point-spread functions for backscattered imaging in the scanning electron microscope. J App Phys 102:123101–123108. https://doi.org/10.1063/1.2817591

31. Wu Y, Whiteus C, Xu CS et al (2017) Contacts between the endoplasmic reticulum and other membranes in neurons. PNAS 114(24): E4859–E4867. https://doi.org/10.1073/pnas.1701078114

32. Hua Y, Laserstein P, Helmstaedter M (2015) Large-volume en-bloc staining for electron microscopy-based connectomics. Nat Commun 6:7923. https://doi.org/10.1038/ncomms8923

Chapter 13

Image Processing for Volume Electron Microscopy

Jörgen Kornfeld, Fabian Svara, and Adrian A. Wanner

Abstract

Today's volume electron microscopy techniques produce large image datasets on the order of thousands of gigabytes. The vast amount of data makes manual analysis almost infeasible, and data storing and processing challenging. Specialized infrastructure and software was therefore developed during the last decade to address these problems, ranging from distributed and versioned 3D image stores to deep neural network architectures optimized for the segmentation of objects of interest. Illustrated by the example of connectomics, the reconstruction of neural circuitry from 3D images of brain tissue, the most common approaches and solutions are discussed.

   Key words Image processing, Automated segmentation, Connectomics, ATUM, TEM, SBEM

---

1   Introduction

In the past decade, volume electron microscopy (vEM) has seen an unprecedented increase in the number and size of datasets acquired, due to automation of physical sample sectioning and increases in imaging throughput. Image stacks at nanometer resolution and on the scale of terabytes with thousands of sections can now be acquired routinely and many large datasets were recently published [1–15]. The size of these datasets requires sophisticated image analysis workflows that integrate storage management systems and high-throughput image analysis (Fig. 1). In the following, we outline such solutions and review recent advances in the field. We will focus on pipelines for connectomics, the reconstruction of neuronal circuits from vEM stacks of brain tissue, since the field usually generates the largest datasets and faces a very difficult image analysis problem: tracing nanometer-thin neurites over millimeters. However, the described workflows, problems, and solutions should generalize well to other vEM applications and tissues types.

   Section 2 introduces how vEM datasets can be organized and stored, and Sect. 3 discusses image normalization and registration techniques, that are used to assemble a seamless and continuous 3D

Irene Wacker et al. (eds.), *Volume Microscopy: Multiscale Imaging with Photons, Electrons, and Ions*, Neuromethods, vol. 155,
https://doi.org/10.1007/978-1-0716-0691-9_13, © Springer Science+Business Media, LLC, part of Springer Nature 2020

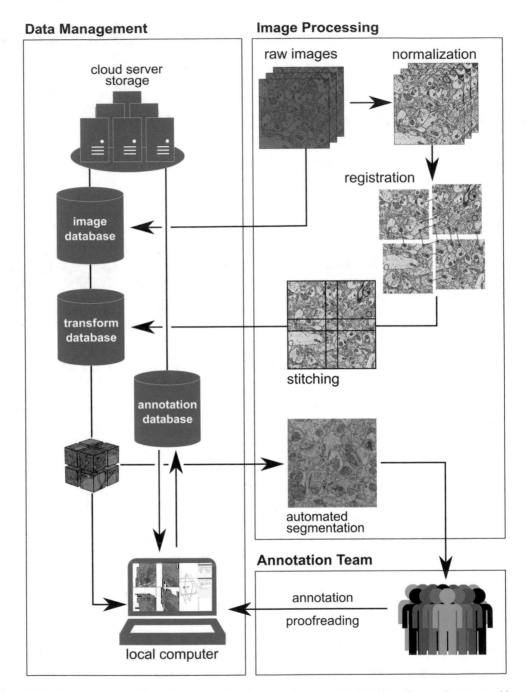

Fig. 1 Generic image processing, storage, and analysis workflow. A server stores the raw images and image transforms which are combined to a 3D image volume that is then split into subvolumes, such that any part of the dataset can be rapidly and randomly accessed. Manual and automated annotations are hosted and managed through an annotation database, which can be accessed concurrently by different clients

volume from many 2D images. The manual reconstruction of even a handful of neurons and their synaptic connections is a very time consuming and challenging task that can easily take hundreds or thousands of human annotation hours. Browsing such large datasets and visualizing and analyzing hundreds or thousands of interconnected neurons requires specialized software tools, some of which are described in Sect. 4. Finally, we review in Sect. 5 recent advances in automated image segmentation and manual proofreading of the algorithmic results, that increase the reconstruction throughput by orders of magnitudes in comparison to purely manual analysis.

2   Data Management and Storage

One cubic millimeter is often considered a landmark volume for vEM, in particular for studying neuronal connectivity in cortex, because it roughly corresponds to the size of a cortical column [16]. To date, the largest published datasets are about 100 times smaller than a cubic millimeter [17]. For the reconstruction of neurites and the identification of chemical synapses, a typical voxel size of between $10 \times 10 \times 30$ nm for serial block-face electron microscopy (SBEM) [18] and $4 \times 4 \times 40$ nm for serial section scanning electron microscopy (SEM) and transmission electron microscopy (TEM) is used [19]. The cubic millimeter volume corresponds therefore to a dataset of $3–16 \times 10^{14}$ voxels requiring about 300–1420 TB of storage space at 8 bit per voxel. Datasets of this size exceed by far the capabilities of desktop computers and have to be redundantly stored and hosted on dedicated (cloud) server infrastructure that supports fast and reliable data access. In comparison, it was estimated that about 2220 petabytes of data are generated worldwide every day [20], showing that it is mainly a matter of available resources and not technological feasibility that makes the handling of vEM data challenging.

2.1   Data Formats

The 3D image data is typically generated by a microscope as a series of 2D image files. Large samples require a division into a mosaic of many, partly overlapping image tiles since the field of view of the electron microscopes is limited. Using these image files directly for inspection and analysis is impractical, since most analyses require a single, global coordinate system. The set of flat images produced by the microscope is therefore consolidated into a 3D volume (*see* Sect. 3.2 below).

Several different specifications have been developed to store the consolidated vEM image data. Since the volumes are often too large to be stored on a user's workstation, most current tools can, at least optionally, retrieve the data from a remote server, in many cases along with different types of annotation data (Fig. 1).

Segmentations, in which each voxel is labeled with an identifier value representing different objects or different object types, are a common annotation type for vEM data. Since segmentation data is, like the raw image data, essentially a 3D matrix of scalar values, many software tools store it in the same fashion as the image data. In the following, a nonexhaustive overview of some of the most commonly used storage solutions and data formats for large vEM datasets is provided.

Volume EM data obtained by serial sectioning methods typically has much higher lateral resolution than section thickness. Consequently, these images are preferably viewed in 2D. To enable rapid zooming and browsing, these datasets can be stored as pyramids of increasingly downsampled flat images, the so-called MIP maps, either locally, as for example in the ImageJ plugin TrakEM2 [21], or on a remote server, such as in the widely used web-based viewer and annotation tool CATMAID [22].

For datasets with (near-)isotropic resolution, such as those generated by SBEM or FIB-SEM, it is often useful to view the data along virtual reslices, that is, slices along planes different from the plane of physical sectioning. In that case, storing the dataset as flat images is suboptimal, since large numbers of these would need to be accessed and combined to build up a resliced image. One data format developed to address that issue is the cube format of the KNOSSOS annotation tool, which was later adopted by its descendants PyKNOSSOS and webKnossos. In that format, the volume is split up into small cubes (e.g., 128^3 8-bit voxels), each of which is stored as a single file in a directory hierarchy on the file system. In order to make it possible to view progressively larger parts of the dataset when zooming out without increasing memory consumption, a pyramid of increasingly downsampled volumes is stored. Extensions of this format allow for reducing the size of the individual cubes using lossy (e.g., JPEG or JPEG2000) or lossless (e.g., PNG) compression to optimize streaming over low bandwidth connections. Additionally, this format makes it easy to perform image processing operations on subvolumes in parallel, since different worker processes can work on different files. Furthermore, it can be extended to additional data channels, such as channels containing segmentation results, which are typically stored with lossless compression.

Another versatile file format is the Hierarchical Data Format 5 (HDF5) that comes with library support for many different programming languages [23]. While it is not currently the native on-disk format of any connectomics viewer/annotation tool, its widespread support by software libraries and viewing/editing tools make it an appealing format for data exchange and programmatic processing. For example, ImageJ [24] can load 3D image data from HDF5 using a plugin [25] and the ElektroNN deep learning library [26, 27] reads ground truth image data provided in the HDF5 format. Data in the HDF5 format can be distributed over multiple files, and recent versions support parallel I/O on single files.

An extensible database solution for storing, editing, and versioning volume image data and annotations is DVID [28], which supports both flat and 3D image data storage. In the 3D case, the data is internally split up into small chunks, similar to the KNOSSOS cubes described above, but can be flexibly stored on different storage backends, such as Google Cloud Storage or local disk drives. DVID features dataset version control such that the user can, for example, revert back to previous states of a segmentation or compare different edited versions derived from the same starting segmentation.

A similar large-scale spatial database service for storing multidimensional neuroimaging data and associated voxel annotations is The Boss [29]. The cloud-based service supports 3D, multichannel, and time series source data and annotations. It supports Amazon Web Services (AWS) as a backend and has a tiered storage architecture that balances costs and performance. The Boss is used by the public hosting service Neurodata.io [30] that provides software tools like KNOSSOS or VAST [31] access to the hosted datasets through a RESTful API.

3   Image Optimization and Registration

The assembly of millions of single image tiles into a seamless 3D volume is usually performed in several steps. Depending on the EM acquisition method used, different problems, such as folds or cracks in the case of serial section collection, or debris on the block face in the case of SBEM, have to be addressed using image processing. In this section, we first discuss different image normalization techniques and how these images can be registered into a single 3D coordinate system.

3.1   Image Normalization and Artifact Correction

Inhomogeneous staining and variations in image acquisition conditions (e.g., beam drift, electron detector bias shifts or sample charging) can cause strong contrast and brightness differences between and within the 2D images acquired by the electron microscope. Various solutions exist to normalize the image data, that is, the intensity values of each image pixel are scaled such that the image histogram, the distribution of pixel intensity values, matches a particular distribution [32]. This kind of histogram equalization is often accompanied by a conversion of the data types of the images, for example, from unsigned 16-bit integers to unsigned 8-bit integers. Most common are global histogram normalization methods that are applied to the entire image (or even the entire set of images), which have the drawback of not being able to address local intensity variations inside of an individual image. Adaptive histogram equalization methods such as contrast limited adaptive histogram equalization (CLAHE) [33] perform histogram equalization in a local image window, making it possible to correct differences that arose during acquisition of a single 2D image but require careful parameter tuning, and invalidate analyses based on local intensity measurements.

While many of these techniques have been used for decades, there has recently been a surge in the development of novel image correction and resolution improvement methods, sometimes based on deep-learning [34, 35], for example to programmatically correct variations in section thickness or other distortions [36]. These approaches, which model a continuous 3D space, bridge between purely 2D image correction and volume registration methods as outlined in the next section.

3.2 Image Registration

The size of a single EM image that can be acquired by a SEM or TEM is typically limited to 10–120 μm edge length. In order to cover larger fields of view, mosaics of overlapping image tiles are acquired, which can easily result in millions of individual images that have to be arranged in a common coordinate system to form a coherent 3D volume. The overlap between neighboring tiles is typically set between 50 and 1000 pixels, depending on the precision of the microscope stage movements. The offsets between neighboring, overlapping tiles are usually calculated by using normalized cross-correlation [37] or by extracting corresponding features, for example SIFT features [38], in both tiles. Matching pairs of patches or features in the overlap between neighboring images are selected using robust sampling methods such as RANSAC [39]. From these, translational offsets between patches are calculated and combined in a global equation system from which the optimal tile positions are calculated in a least square displacement sense [40, 41].

This process is complicated by nonlinear distortions within the individual images caused by variations in the image acquisition conditions (e.g., sample charging) and lens distortions of the electromagnetic, and in the case of TEM-imaging, optical lenses of the microscope. Lens distortions are typically stationary and can therefore be corrected by calculating a static lens-distortion model [42]. The former, however, are usually nonstationary.

Similarly, the section collection process in serial section TEM and SEM introduces a combination of rotations, folds, cracks, compressions and dilations that result in nonstationary, nonlinear distortions between subsequent sections and planes. In those cases, linear alignment corrections such as rotations and other rigid or affine transforms can only deliver approximative results. Subsequent automated annotation and segmentation pipelines are typically very sensitive to alignment errors and therefore more sophisticated methods have recently been introduced that combine registration and artifact correction [35, 36].

For small datasets, consisting of a few hundreds or thousands of tiles with little or no nonlinear distortions in the overlapping regions, the following sequential alignment workflow, for example using the ImageJ plugin TrakEM2 [21, 43], has proven to give sufficient results, at least for manual data analysis:

1. Contrast normalization of individual tiles.

2. In-plane alignment of overlapping tiles by translational offset correction.

3. In-plane montage: stitching of overlapping tiles.

4. Across-planes registration: Sequential (possibly nonlinear) alignment of subsequent sections using the previous, already aligned section as a reference.

However, this heuristic, sequential alignment workflow does not scale to larger datasets with tens of thousands or millions of images because it is lacking the registration version and quality control that allow the user to locally correct and optimize alignment parameters efficiently without having to rerun the entire alignment pipeline. The registration of millions of images in terabyte-sized datasets requires more sophisticated infrastructure and processing pipelines. These pipelines typically keep the raw image data and the corresponding image metadata and image transforms in separate databases [44], in order to allow for an iterative alignment procedure, in which the image transforms are recalculated and updated without applying them each time to the actual image data. This design saves computation and storage space, avoids the problem of artifact propagation and amplification due to multiple interpolation iterations and it makes it possible to use version control in order to keep track of the alignment history. The typical workflow for these database-based pipelines is applied iteratively by starting with a coarse registration that gets subsequently refined:

1. New raw images are added to the database and a set of landmarks and parameters are extracted for each new image (e.g., SIFT features) and are stored in the metadata database.

2. These landmarks are then used to iteratively calculate and optimize the local and global image transforms.

3. Manually or automatically identify and inspect problematic regions such as images with large shifts or strong distortions and optimize the alignment parameters locally.

4. Recalculate the global image transforms with the new set of parameters.

4 Manual Image Annotation and Software Tools

vEM data can be annotated and analyzed manually, automatically, or semiautomatically, for example by manually proofreading an automatically generated proposal segmentation (*see* also Sect. 5 below). Table 1 gives an overview of the most commonly used noncommercial software packages and their key annotation tools.

Fully manual annotation can in many cases be the most efficient choice. This is particularly the case when the goal is to reconstruct the morphologies of a relatively limited number of cells, since the up-front effort required to obtain an automatic segmentation from scratch can be large. Neuron morphology in particular can be reconstructed efficiently by skeletonization [21, 45]. Skeletonization refers to the creation of a spatial graph that represents the center lines of the processes of a cell, while ignoring their precise three-dimensional extent (Fig. 2a). Skeletons can be sufficient, for example when the scientific questions pertain mainly to the connectivity between cells. Crucially, obtaining complete three-dimensional volumes (segmentations) of neurons manually by painting, that is, by assigning a label to each voxel (Fig. 2b), is substantially more time-consuming, typically by an order of magnitude [45]. Skeletonization has therefore been used to reconstruct large sets of neurons by teams of human annotators [8, 10]. However, painting is required when detailed morphologies of the cells or parts of the cells are sought (e.g., when measuring synaptic contact areas or spine volumes), or when the objects of interest do not primarily have an elongated, tubular structure (e.g., organelles or cells other than neurons).

Other manual annotation approaches are used when a proposal segmentation already exist, where voxels have been painted automatically. Manual labor is then used to correct that segmentation by merging or splitting objects. In that scenario, a distinction can be made between actions that operate on the level of objects (Fig. 2c) and actions that operate on the underlying voxels (Fig. 2a). The former case is computationally cheap, because the annotation software only needs to keep track of a list of objects that belong together. By contrast, in the latter case, labels need to be changed for all the voxels in one of the objects.

5 Automatic Segmentation and Proofreading Strategies

While optimized manual annotation software [45, 46] used by large teams of human annotators enabled the dense reconstruction of neurons in cubic volumes up to 100 μm edge length [8–10], automation is required for larger volumes. This becomes evident when considering that the total manual neuron skeletonization time for a larval zebrafish brain would be close to 300 person-years [26].

Table 1
Overview of different vEM analysis and annotation software tools

Software	Highlights	Stand-alone/browser	Max dataset size with 8GB RAM	Skeleton tracing	Voxel painting	Segmentation proofreading	Arbitrary reslices	Mesh rendering	Scripting/plugin interface
BigData viewer BigCat https://imagej.net/BigDataViewer		s			×	×	×		
CATMAID https://github.com/catmaid/CATMAID	Collaborative environment, many analysis tools, audit and provenance tracking, landmark-based skeleton transformations, NBLAST support.	b	Unlimited[a]	×				×	×
IMOD http://bio3d.colorado.edu/imod/	3D reconstruction from tilt series and serial sections, alignment, image processing, volume viewing, and contour tracing	s	Unlimited[a]				×	×	×
KNOSSOS https://knossostool.org https://github.com/knossos-project	Very fast data and annotation online streaming, on-the-fly 3D mesh generation and high-performance C++ OpenGL rendering, many different annotation modes available, Python plugin interface, tested and used in many large-scale annotation projects. Extensive open-source libraries to work programmatically with Knossos datasets, for example *knossos-utils*	s	>50TB[b] Unlimited[a]	×	×	×	×	×	×

(continued)

Table 1
(continued)

Software	Highlights	Stand-alone/browser	Max dataset size with 8GB RAM	Skeleton tracing	Voxel painting	Segmentation proofreading	Arbitrary reslices	Mesh rendering	Scripting/plugin interface
PyKNOSSOS https://github.com/adwanner/PyKNOSSOS	Flight mode for skeleton proofreading and synapse annotation and arbitrary reslices, fast multiresolution data streaming, extensive python API and 3D rendering capabilities through the Visualization Toolkit (VTK)	s	>12TB[b] Unlimited[a]	×			×	×	×
webKnossos https://webknossos.org/	Flight mode, collaborative environment, task management features, versioned annotations	b	>50TB[b] Unlimited[a]	×	×	×	×	×	×
Microscopy Image Browser http://mib.helsinki.fi/	Image processing, alignment, segmentation and quantification for 2D–5D datasets	s	In RAM: 3GB On HDD: unlimited[a]	×	In RAM only	×		×	×
Neuroglancer https://github.com/google/neuroglancer		b	Unlimited[a]			×	×	×	×
NeuTu https://github.com/janelia-flyem/NeuTu		s	>50TB[b] Unlimited[a]			×		×	
Nornir http://nornir.github.io	Nornir takes large sets of overlapping images in 2D and produces registered 2D and 3D volumes of any size and scale	s	Unlimited[a]						
Reconstruct https://synapseweb.clm.utexas.edu/software-0	Many annotation tools for manual 3D reconstruction. Extensive documentation	s	Unlimited[a]	×	×			×	

Name / URL	Description						
TrakEM2 https://imagej.net/TrakEM2	Various image normalization, registration and annotation tools, seamless browsing in 2D	s		×	×	×	3D viewer plugin	×
Vast Lite https://software.rc.fas.harvard.edu/lichtman/vast/	Local and online datasets, works with EM and LM data, voxel painting and filling, segment hierarchies, layers, trans-layer masking, 3D viewer	s	$1.3\ PB^b$ $(2^{31\times3})^a$	×	×	×	Export via API	×
Viking http://connectomes.utah.edu	Viking: Online multiuser annotation with arbitrary OpenGIS shapes. Spatial SQL database. Data export via online web services in standard file formats	s	Unlimited[a]	×	×		×	×

Note that many of these tools are still being developed and extended very actively

[a]Theoretical

[b]Tested

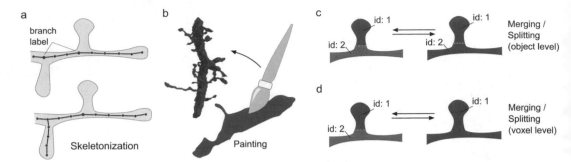

Fig. 2 Manual annotation operations. (**a**) In skeletonization, a user creates a sparse, 3D spatial graph representing the morphology of a cell, without detailed volume information. Branch points can be labeled for reinspection at a later point in time (blue nodes in the drawing), so that branches can be traced once the current branch is complete. Top/bottom: Before and after an annotator handles the leftmost branch label, respectively. (**b**) Painting refers to the operation where the user assigns a label (e.g., a 64-bit integer) to individual voxels, in order to reconstruct the volume of complete cells (or sometimes other structures). (**c, d**) If there are preexisting voxel labels, the annotator can operate on these by defining certain labels as equivalent (merging), or by breaking them apart into subvolumes (splitting). These operations can occur either on the level of entire objects (**c**) or can be propagated down to the individual voxel level (**d**)

5.1 Automatic Segmentation

Automatic reconstruction pipelines usually attempt to assign a unique integer identifier (ID) to every voxel in the volume, with the goal of assigning all voxels that belong to the same neuron the same ID (often a 64 bit value). Errors in these assignments can broadly be split into two categories (Fig. 3a):

1. False merge errors, meaning that the voxels belonging to two different cells were assigned the same ID.

2. False split errors, meaning that the voxels belonging to a single cell were assigned different IDs.

It is important to note that the first error category is harder to undo later, since correcting a false merger requires changing all affected voxel IDs, while the different IDs of fragments belonging to the same cell can be mapped to a common ID using a lookup table (Fig. 2c, d). Therefore, the goal in automatic segmentation is typically to first generate an oversegmentation, that is, a segmentation in which false merge errors are rare or absent, but false split errors may still be present, and fix the remaining split errors in subsequent steps.

This is commonly achieved using a multistep approach (Fig. 3b), which starts with the generation of a boundary map or affinity map which describes which neighboring voxels belong together. Ideally, all cells would be correctly separated in this representation after applying a connected components algorithm. Boundary or affinity maps are usually generated using computer vision methods, either based on classifiers with hand-selected features, such as edge (e.g., Gaussian Gradient Magnitude) or texture

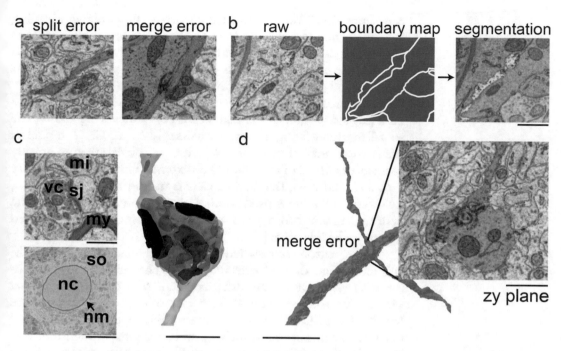

Fig. 3 Automatic segmentation of volume electron microscopy data. (**a**) Left: False split error. Two segments were generated (blue and red), despite belonging to the same cell. Right: False merge error. A single segment extends over a myelin sheath over two neurites. (**b**) Steps of a common segmentation pipeline: a raw image (left) is automatically converted into a boundary representation (middle), which is then used as input for a segmentation algorithm, which generates labeled regions (right). (**c**) Some local segmentation problems in volume electron microscopy datasets of nervous tissue: *vc* vesicle clouds, *sj* synaptic junctions, *mi* mitochondria, *my* myelin, *nc* nucleus, *nm* nuclear membrane, *so* soma. (**d**) A typical false merger error, resulting in an X-crossing, which is easy to recognize in the 3D representation but hard to notice by inspection of the cutting planes. Scale bars in (**a**), (**b**) and (**c**) top: 850 nm; (**c**) bottom: 7 μm; (**c**) surface rendering: 1.5 μm; (**d**) surface rendering: 6.5 μm; (**d**) inset: 1 μm

features (e.g., Hessian of Gaussian Eigenvalues) [47] or by using neural networks, in particular convolutional neural network (CNN) architectures [48–50]. Currently, CNNs can be considered the best choice, winning most segmentation competitions [51, 52].

The next step is the generation of a base segmentation, for example by using a seeded watershed algorithm that is applied to the boundary map [48, 53, 54]. To reduce the number of false mergers, parameters are tuned at this stage to ensure an oversegmentation. Importantly, the oversegmentation reduces the amount of data significantly, since the resulting "supervoxels" (collection of voxels/fragments with the same IDs) usually cover thousands or even millions of raw data voxels, making further agglomeration approaches, such as the greedy GALA algorithm [55], possible. In GALA, local features between supervoxels are extracted and a feature-based classifier is trained to predict the probability of two supervoxels or regions belonging to the same true object.

Agglomeration is then performed hierarchically, potentially leading to globally nonoptimal solutions. Globally optimal agglomerations can also be calculated by formulating the segmentation problem as a graph cut problem, that can then be solved as an integer linear program [53], albeit at high computational cost(but *see* [56]).

A recent alternative approach to neuron segmentation is based on a learned version of the connected-components algorithm, the so-called flood-filling neural networks [57]. A CNN model is trained on predicting locally whether an object that is already partially in its field of view should be extended into unseen territory [58, 59], and if so, the field of view is moved into this direction, followed by the next prediction. Using that approach, complete neurons can be reconstructed serially, and the method can be scaled to entire vEM datasets [57].

While neuron reconstruction is currently one of the most difficult reconstruction problems that can be encountered when working with vEM data, many other segmentation problems exist (*see* Fig. 3c for some examples), that can often be addressed more locally, that is, the objects of interest rarely extend over an entire dataset. In connectomics, such a problem is synapse identification, in which synaptic densities and vesicles have to be located to reliably determine the connectivity between neurons. Similar to the neuron reconstruction problem, voxel-wise probability maps are usually inferred [60–62], that are then segmented into separate objects using connected-components on a thresholded map and mapped to neuron reconstructions [26, 63]. Fundamentally, there is no difference between segmenting synapses, mitochondria, the Golgi apparatus or endoplasmic reticulum, and, depending on dataset quality, available training data and classifier, these problems can be solved fully automatically at excellent error rates nowadays. As a rule of thumb, the achievable automatic segmentation quality is correlated to how easy the structures of interest can be identified by the human observer, at least given a highly optimized machine learning model.

5.2 Segmentation Proofreading

Despite the recent progress in automated segmentation, the achievable segmentation error rates are often insufficient to address a specific biological question in a vEM dataset, and therefore manual proofreading of the results is necessary. The simplest approach is to use the same tools as for purely manual segmentation, that is, the re-labeling of voxels with, for example, paint brushes of various sizes or flood-fill tools, based on the visual identification of automated segmentation errors by browsing the dataset. This method usually works well for small volumes (Megavoxels to Gigavoxels) but becomes problematic for larger volumes not only because of the additional annotation time but also because without a guidance mechanism, a human annotator easily loses track of areas or segmentation objects that require special attention. An often-used solution consists of splitting up a larger volume into smaller chunks for inspection and proofreading, with the downside that the chunks have to be unified afterward. While seemingly trivial, this should not be underestimated, given that manual annotations are rarely perfect and small local errors may propagate upon stitching of proofread subvolumes. Especially for neuron reconstructions, the so-called mesh proofreading, the 3D visualization of the shape of individual neurites, appears to be necessary, allowing a human annotator to identify errors without the cumbersome and error prone inspection of cutting planes through the volumes (*see* Fig. 3d). Errors identified in the 3D view can then be corrected on the slice plane data, or depending on the segmentation task and capabilities of the used software directly in the 3D projection.

While proofreading can be performed on the level of individual voxels, it can be sufficient to restrict proofreading to a coarser level. As outlined before, a (dense) segmentation of a vEM dataset is usually performed by first identifying sets of voxels that have a very high probability of being part of the same object, with the goal of preventing false merge errors (oversegmentation). A graph of supervoxels can then be generated, where every edge represents a potential merge decision. Proofreading can then proceed by either removing edges, confirming edges or by adding new edges, which requires significant manual input. While conceptually easy, the user interfaces to perform these tasks optimally are still being actively researched. It is particularly challenging to proofread large and visually complex structures, such as neurons found in *Drosophila melanogaster* [12] or spatially dense vertebrate neurons with many branches, which make it very difficult for an annotator to spot missing branches or small false mergers. Neural network models were recently developed that operate on a larger fragment or super-voxel scale [64–66]. These could be used to obtain improved merge decision predictions on the supervoxel graph, which might speed up human proofreading and eventually eliminate the need for human input entirely by performing large-scale shape plausibility analysis that could then change the parameters of the underlying local segmentation algorithms to correct the segmentation errors.

Acknowledgments

We would like to thank the authors of the software packages listed in Table 1 for providing details on their software.

References

1. Briggman KL, Helmstaedter M, Denk W (2011) Wiring specificity in the direction-selectivity circuit of the retina. Nature 471:183–188

2. Ohyama T, Schneider-Mizell CM, Fetter RD et al (2015) A multilevel multimodal circuit enhances action selection in Drosophila. Nature 520:633–639

3. Takemura S-Y, Bharioke A, Lu Z et al (2013) A visual motion detection circuit suggested by Drosophila connectomics. Nature 500:175–181

4. Morgan JL, Berger DR, Wetzel AW, Lichtman JW (2016) The fuzzy logic of network connectivity in mouse visual thalamus. Cell 165:192–206

5. Kasthuri N, Hayworth KJ, Berger DR et al (2015) Saturated reconstruction of a volume of neocortex. Cell 162:648–661

6. Lee W-CA, Bonin V, Reed M et al (2016) Anatomy and function of an excitatory network in the visual cortex. Nature 532:370–374

7. Kornfeld J, Benezra SE, Narayanan RT et al (2017) EM connectomics reveals axonal target variation in a sequence-generating network. eLife 6

8. Helmstaedter M, Briggman KL, Turaga SC et al (2013) Connectomic reconstruction of the inner plexiform layer in the mouse retina. Nature 500:168–174

9. Kim JS, Greene MJ, Zlateski A et al (2014) Space-time wiring specificity supports direction selectivity in the retina. Nature 509:331–336

10. Wanner AA, Genoud C, Masudi T et al (2016) Dense EM-based reconstruction of the interglomerular projectome in the zebrafish olfactory bulb. Nat Neurosci 19:816–825

11. Schmidt H, Gour A, Straehle J et al (2017) Axonal synapse sorting in medial entorhinal cortex. Nature 549:469–475

12. Zheng Z, Lauritzen JS, Perlman E et al (2018) A complete electron microscopy volume of the brain of adult Drosophila melanogaster. Cell 174:730–743.e22

13. Bock DD, Lee W-CA, Kerlin AM et al (2011) Network anatomy and in vivo physiology of visual cortical neurons. Nature 471:177–182

14. Svara FN, Kornfeld J, Denk W, Bollmann JH (2018) Volume EM reconstruction of spinal cord reveals wiring specificity in speed-related motor circuits. Cell Rep 23:2942–2954

15. Vishwanathan A, Daie K, Ramirez AD et al (2017) Electron microscopic reconstruction of functionally identified cells in a neural integrator. Curr Biol 27:2137–2147.e3

16. Wanner AA, Kirschmann MA, Genoud C (2015) Challenges of microtome-based serial block-face scanning electron microscopy in neuroscience. J Microsc 259:137–142

17. Kornfeld J, Denk W (2018) Progress and remaining challenges in high-throughput volume electron microscopy. Curr Opin Neurobiol 50:261–267

18. Denk W, Horstmann H (2004) Serial block-face scanning electron microscopy to reconstruct three-dimensional tissue nanostructure. PLoS Biol 2:e329

19. Briggman KL, Bock DD (2012) Volume electron microscopy for neuronal circuit reconstruction. Curr Opin Neurobiol 22:154–161

20. Marr B (2018) How much data do we create every day? The mind-blowing stats everyone should read. Forbes. https://www.forbes.com/sites/bernardmarr/2018/05/21/how-much-data-do-we-create-every-day-the-mind-blowing-stats-everyone-should-read/. Accessed 15 Sep 2018

21. Cardona A, Saalfeld S, Schindelin J et al (2012) TrakEM2 software for neural circuit reconstruction. PLoS One 7:e38011

22. Saalfeld S, Cardona A, Hartenstein V, Tomancak P (2009) CATMAID: collaborative annotation toolkit for massive amounts of image data. Bioinformatics 25:1984–1986

23. The HDF5® library & file format – the HDF Group. The HDF Group. https://www.hdfgroup.org/solutions/hdf5/. Accessed 3 Mar 2019

24. Schneider CA, Rasband WS, Eliceiri KW (2012) NIH Image to ImageJ: 25 years of image analysis. Nat Methods 9:671–675

25. HDF5 plugin for ImageJ. https://lmb.informatik.uni-freiburg.de/resources/opensource/imagej_plugins/hdf5.html. Accessed 3 Mar 2019

26. Dorkenwald S, Schubert PJ, Killinger MF et al (2017) Automated synaptic connectivity inference for volume electron microscopy. Nat Methods 14:435–442

27. ELEKTRONN – Convolutional neural network toolkit in python. Fast GPU acceleration and easy usage. http://elektronn.org. Accessed 3 Mar 2019

28. Katz WT, Plaza SM (2019) DVID: distributed versioned image-oriented dataservice. Front Neural Circuits 13:5

29. Kleissas D, Hider R, Pryor D et al (2017) The block object storage service (bossDB): a cloud-native approach for petascale neuroscience discovery. bioRxiv 2017:217745

30. Burns R, Perlman E, Baden A et al (2018) A community-developed open-source computational ecosystem for big neuro data. Nat Methods 15(11):846–847

31. Berger DR, Seung HS, Lichtman JW (2018) VAST (volume annotation and segmentation tool): efficient manual and semi-automatic labeling of large 3D image stacks. Front Neural Circuits 12:88

32. Gonzalez RC, Woods RE (2008) Digital image processing. Prentice Hall, Upper Saddle River, NJ

33. Pizer SM, Philip Amburn E, Austin JD et al (1987) Adaptive histogram equalization and its variations. Comput Vis Graph Image Process 39:355–368

34. Heinrich L, Bogovic JA, Saalfeld S (2017) Deep learning for isotropic super-resolution from non-isotropic 3D electron microscopy. Lect Notes Comput Sci 2017:135–143

35. Jain V (2017) Adversarial image alignment and interpolation. arXiv:1707.00067

36. Hanslovsky P, Bogovic JA, Saalfeld S (2017) Image-based correction of continuous and discontinuous non-planar axial distortion in serial section microscopy. Bioinformatics 33:1379–1386

37. Buniatyan D, Macrina T, Ih D et al (2017) Deep learning improves template matching by normalized cross correlation. arXiv:1705.08593

38. Lowe DG (1999) Object recognition from local scale-invariant features. In: Proceedings of the seventh IEEE international conference on computer vision

39. SRI International. Artificial Intelligence Center, Fischler MA, Bolles RC (1980) Random sample consensus: a paradigm for model fitting with applications to image analysis and automated cartography. Commun ACM 24(6):381–395

40. Sun C, Beare R, Hilsenstein V, Jackway P (2006) Mosaicing of microscope images with global geometric and radiometric corrections. J Microsc 224:158–165

41. Saalfeld S, Fetter R, Cardona A, Tomancak P (2012) Elastic volume reconstruction from series of ultra-thin microscopy sections. Nat Methods 9:717–720

42. Kaynig V, Fischer B, Müller E, Buhmann JM (2010) Fully automatic stitching and distortion correction of transmission electron microscope images. J Struct Biol 171:163–173

43. TrakEM2. ImageJ. https://imagej.net/TrakEM2. Accessed 3 Mar 2019

44. Image Transformation Web Services. https://www.janelia.org/image-transformation-web-services. Accessed 3 Mar 2019

45. Helmstaedter M, Briggman KL, Denk W (2011) High-accuracy neurite reconstruction for high-throughput neuroanatomy. Nat Neurosci 14:1081–1088

46. Boergens KM, Berning M, Bocklisch T et al (2017) webKnossos: efficient online 3D data annotation for connectomics. Nat Methods 14:691–694

47. Sommer C, Straehle C, Kothe U, Hamprecht FA (2011) Ilastik: interactive learning and segmentation toolkit. In: 2011 IEEE international symposium on biomedical imaging: from nano to macro

48. Turaga SC, Murray JF, Jain V et al (2010) Convolutional networks can learn to generate affinity graphs for image segmentation. Neural Comput 22:511–538

49. Ronneberger O, Fischer P, Brox T (2015) U-Net: convolutional networks for biomedical image segmentation. arXiv [cs.CV]. arXiv:1505.04597

50. Ciresan D, Giusti A, Gambardella LM (2012) Deep neural networks segment neuronal membranes in electron microscopy images. Adv Neural Inf Process Syst 2012:1–9

51. SNEMI3D. http://brainiac2.mit.edu/SNEMI3D/. Accessed 3 Mar 2019

52. CREMI. https://cremi.org/. Accessed 3 Mar 2019

53. Beier T, Pape C, Rahaman N et al (2017) Multicut brings automated neurite segmentation closer to human performance. Nat Methods 14:101

54. Berning M, Boergens KM, Helmstaedter M (2015) SegEM: efficient image analysis for high-resolution connectomics. Neuron 87:1193–1206

55. Nunez-Iglesias J, Kennedy R, Parag T et al (2013) Machine learning of hierarchical clustering to segment 2D and 3D images. PLoS One 8:e71715

56. Pape C, Beier T, Li P et al (2017) Solving large multicut problems for connectomics via domain decomposition. In: 2017 IEEE international conference on computer vision workshops (ICCVW)

57. Januszewski M, Kornfeld J, Li PH et al (2018) High-precision automated reconstruction of neurons with flood-filling networks. Nat Methods 15(8):605–610

58. Meirovitch Y, Matveev A, Saribekyan H et al (2016) A multi-pass approach to large-scale connectomics. arXiv [q-bio.QM]

59. Januszewski M, Maitin-Shepard J, Li P et al (2016) Flood-filling networks. arXiv [cs.CV]

60. Kreshuk A, Straehle CN, Sommer C et al (2011) Automated detection and segmentation of synaptic contacts in nearly isotropic serial electron microscopy images. PLoS One 6:e24899

61. Kreshuk A, Koethe U, Pax E et al (2014) Automated detection of synapses in serial section transmission electron microscopy image stacks. PLoS One 9:e87351

62. Roncal WG, Pekala M, Kaynig-Fittkau V et al (2015) VESICLE: volumetric evaluation of synaptic interfaces using computer vision at large scale. Proceedings of the British machine vision conference 2015

63. Staffler B, Berning M, Boergens KM et al (2017) SynEM, automated synapse detection for connectomics. eLife 6:e26414

64. Schubert P, Dorkenwald S, Januszewski M et al (2019) Learning cellular morphology with neural networks. Nat Commun 10:2736

65. Zung J, Tartavull I, Lee K, Seung HS (2017) An error detection and correction framework for connectomics. Advances in Neural Information Processing Systems 30 (NIPS 2018)

66. Rolnick D, Meirovitch Y, Parag T et al (2017) Morphological error detection in 3D segmentations. arXiv:1705.10882

Chapter 14

Forget About Electron Micrographs: A Novel Guide for Using 3D Models for Quantitative Analysis of Dense Reconstructions

Daniya J. Boges, Marco Agus, Pierre Julius Magistretti, and Corrado Calì

Abstract

With the rapid evolvement in the automation of serial micrographs, acquiring fast and reliably giga- to terabytes of data is becoming increasingly common. Optical, or physical sectioning, and subsequent imaging of biological tissue at high resolution, offers the chance to postprocess, segment, and reconstruct micro- and nanoscopical structures, and then reveal spatial arrangements previously inaccessible or hardly imaginable with simple, single section, two-dimensional images. In some cases, three-dimensional models highlighted peculiar morphologies in a way that two-dimensional representations cannot be considered representative of that particular object morphology anymore, like mitochondria for instance. Observations like these are taking scientists toward a more common use of 3D models to formulate functional hypothesis, based on morphology. Because such models are so rich in details, we developed tools allowing for performing qualitative, visual assessments, as well as quantification directly in 3D. In this chapter we will revise our working pipeline and show a step-by-step guide to analyze our dataset.

Key words 3DEM, 3D models, 3D reconstruction, 3D analysis, Virtual reality, Morphology

1 Introduction

The importance of analyzing brain cells morphology to understand brain function has been recognized for over a century by neuroscientists and is considered one of the biggest scientific challenges of this century. Since last decade, an increasing need for digital reconstructions of neuronal morphology has stimulated the rapid development of numerous synergistic tools for data acquisition, anatomical analysis, three-dimensional rendering, growth models, physical and functional models, and data sharing [1, 2].

However, a number of technical issues have been faced, like scalability, since modern digital acquisition systems [3] produce "big data," unprecedented quantities of digital information at

Irene Wacker et al. (eds.), *Volume Microscopy: Multiscale Imaging with Photons, Electrons, and Ions*, Neuromethods, vol. 155, https://doi.org/10.1007/978-1-0716-0691-9_14, © Springer Science+Business Media, LLC, part of Springer Nature 2020

unprecedented rates, and require, as with genomics at the time, breakthrough algorithmic, processing, and computational solutions [4, 5]. Worldwide, research and political institutions have put effort into the common goal of understanding and characterizing the mammalian brain functioning, connecting the many information coming from anatomy, biochemistry, connectivity, development, and gene expression (ABCDE) [6]. This chapter will focus on neuroanatomy and the state of the art methods for visual and quantitative analysis of micron- and at nanoscale resolution brain cell and their processes and intracellular apparatus. To this end, we will overview 3D reconstruction, analysis, and visualization techniques applied to EM datasets, as well as the pioneering application of emerging virtual reality technologies in the context of visual morphometric analysis of 3D reconstruction from EM stacks [7–10].

The methods and tools described in this chapter have been designed for qualitative and quantitative analysis of 3D data from rodent EM image stacks, to understand the role of glycogen in the central nervous systems [11,12]. We will give first a brief overview of the currently available pipelines for morphometric analysis targeted to discover peculiarities on cellular structures and perform statistical computations. In addition, we will describe systems and tools for immersive and natural visual exploration, like virtual reality (VR). Benefits of immersive visualization methods compared to desktop solutions include the possibility of collaborative discussions, and an improved visual experience of complex dataset, allowing for a more rich understanding and characterization of morphological features not accessible with 2D images, or on a flat screen. Future systems are aimed at embedding of quantitative and measurement tools within virtual reality environments [7, 13].

1.1 State of the Art of 3DEM Imaging, Segmentation, Reconstruction, and Analysis

1.1.1 Electron Microscopy Imaging

Automated volume SEM techniques have substantially improved the acquisition of biological tissue in three dimensions with regard to reliability, z-resolution, and speed. Nowadays, the mainly used systems are the following:

- *Serial block-face electron microscopy (SBEM)*, in which a diamond knife iteratively removes a thin surface layer of the sample. After each cut, the exposed smooth block face is imaged [14]. For applications where one is interested in imaging big fields of view (>20 μm), this approach is preferable, although the z-resolution will be limited by the precision of the cut with an ultramicrotome, which is highly influenced by the quality of embedding, the type of resin, and other environmental factors.

- *Focused Ion beam SEM (FIB-SEM)*, in which, instead of a diamond knife, a focused beam of gallium ions removes thin layers of material from the sample block face [15]. This technique offers the advantage of a better resolution on the z-axis, resulting eventually on isotropic voxel resolution, down to 5 nm/voxel in optimal conditions. Samples prepared for conventional TEM can be used, as in these setups the distance between the sample block face and the pole piece is small enough to limit the charge of the sample, making it stable, with a very good contrast. On the contrary, FIBSEM has the disadvantage of showing artifacts on the boundaries when the field of view is starting to exceed 15–20 μm.

- *Automated tape-collecting ultramicrotome SEM (ATUM-SEM)*, in which serial thin sections are automatically on tape after they are cut off a sample block with a diamond knife. The tape holding the sections is then manually transferred onto wafers for SEM imaging [16].

Since a growing community of scientists is using these techniques, sample preparation protocols are continually being developed and improved, and software tools for data acquisition, processing, and analysis are becoming more effective an d user friendly. As a consequence, the volume SEM workflow is expected to become more robust and routine in the coming years [17].

1.1.2 3D Reconstruction

Volume SEM, or 3D EM, has successfully filled the "imaging gap" that had existed in biology. Ultrastructural 3D datasets covering distances of tens or even hundreds of micrometers are becoming readily available in many laboratories. However, as datasets have grown in size, image processing and analysis have become the bottleneck for most studies [3]. For this reason, large-scale automatic methods will become available, but as for now, most efficiently used processing pipeline are semiautomated, and still requiring manual time-consuming efforts, especially for proofreading [18]. To this end, large-scale community efforts like the BigNeuron project [19] are being carried out to bench-test a large set of open-source, automated neuron reconstruction algorithms, in order to produce large, community-generated databases of single-neuron morphologies, open-source tools for neuroscience, and community-driven protocols intended to serve as the standard for digital reconstruction of single neurons. With respect to reconstruction methods, they can be subdivided in sparse and dense approaches: for a dense reconstruction, the aim is to capture all visible structures of the acquired volume, whereas a sparse reconstruction targets a small subset [20]. Since the time needed to digitally reconstruct a tissue volume is usually much longer than the time needed to acquire it, one must carefully set realistic goals for reconstruction efforts, since (semi)automated reconstruction tools are not yet ready for routine use and manual reconstruction is often the only option [21]. In the category of dense reconstruction, Liu et al. [22] recently proposed a fully automatic approach for intrasection segmentation and intersection reconstruction of neurons using EM images. They also developed a semiautomatic method that utilizes the intermediate outputs of our automatic algorithm and achieves intrasegmentation with minimal user intervention. Similarly, Kaynig et al. [23] developed a semiautomatic pipeline that provides state-of-the-art reconstruction performance while scaling to data sets in the GB-TB range, employing random forest classifier on interactive sparse user annotations.

The classifier output is combined with an anisotropic smoothing prior in a Conditional Random Field framework to generate multiple segmentation hypotheses per image. These segmentations are then combined into geometrically consistent 3D objects by segmentation fusion. Berning et al. [24] introduced SegEM, a toolset for efficient semiautomated analysis of large-scale fully stained 3D-EM datasets for the reconstruction of neuronal circuits. SegEM provides a robust classifier selection procedure for finding the best automated image classifier for different types of nerve tissue. SegEM resolves the tradeoff between synapse detection and semiautomated reconstruction performance in high-resolution connectomics and makes efficient circuit reconstruction in fully stained EM datasets a ready-to-use technique for neuroscience. Very recently, Kasthuri et al. [25] applied automated technologies to probe the structure of neural tissue at nanometer resolution and use them to generate a saturated reconstruction of a subvolume of mouse neocortex in which all cellular objects (axons, dendrites, and glia) and many subcellular components (synapses, synaptic vesicles, spines, spine apparati, postsynaptic densities, and mitochondria) are rendered and itemized in a database. As result of analysis of the trajectories of all excitatory axons with respect to dendritic spines, they were able to refute the idea that physical proximity is sufficient to predict synaptic connectivity (the so-called Peters's rule). In order to reduce the human efforts, Templier et al. [26] introduced a promising image annotation approach for the analysis of volumetric electron microscopic imagery of brain tissue, consisting of exploiting eye tracking in a way to let the operator navigate through the 3D data with gamepad controller at a high speed while keeping eye gaze focus on a single neuronal fiber, and automatically annotate it.

1.1.3 Visual and Morphometric Analysis

Following registration and digital reconstruction, the 3D dataset can be explored and analyzed with a variety of tools that offer visualization and annotation. In the field of cell and tissue biology, a large number of studies in recent years have used SBEM or FIB-SEM to characterize organelles and perform quantitative and visual analysis in cell cultures and tissues.

- *Quantitative morphometric analysis*: in this category, Scorcioni et al. [27] released L-Measure (LM), a freely available software tool for the quantitative characterization of neuronal morphology. The system computes a large number of neuroanatomical parameters from 3D digital reconstruction files starting from and combining a set of core metrics, and allowing for a set of operations, ranging from the extraction of basic morphological parameters to filtered selections and searches from collections of neurons based on any Boolean combination of the available morphometric measures. Similarly, Billeci et al. [28] developed a NEuronMOrphological analysis tool called NEMO capable of handling and processing large numbers of optical microscopy image files of neurons in culture or slices in order to automatically run batch routines, store data, and apply multivariate classification and feature extraction using three-way principal component analysis (PCA).

- *Mesh-based visualization and analysis:* currently many neuroscientists employ commercial, like Amira or Imaris, or free software solutions, like KNOSSOS [29], TrakEM2 [30], CATMAID [31], ilastik [32], and VAST [25], for 3D visual analysis of segmented data, and develop custom plug-ins for specific statistical and morphometric analysis on top of popular 3D modeling software [7, 33, 34]. In this category, Aguiar et al. [35] released Py3DN, an open-source Python solution for analyzing and visualizing 3D data collected with the widely used Neurolucida (MBF) system and integrated with Blender, allowing for the construction of mathematical representations of neuronal topology, detailed visualization and the possibility to define nonstandard morphometric analysis on the neuronal structures. Similarly, Jorstad et al. [36] customized the 3D modeling environment Blender with NeuroMorph (neuromorph.epfl.ch), a collection of semiautomatic software tools, with which users can view the segmentation results, in conjunction with the original image stack, manipulate these objects in 3D, and make measurements of any region. This approach to collecting morphometric data provides a faster means of analyzing the geometry of structures, such as dendritic spines and axonal boutons. Asadulina et al. [34] extended Blender to visualize and analyze anatomical atlases from larval stages of the marine annelid *Platynereis dumerilii*. Specifically, they developed tools for annotation and coexpression analysis, also representing and analyzing connectome data including neuronal reconstructions and underlying synaptic connectivity. Other custom solutions include the Filament editor [37], which is an integrated set of tools for creating reliable neuron tracings from sparsely labeled in vivo datasets. With respect to histology imaging, very recently Van Den Berghe et al. [38] proposed a ready-to-use, automated, and scalable method to thoroughly quantify histopathological markers in 3D in rodent whole brains, relying on block-face photography, serial histology and 3D-HAPi (Three-Dimensional Histology Analysis Pipeline), an open source image analysis software. They illustrate our method in studies involving mouse models of Alzheimer's disease and show that it can be broadly applied to characterize animal models of brain diseases, to evaluate therapeutic interventions, to anatomically correlate cellular and pathological markers throughout the entire brain, and to validate in vivo imaging techniques.

- *Volume-based visualization systems:* tools for visual analysis directly relying on segmented volume data are becoming available for usage especially for the reconstruction of nanoscale neuronal connectivity [13]. These systems are expected to be extensively used in future, given the quality of reconstructed images in semitransparency and the potentialities for creating sophisticated visual selection queries and focus-in-context interactive explorations.

2   Methods and Tools

In [7], we have created an analysis pipeline for visualizing and analyzing EM data (Fig. 1). Since then, we have upgraded a number of our analysis tools as well as developed new ones, and incorporated all with our existing tool chain. As a result, our improved analysis pipeline performs much better in terms of accuracy when creating 3D models, and in efficiency when executing analysis tasks. It also includes enhanced output, which represents information as a product of the analysis operation. The contents of the end product of the pipeline are variable and depend on the analysis operations that took place during processing.

Our pipeline follows different pathways; each stage is dependent on the output of the previous stage. Domain experts can proofread the results of each stage and perform the corrections needed. Any refinements made to the output will result in the pipeline to follow a new pathway.

One exciting addition to our analysis pipeline is the element of virtual reality (VR) via head-mounted displays (HMD), where we invest a lot of effort in developing data visualization applications in VR. Furthermore, a user case study has been conducted for the aim of evaluating usability of the visual setups of our systems. For this, we perform a comparative study of conducting visualization and analysis tasks in a stereoscopic setup (VR via HMD) and a monoscopic setup (3D via desktop applications). Subheading 2.5 will include more information regarding the technical side of this development work and how it is integrated with our analysis pipeline.

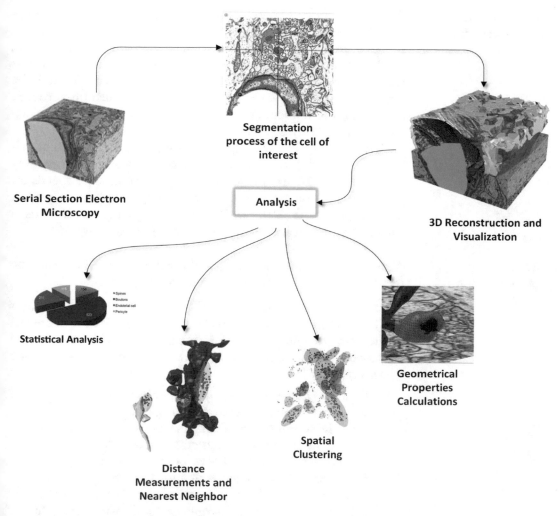

Fig. 1 Scheme of the entire pipeline, from image acquisition to analysis of 3D models

In the next sections we will demonstrate how our improved analysis pipeline is implemented and utilized for (1) analyzing astrocytic glycogen and (2) analyzing morphological features within one astrocytic process, in particular mitochondria and ER distribution compared to synapse, explaining in detail each step within the pipeline.

2.1 Serial Section EM The choice of the imaging technique depends on the type of structure one wishes to resolve and the field of view he or she wants to image. In our case, we were interested in observing a limited volume of an astrocytic process, its presynaptic extensions, and glycogen granules. For these reasons, we decided FIBSEM was appropriate, as a limited imaged volume and high z-resolution are required.

2.2 Cell of Interest and Segmentation

We were interested to reconstruct all neurites (axons and dendrites), as well as their connection points, the synaptic densities, the glial processes (processes belonging to astrocytes, microglia, NG2, oligodendrocytes, pericytes) and vascular processes such as epithelial cells in our stack. In addition, we reconstructed the intracellular machinery; we were focusing in particular on astrocytic machinery, ER, mitochondria and glycogen, and their relationship with synapses (Figs. 2, 3, and 4).

We took largely advantage of iLastik, a software which allows for fast and reliable segmentation and can provide high-quality image stacks [32]. Its semiautomated, carving module speeds up the reconstruction process by several fold, compared to classic, manual approaches [39]. For PSD and glycogen granules we used a manual approach and reconstructed them using TrakEM2.

2.2.1 Cellular Segmentation: Neurites and Glia

Axons and dendrites (Fig. 2, top and insets 1 and 2; Fig. 3a) are the output and the input stations of neurons, respectively.

- Axons (Fig. 2, light yellow; Fig. 3a) have relatively straight tubular structures, with a narrow, circular/elliptical diameter; they can swell at specialized structures filled with clear, round vesicles, and usually (but not always) facing a post-synaptic density, or PSD.
    - Boutons (Fig. 2, inset 1, and Fig. 3c) can be noted in a later step, directly on the 3D mesh, using NeuroMorph tool [36]; from the morphology of the vesicles (round, regular in size, or flattened), we can recognize supposedly excitatory or inhibitory axons, respectively. Boutons containing pleomorphic vesicles are classified as unknown, as their nature cannot be determined solely by morphology; they are supposedly belonging to monoaminergic or catecholaminergic fibers. Occasionally, these boutons are "en passant," not facing a postsynaptic density. Vesicles have been as well reconstructed, using a manual tool embedded within Blender [40].
- Dendrites (Fig. 2, blue; Fig. 3a) have a straight and tubular structure, with a circular diameter, and they are usually bigger than axons of several orders of magnitude, and contain actin bundles clearly visible within their cytosol, whose diameter is roughly in the order of 10 nm. They present protrusions called "spines," which could be of different size, and are connected to the tubular part of the dendrite, called "shaft," through a small narrow neck.
    - Spines (Fig. 2, inset 2, and Fig. 3b) can be noted in a later step, similarly to boutons for axons; they are usually showing a PSD.
- Astrocytes (Fig. 2, top, and insets 3 and 4; Fig. 4) are located in a strategic position, between vasculature and neurites, with

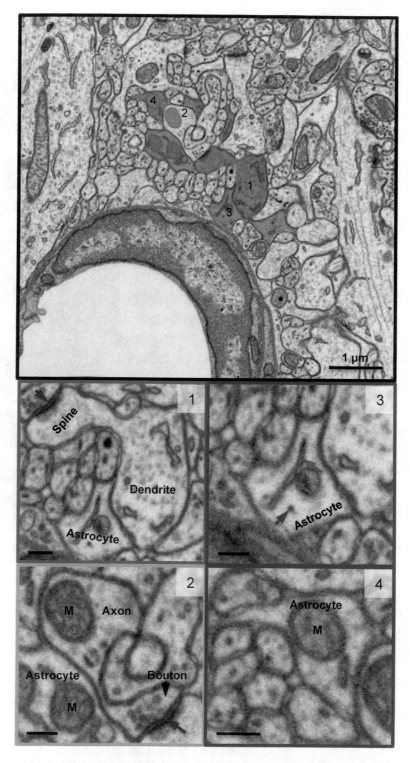

Fig. 2 EM micrographs highlighting neurites, glial and subcellular elements of interest for subsequent 3D reconstruction and analysis. Top: entire field of view with examples of segmented structures. (1) Blue: dendrite. Red; PSD. (2) Light

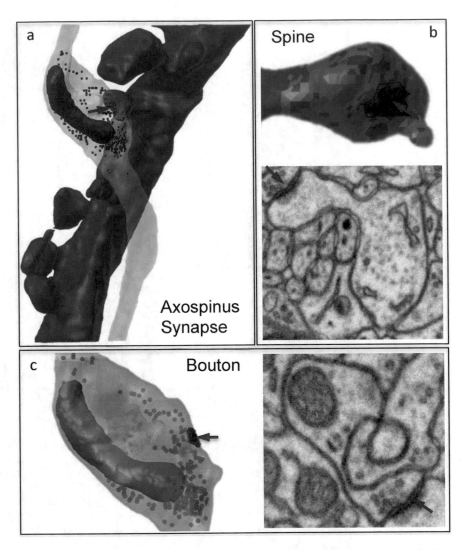

Fig. 3 3D reconstructions of neurites in Fig. 2. (**a**) Dendrite 1 (blue) and axon 2 (grey) establishing a synapse (red arrow). Axon semitransparent to highlight vesicles (yellow spheres) and mitochondrial tubule (purple). (**b**) 3D Magnification of the spine (blue) establishing the synapse in (**a**) and its correspondent 2D micrograph on the bottom. (**c**) 3D Magnification of the bouton (light green to highlight the bouton within the axon) containing synaptic vesicles and one mitochondrion, establishing the synapse in (**a**) and its correspondent 2D micrograph on the right

Fig. 2 (continued) yellow: axon, containing one mitochondria (purple). (3 and 4) Green; astrocytic processes, containing ER (yellow) and mitochondria (purple). Bottom, inset 1: dendrite 1, and its spine. Red arrow: PSD. Inset 2: axon 2 and its bouton (note synaptic vesicles). Red arrow: PSD. Insets 3 and 4: astrocytic processes 3 and 4. Yellow arrows: ER. Green arrow: glycogen granule. Scale bar in insets: 200 nm

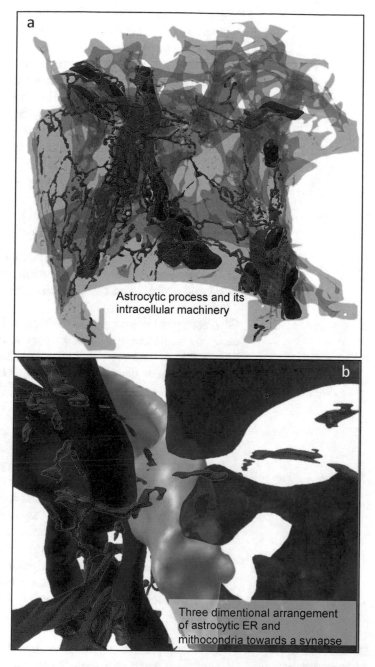

Fig. 4 3D reconstructions of astrocytic process and its intracellular machinery in Fig. 2. (**a**) Entire astrocyte (green, semitransparent) with its complex network of ER (yellow) and mitochondria (purple). (**b**) Detail of the synapse in Fig. 3 surrounded by astrocytic ER (yellow) and mitochondria (purple) to highlight the tight and intimate three-dimensional spatial arrangement between these structures

smaller, lamelliform processes interfacing synapses. Their cytosol appears relatively clear and little electron-dense, and their cross-sectional morphology is not regular by nature, as astrocytes tend to "fill" space in the neuropil; on a single section might be confused for any of the neurites that has been listed previously, as depending on the section they have been imaged, they appear tubular, whereas in other cases they could show flattenings of few nanometers thick, that appears like sheets in 3D. While bigger processes can host more intracellular apparatus, like mitochondria, ER, and actin tubules, astrocytic perisynaptic processes might be relatively empty, with portions of ER reaching where the cross-sectional size allows for it. We used a pipeline involving an improved version of iLastik for the reconstruction, proofread using TrakEM2 for the smaller processes.

The stack contained two portions of pericytes, recognized by their darker cytosol and their location, at the interface between the astrocytic process and the endothelium of the blood vessel.

2.2.2   Glycogen and PSD

Glycogen granules appear as dark dots (Fig. 2, inset 3); they can be easily recognized within the astrocytic cytoplasm. They can be of different sizes and are reconstructed individually using TrakEM2, as spheres whose diameter corresponds to the size of the individual granules.

- PSD (postsynaptic densities; Figs. 2 and 3) are highly stained, dark specializations appearing on part of the membrane of spines, or occasionally on shafts, and facing boutons. They represent the interface between a presynaptic and a postsynaptic element, establishing a synapse, the chemical transductor of an electric signal that needs to cross from one neuron to another one. Thicker ones are called "asymmetric" and are supposedly excitatory, whereas thinner ones are called "symmetric" and supposedly inhibitory, although depending on the direction and the angle of the cut the thickness might create confusion. Therefore, using the presynaptic terminal vesicles is always a safer way to distinguish the PSD. They can also be segmented manually, using TrakEM2.

2.2.3   Astrocytic ER and Mitochondria

- ER (endoplasmic reticulum; Figs. 2 and 4) exerts a number of functions, such as a membrane reservoir, transport of proteins, and calcium storage and release. Depending on its morphology and cross section, it could either appear as a round, circular structure with a clear lumen or an elongated pipe. In the first case, it could be confused for a vesicle, but its cross section would remain constant over hundreds of nanometers. In the second case, although the structure would appear as a tubule, its length and cross section might remain constant over a tenth of serial sections, meaning it is not a tubule but a so-called cistern. The two types of structures appear similar on single sections but could be easily distinguished in 3D.

- Mitochondria (Figs. 2, 3, and 4) have a peculiar, stereotyped, morphology, with the so-called cristae within their inner lumen. They appear as electron-dense, round/oval shaped organelle of a hundred-nanometer cross section; once reconstructed, these structures form long tubules that can travel all along the process containing them. We reconstructed all the mitochondria within the volume, as their features and textures make them easily recognizable and reconstructable.

2.3 Segmentation Proofread and 3D Mesh Generation

After completing the segmentation process of our structures of interest, a proofread of the result is needed before exporting the segmentation into a 3D mesh, a geometry definition wavefront file. It is important that we obtain a 3D mesh that represents its corresponding neuronal structures as accurate as possible. Any defect or deformation on the mesh will reflect on the analyses results, which can be a challenging task for highly complex morphologies. This is where proofreading is needed. The process of proofreading involves a visual inspection of the segmentation at first, which could highlight gross mistakes in the structure. This pre-export proofreading stage is accomplished often using the same segmentation software (e.g., TrakEM2 in ImageJ) and is usually conducted by, or in the presence of a domain expert. TrakEM2 is equipped with a hardware-accelerated add-on that offers visualization of the segmented stack called the 3D viewer. The 3D viewer displays the segmentation as texture-based volume renderings. This feature is essential for proofreading the segmentation prior to exporting it as an obj 3D model (Fig. 5). A visual inspection of the 3D volume can help domain experts recognize any structural defects. Proofreading also involves retracing the reconstructed structure in the 3D volume along its 2D corresponding slice from the EM stack.

Once the segmentation is checked and approved by the domain scientist, it gets exported as an obj file. The obj file represents the 3D model of our targeted cell/organelle.

In computer graphics, our obtained 3D models are called "polygon meshes." A polygon mesh is a collection of edges, vertices, and polygons; each item in the collection is connected based on the following system [41]:

- Each vertex is shared by at least two edges.
- Each edge is shared by at most two polygons.
- Each polygon is a closed sequence of edges.

Therefore, each object in 3D space is made up of a set of connected polygonal bounded planar surfaces [41]. Different

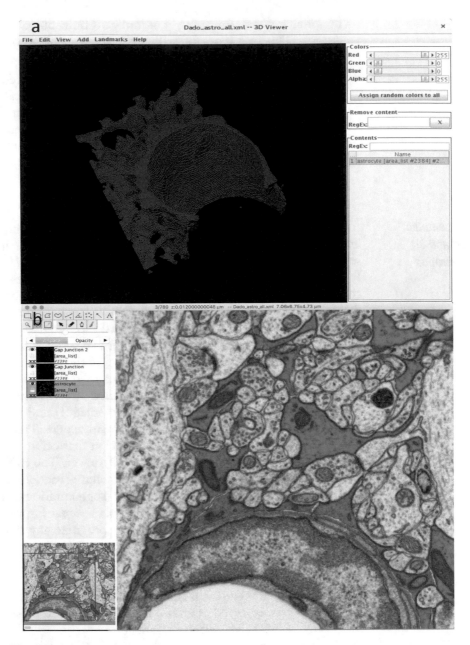

Fig. 5 The 3D viewer in Fiji displaying the segmentation of an astrocyte as texture-based volume renderings to be used for proofreading

literature and contexts may refer to polygon surfaces as faces, as for example in Blender [42]. Polygons come together to form a mesh.

Vertices, edges, and faces, are fundamental elements in mesh modeling. Their arrangement reflects on a number of features concerning 3D meshes, such as topology and geometry [42].

Usually segmentation software, such as TrakEM2 and iLastik, can handle mesh generation and produce accurate 3D models that are processed smoothly through our analysis pipeline. However, when it comes to generating meshes of extremely complex structures such as ER and astrocytes; a combination of two or more processing stages is necessary to obtain a correct and accurate 3D mesh. Different software tools use different algorithms for generating meshes. For example, TrakEM2 implements the marching cubes algorithm [43] to extract its polymesh data [30]. As a result, meshes may suffer from several defects within their topology, like in the case of non-manifold meshes also known as not solid or not watertight.

Preparing a mesh for analysis is crucial and usually done using mesh cleanup functions. Most 3D modeling software tools (e.g., Blender) are equipped with functions that provide non-manifold geometry detection and fixing (Fig. 6).

In addition, a combination of two or more of the following software tools and techniques can be used to handle mesh defects:

- By using MeshLab or Blender, we perform visual checkups and manual cleanups on the defected obj file. This is usually a good option if the number of vertices in the mesh was small.

- Regenerating the mesh using the segmented masks by exporting them into tiff images or an Amira file which are two popular file formats that most 3D visualization software such as Avizo accept as input. TrakEM2 can handle masks exports. We then use Avizo to regenerate surface meshes from 3D voxel data. We can also apply mesh fixes from within Avizo.

- Using Blender's 3D printing add-on. This add-on performs non-manifold tests on a mesh topology. It also performs fixes.

To summarize, our cell 3D mesh acquisition stage in our analyses pipeline is as follows:

1. Domain experts proofread the segmentation data. This is where neuroscientists review the segmentation process by going through the whole stack and check if the targeted cell is equal to the assumed ground truth, that is, the EM image. This process is done usually with the same software tool that handles the segmentation.

2. Exporting 3D mesh/masks. Exporting a mesh allows us to proceed with the analysis stage. Segmentation masks can be exported when further refinements are needed. As the case with a non-manifold mesh, a step backward in the mesh generating process is required until we obtain a valid 3D model. Masks are exported in different file formats. A sequence of tiff images is one example.

3. Mesh cleanup:

   (a) Noise and artifacts: Manually delete any bits or sections that are identified as noise or isolated components during

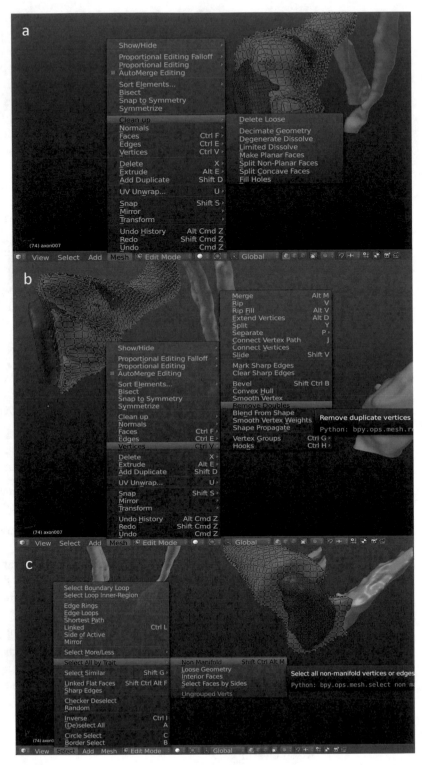

Fig. 6 Screenshots from Blender software showing options for fixing a defected mesh. Features are available one entering Edit Mode (by selecting the target object with right mouse click, then, press tab key to enter Edit Mode). (**a**) Showing available options under mesh cleanup. Delete Loose is most common. (**b**) Showing mesh topology issues fixes concerning vertices. (**c**) Showing non-manifold detection feature

the segmentation process. They are tedious to get rid of by the final stage of the segmentation, but could also be very small bits not easily identifiable by visual inspection in the pre-export stage; 3D modeling software tools, such as Blender and MeshLab can easily get rid of these bits once identified.

(b) Non-manifold and topology issues: Blender and MeshLab come with mesh cleaning functions that identifies non-manifold vertices and edges. With proper examination a mesh can be treated and recreated as a new modified model.

2.4 3D Models Analyses

Once 3D models are obtained from our EM stack, we proceed with annotating each neuronal object. This is done using NeuroMorph add-on with Blender [36].

The analysis context dictates the type of computational operations to execute against our objects. Under our two analysis examples: astrocytic glycogen and ER/mitochondria analysis, we quantify each object by executing a series of computational functions (Fig. 7). These computations include the following:

1. Geometrical properties (volume, surface area, cross-sectional area, dimension, and perimeter).

2. Statistics, such as a simple count on each organelle/cell type followed by assigning each group of counted objects to a category.

3. From 1 and 2, we extract more statistical measures such as mean, variance, and standard deviation.

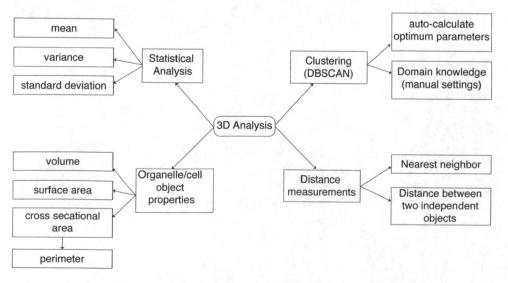

Fig. 7 Mind map of all analysis computations executed on 3D models

Such computations performed on the obtained mesh data requires to be automated and this is done via custom analysis tools. These tools should be made to handle such data and are optimized for faster processing time. With this intention, we seek to perform calculations where we have two types of measurements, each with six classes per type (Table 2). We also need to implement spatial clustering algorithms on targeted cellular organelle models, for example, Glycogen granules, and vesicles.

Table 1
List of technical resources

Resource/software	Web address
Blender	https://www.blender.org/
TrakEM2 (ImageJ)	http://imagej.net/TrakEM2
Scikit learn (DBSCAN)	http://scikit-learn.org/stable/modules/generated/sklearn.cluster.DBSCAN.html
Scipy	https://www.scipy.org/
Blender Wiki	https://wiki.blender.org/
Cross section script	https://wiki.blender.org/index.php/Extensions:2.4/Py/Scripts/System/CrossSection
MathWorks	https://www.mathworks.com/matlabcentral/fileexchange/9542-minimum-volume-enclosing-ellipsoid
NueroMorph	https://github.com/ajorstad/NeuroMorph
Glycogen Analysis	https://github.com/daniJb/glyco-analysis
Cython	https://cython.org
MeshLab	http://meshlab.sourceforge.net/
Anaconda 3	https://www.continuum.io/downloads
ER-Mito Analysis	https://bitbucket.org/garfy7/er-analysis
Java SE	http://www.oracle.com/technetwork/java/javase/downloads/index.html
ImageMagick	https://www.imagemagick.org/script/index.php
Avizo	https://www.fei.com/software/amira-avizo/
SteamVR	https://assetstore.unity.com/packages/templates/systems/steamvr-plugin-32647
VRTK	https://github.com/thestonefox/VRTK
Unity3D	https://unity3d.com/get-unity/download
AsImpL	https://github.com/gpvigano/AsImpL
HSV-Color-Picker	https://github.com/judah4/HSV-Color-Picker-Unity

For this reason, we exploit three analysis tools: *NeuroMorph* [36], *Glycogen Analysis* [7], and *ER-Mito Analysis* (Table 1). Detailed user case scenarios of Glycogen Analysis and ER-Mito Analysis will be demonstrated in Subheadings 3 and 4.

2.4.1 Distance Measurements and Nearest Neighbor Lookup

Obtaining intracellular spatial information of organelles from our 3D models requires the implementation of precise distance functions between two objects of interest. An efficient nearest neighbor search in multidimensional space is necessary to obtain such information. A brute force strategy would be enough if we were to deal with simple data morphologies; however, neuronal data have demonstrated to be quite complex when evaluated in 3D meshes. For instance, a single astrocyte would comprise more than 1,000,000 vertices in 3D space. For an average computer, the processing of such numbers takes up more than 20 min to loop through every vertex in the mesh. Our need for optimization led us to make use of binary trees spatial algorithms. For our distance functions, the K-d (K-dimensional) tree data structure created by Bentley [44, 45] has demonstrated in all our use cases to be a powerful structure that serves our need.

Python library packages such as scipy-spatial [46] and scikit-learn [47] has nearest neighbor lookup algorithm libraries with KDTree data structure implementation. The KDTree under scipy-spatial has an average complexity of $O(n \log n)$ with respect to sample size [48] when computing distances in 3D for large numbers of vertices. Furthermore, the cKDTree, which is a C implementation of the KDTree for quick nearest-neighbor lookup, has demonstrated to be faster when implemented against our data. By using the cKDTree class library, and incorporating that into Blender software by converting them to add-ons [7], we were able to efficiently obtain measurement figures from our reconstructed 3D models.

We perform distance calculations based on either the centroid of the target mesh or all its vertices.

With centroid-based measurements, our add-on reads-in all vertices that comprise each cell/organelle mesh object. Next, it computes the median in 3D space, which gives us the centroid of that individual mesh. Centroids are then carried on to the nearest neighbor lookup function, which in turn gives the resulting output as a list containing: a floating point number representing the distance in microns, and the names of both, the cell/organelle of interest and its nearest neighbor.

The all-vertices-based measurements reads-in all vertices that comprise each cell/organelle mesh object into a single matrix and then sends the constructed matrix to the nearest-neighbor function. The output is a list of nearest neighbor vertices where each vertex from the cell/organelle mesh of interest has a nearest neighbor vertex from the other targeted cell/organelle. Furthermore,

Table 2
ER-Mito analysis nearest neighbor lookups

Object/s of interest		
A single group of cells/organelles:	*Against*	Everything else
A single cell/organelle:		Another specific group/cell/organelle
		Multiselected groups/cells/organelles

the minimum distance is then evaluated among all vertices per mesh object of interest resulting in one single value per mesh to represent the closest distance in microns. The final output will be with similar headings to the centroid-based measurements.

The ER-Mito analysis add-on offers a number of options for customizing the parameters for nearest neighbor lookup (Table 2).

2.4.2 Spatial Clustering with DBSCAN

DBSCAN (density based spatial clustering algorithm with noise) is an elegant approach for discovering arbitrary shaped clusters within a finite set of spatial data. It was invented by Ester [49] and is one of the most cited in research and most popular method for clustering objects in Euclidean space. It was created to handle spatial data, especially with space being multidimensional where $d \geq 3$. Density based clustering algorithms form clusters with arbitrary shapes, unlike other k-means methods that are built to return ball-like clusters [50]. The DBSCAN implementation requires some domain knowledge of the data and space. Hence, controlling it is achieved with the value of two main parameters:

- ε: a positive real value representing the radius of an area relative to data space.

- *MinPts*: a small constant positive integer representing the minimum number of points relative to the data.

The main idea in DBSCAN is that for each point of a cluster, the area of a given radius ε has to contain a number of points equal or bigger than *MinPts*. This is so that the density in that area has to exceed some threshold. Moreover, if we have a ball shaped d-dimensional area with a radius ε and centered at point p; hence, the area is denoted by $B(p, \varepsilon)$. $B(p, \varepsilon)$ is considered "dense," if it contains a number of points P, where $P \geq MinPts$. Therefore, with DBSCAN, if $B(p, \varepsilon)$ is dense, then all points in $B(p, \varepsilon)$ belong to the same cluster as p, which is the centroid. The cluster of p will continue to grow accordingly [50].

Our Glycogen granules dataset is a good use case for DBSCAN, where we have developed a specialized analysis tool for that purpose. The Glycogen Analysis add-on implements DBSCAN spatial clustering via the sklearn.cluster python library. Subheading 3 demonstrates Glycogen Analysis user interface, as well as a step-by-step guide on how to obtain analysis results based on domain knowledge of the data, such as size and nature.

Automating Parameter Calculation

One of the main features of Glycogen Analysis is that it calculates the optimum value of ε, for a given set of points against their generated clusters. This is achieved by calculating the *silhouettes coefficient* [47, 51], a number whose value represents a metric for the goodness-of-fit for the DBSCAN parameters p and ε. The higher the silhouettes score, the better the resulting clusters. We can define this operation by the following:

For a range of ε values specified by the domain expert user, and for a given set of points with their clusters for each ε value, we calculate the silhouettes coefficient value.

Subheading 3 will demonstrate how Glycogen Analysis implements DBSCAN parameter calculation for an optimum ε value.

2.4.3 3D World Objects Geometrical Properties

We get to define our data in 3D by evaluating them geometrically. Glycogen Analysis and NeuroMorph are tools that allow us to apply such operations.

We quantitatively analyze every model, in terms of 3D space geometry by evaluating the following metric properties:

- Cross-sectional area obtained with ER-Mito Analysis.
- Perimeter of a cross section obtained with ER-Mito Analysis.
- Volume, obtained with NeuroMorph.
- Surface Area, obtained with NeuroMorph.
- Dimensions of targeted cells, automated in Glycogen Analysis.

2.5 Immersive VR and Interactive 3D

Virtual reality (VR) is an expanding field, which is rapidly increasing within the gaming industry. At the beginning of 2016, we published a pioneering approach combining VR to neuroscience, for the purpose of enhancing accessibility and visual assessment of complex 3D models from brain dense reconstructions [7]. In particular, we describe how the use of a CAVE system was helpful to formulate the hypothesis of nonrandom distribution of glycogen granules, by staring at the cloud of points floating around a user immersed within a 3D space filled with a sample whose original size is of less than 10 μm length.

This approach was very powerful, and its usefulness extends to any kind of study where the accessibility of a sample is limited by its size or complexity, but its structure can be reconstructed and rendered by using 3D software; nevertheless, the use of CAVE requires the availability of the system, which is expensive and available in only few institutions all over the world [52]. For this reason, the use of more portable lightweight systems and head mounted displays (HMD), for example, Oculus Rift and HTC Vive, are two excellent alternatives, allowing visualization scientists, computer

Fig. 8 Screenshot of the main scene from Virtual Mente VR application

scientists, and neuroscientists to collaborate in developing software environment for improved visualization, and more importantly interactivity in the 3D space. Cross-platform software like Unity allow for immediate feedback and proofreading of codes using portable VR goggles, and the same software could be later built and run on other platforms including CAVE but also mobile devices.

While VR is already being exploited for educational purposes by our and other institutions and developers, we are set to extend the analysis tools to perform measurements not only in 3D but also in a VR environment, by improving the interactivity with the reconstructed model. In light of this, we developed two projects, specifically for virtual reality: *VirtualMente*, and *VR Data Interact*.

VirtualMente was developed with the idea of animating the ANLS (Astrocyte Neuron Lactate Shuttle) [53] process using models reconstructed from an EM stack. The 3D models included are five (a blood vessel, three astrocytes, and one neuron) (Fig. 8) that are loaded as constant assets in the main scene. The user is guided through a storyboard explaining the role of synapses, astrocytes, and neurons during learning and memory formation. In addition, with the aid of a joystick controller, and an Oculus Rift HMD, users can freely explore the entire scene and, most importantly, examine our reconstructed 3D models in an immersive VR space.

VR Data Interact is another virtual reality application that was created using Unity development environment and supported by SteamVR, and VRTK libraries (Table 1). Much like the typical analysis framework practiced by neuroscientists in a standard desktop setup [7], VR Data Interact allows for explorative analysis tasks to take place, such as proofreading and driving hypotheses concerning possible correlations between cellular structures in the reconstructed EM data. The application requires OBJ files as 3D models, to be visualized in an immersive VR space. Developing a VR system dedicated for neuronal data analysis required then a user case study, to test the usability of the application from the perspective of domain expert users. We will present ours, as an example, in Subheading 2.5.1. Moreover, collaborative analysis can be achieved with VR Data Interact, by incorporating a High Level API (HLAPI) in the application, that allows for multiuser networking setups to take place. With this feature, multiple instances of the application can run on different machines, in sync, and free from any physical space constrains.

To this end, autostereoscopic light field displays are an exciting and promising alternative for collaborative analysis, since they enable multiple untracked naked-eye users in a sufficiently large interaction area to coherently perceive rendered scenes as real objects, with stereo and motion parallax cues [54, 55]. They have been successfully tested for natural real-time interaction with high-resolution surface models [56], and CT-based volumetric data, and we plan to evaluate their depth discrimination capabilities in the context of morphometric analysis of EM high-resolution data.

2.5.1 VR Data Interact Usability Study

VR Data Interact is a virtual reality software application that was developed using Unity game engine. It allows the end user to load EM reconstructed data in OBJ formats. Data interaction is represented in selecting 3D neurite models while interactively scrolling through the corresponding 2D EM image from the original stack (Fig. 9). This method of interaction with the data is useful for data validation as well as performing correlative visual analysis. Domain scientists already practice this approach using typical desktop setups in monoscopic view [36]. On the other hand, our case study investigates the impact of introducing the VR element into this

Fig. 9 A screenshot from VR Data Interact application showing a 3D EM image super imposed on a transparent glia 3D model. The EM slice can be interacted with using the Vive controllers by moving it up and down

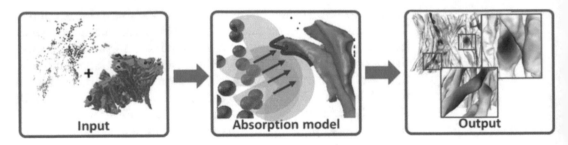

Fig. 10 Block diagram of the absorption model representing stages to obtain a glycogen derived color map

analysis approach. In other words, we are looking to compare monoscopic and stereoscopic setups and to conclude if the latter supports correlative visual analysis better or not.

In [57], a visualization technique was proposed that supports hypothesis formation on astrocytic glycogen with respect to its surrounding neuritis, specifically, during synaptic development. This technique is based on energy absorption mechanisms computed out of its radiance transfer, where energy sources come in place of glycogen granules (Fig. 10).

In the light of this, we employed our VR Data Interact application to serve the visual analysis of absorption maps and energy peaks recognition in the final computed model (the absorption model). Hence, evaluating this methodology of explorative analysis and visualization is carried out via the proposed comparative study.

Moreover, planning this study was based on four directives that will be explained in the next subheadings.

Study Objective and Method

Considering the early stages where VR technologies exist in this domain of analysis, the user study was meant to be subjective at evaluating the effectiveness of VR in performing visual analysis of glycogen derived absorption models [57]. We achieve this by comparing two setups: a desktop system with monoscopic view and a head-mounted display setup with stereoscopic view.

We aim to measure user satisfaction upon using both of our systems, via a hybrid of two evaluation methods: questionnaires (Table 3) and open-ended questions. We adapted our first evaluation method from the System Usability Questionnaire (SUS) [58]. As with open-ended questions, they were found to be quite useful for collecting user's feedback and have them comfortably talk about their experience in each session.

Participants Group

The study was performed with a group of 27 staff members who were potential users of our system that volunteered to participate in the study. Participants were a mixture of domain scientists and research staff with varying experience in neuroscience, technology and VR. Our user selection criteria were put based on the application context, time availability and study objective. In addition, any

Table 3
VR Data Interaction SUS derived questionnaire [58]

Questions	Likert scale: 1 = Strongly Disagree, 2 = Disagree, 3 = Neutral, 4 = Agree, 5 = Strongly agree				
Q1: I think I would use this application frequently	1	2	3	4	5
Q2: I found the application unnecessarily complex	1	2	3	4	5
Q3: I think I would need the support of a technical person to be able to use this application	1	2	3	4	5
Q4: I like using the interface of this application	1	2	3	4	5
Q5: I believe most people would learn to use this application very quickly	1	2	3	4	5
Q6: I felt very confident using the application	1	2	3	4	5
Q7: I needed to learn a lot of things before I could get going with this application	1	2	3	4	5
Q8: The information (e.g., menu) provided by the application is clear and helpful	1	2	3	4	5
Q9: I felt difficult interacting and controlling the system	1	2	3	4	5

type of impairment that could prevent users from experiencing the nature of VR is one criterion for user exclusion. Furthermore, collecting user demographics data can play a huge role when concluding the study results, which can be achieved via a demographics questionnaire given at the beginning of the trial.

Ethics

Prior to the study, participants were given a brief explanation on how the VR system functions, highlighting possible complications or potential hazards that could rise from interacting with virtual reality space. This includes nausea, motion (cyber) sickness, and fatigue. During the trial session, it was essential that participants were constantly checked upon, by repeatedly asking them if they were okay. Upon any feeling of discomfort, the participant was asked to stop. Furthermore, participants were assured with their personal rights and the confidentiality of their data, via a consent letter that was obtained via an online form, a copy which is forwarded to each participant.

Pilot Study

We started off with a preliminary pilot study to assure consistency of the design for the final formal study. During which we made sure of the availability of resources, technical support and proper functionality of all hardware equipment. The pilot study involved six participants [59]. Conducting a pilot study helped in selecting proper data samples (3D neural models) and to evaluate important factors concerning data such as quantity, level of complexity, and

Fig. 11 VR Data Interact usability study space setup

possible artifacts. The results from the pilot study helped in reshaping our method and performing alterations to the study design.

Instrumentation and Protocol

For preparation, we set up our VR system on a Supermicro desktop machine running Windows 10 Professional (hardware specifications are listed in Subheading 5). In addition, we took advantage of a portable immersive environment setup, represented in the HTC Vive system. This study took place in the visualization lab facility of King Abdullah University of Science and Technology (Fig. 11).

For the formal study, we implemented the following protocol:

As a first step, participants were given instructions on the procedure they would need to follow in order to successfully complete all tasks. Each participant was given a demographics questionnaire to fill, accompanied by knowledge questions asking them to rate their own acquaintance with technology in general and virtual reality and neuroscience in specific. Questions were on a scale from 1 to 5.

We followed with a short introduction explaining the absorption model, including the definition of what a peak is and how it is recognized on an absorption model. Following the introduction was a small demo serving as a guide for using our software tools. The demo here served as a simple use case scenario. During which participants were asked to watch and learn as the person who is conducting the study demonstrates software and hardware usage.

Participants were required to go through two designed sessions corresponding with each of our software setups: a desktop application represented in Blender and NeuroMorph running in monoscopic view, and a desktop VR application represented in VR Data Interact incorporating an HMD system (HTC Vive) running in stereoscopic view. Considering the lack of familiarity with VR as the case for most participants, extra care was taking into account when initiating VR sessions. Before the VR session, participants were given the chance to get familiar with the Vive hardware, the two controllers and a headset. They were encouraged to try them on for fitting as well as get their hand adjusted to the grip of the Vive controllers, for example, to properly position their fingers to reach all buttons. The Vive is equipped with two identical controllers, one for each hand. As part of VR Data Interact design, each Vive controller is programmed differently. One controller is meant for interacting with objects, such as moving items around or clicking on menu buttons, and the other dedicated for user navigation, that is, the virtual physical movement around the entire virtual scene. For this reason, participants were given some time to feel comfortable on to which controller they want to assign to which hand; for example, right-handed users tend to prefer navigating with their left hand while interacting with objects using the other.

Each session required the completion of two tasks involving the absorption model: Task 1 involved counting the peaks on a randomly selected absorption model, whereas task 2 required conducting preliminary visual analysis on the same 3D model by correlating one of the peaks they identified from task 1, with the surrounding cellular structures identified on the corresponding 2D EM image.

The main data source used in this study was from a stack acquired from rat hippocampus, from which we created a set of five 3D neurite models. At each round session for each participant, 3D models were picked randomly.

At the end of each session, participants were asked to fill a 9-item questionnaire derived from [58]. The questionnaire contained 9 statements in which participants are required to select from a Likert scale of 1–5, with 1 = strongly disagree, and 5 = strongly agree (Table 3). At the end of each session, we asked open-ended questions to the participants, in order to identify the best and worse features they encountered with each system as well as identify possible missing features they strongly felt that should be there. The last user feedback is considered beneficial for future software developments and enhancements. Moreover, user feedbacks were recorded constantly. As they engage with each task, they were encouraged to think out loud and express their thoughts. The total time required for completion was approximately 30–40 min per participant.

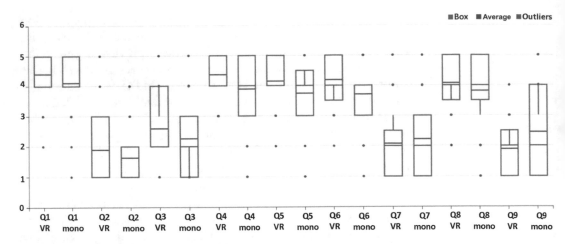

Fig. 12 Results from the comparison study between stereo and mono setups represented in a box-plot chart. The chart shows comparison between Likert scores of the usability questionnaire

Considerations

The results obtained from the nine-item questionnaire can be represented as a compact box plots. As an example, the results of our case study are presented in Fig. 12. Usability scores are similar between the two setups, with a slight leaning toward the virtual reality setup. All users found that the immersiveness that virtual reality has to offer served them best when handling exploratory analysis tasks.

On the negative side, out-loud comments highlighted expected issues that normally rise with VR, in particular lost in space sensation, cyber sickness, and fatigue. Importantly, domain experts preferred to use VR only on specific targeted models, because of the higher resolution of the VR setup combined with the 3D models, compared to the 2D EM image, with better spatial discrimination.

As a conclusion, the user study results show that VR can help in visual analysis and serves slightly better in highlighting certain details in specific 3D neural models.

3   Use Case Scenario I: Glycogen Granule Analysis

From an energy perspective, distribution of glycogen granules has been observed as an area of interest. This is because granules store glucose which is considered a precursor of lactate, which neurons use as an energy substrate. We hypothesized that the 3D localization of the granules is not random; investigating their distribution within astrocytes is a way to determine whether they are more likely to interact with pre- or postsynaptic elements, boutons or spines, respectively. With the aid of Glycogen Analysis and NeuroMorph add-ons within Blender environment, we execute the DBSCAN clustering algorithm on our granules dataset and then visualize the results.

For Glycogen Analysis to work properly, there are few prein-stallation steps that need to be implemented beforehand. That involves making sure dependencies are properly installed in the system. In addition, software and hardware specifications mentioned in Subheadings 5 and 6, are as important in order for the analysis functions to execute smoothly.

The following steps summarize the whole process of dependencies and environment set up on a Mac OS or Linux system:

1. Install Anaconda3 with python3.4. *See* Table 1 for download links.

2. Once installation is successful, start the system's command line terminal, and update environment variables paths for $PATH and $PYTHONPATH, as follows:

   $ export PATH = $anaconda_installation_directory/bin: $PATH.

   $ export PYTHONPATH = $anaconda_installation_directory/lib/python3.4/site-packages:$PYTHONPATH.

3. Install Blender 2.76 (Table 1).

4. Perform a small test to see if the external anaconda python libraries can be reached correctly from within Blender with the following:

   From command line launch Blender's python environment by running the following:

   $. /blender --python-console

   Then, type the following commands below to import the required libraries, which should not result in any error message:

   >>>import scipy

   >>>import sklearn

   >>>from scipy.spatial import cKDTree, distance

   >>>from scipy.stats import itemfreq

   >>>from sklearn import metrics

   >>>from sklearn.cluster import DBSCAN

5. When all libraries imports are successful, start Blender from terminal by navigating to the Blender Unix executable and launching it with:
   $. /blender.

6. Finally, to install Glycogen Analysis and NeuroMorph, download their python scripts from the links in Table 1. Then install them from Blender by lunching the user preferences window from File → User Preferences, select the add-ons tab, and then click on "Install from File." A file navigation window will open.

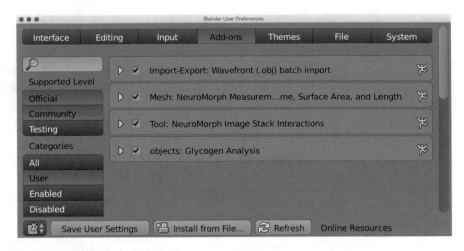

Fig. 13 Blender user preferences window

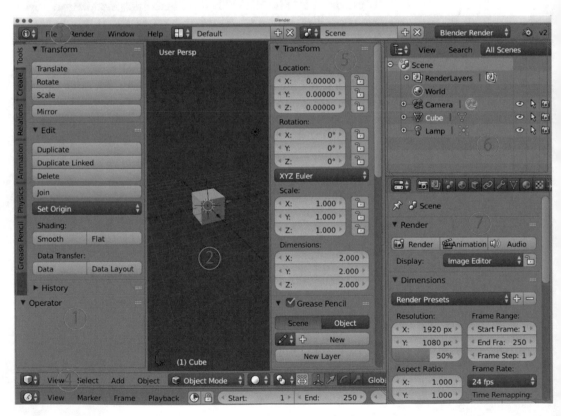

Fig. 14 Blender user interface elements. (1) Tool shelf, (2) 3D view main region, (3) info, (4) Header, (5) Properties region, (6) Outliner, (7) Properties

Navigate to where the two scripts are located, and then click on "install from file." Select the checkbox for each one and then exit Blender preferences (Fig. 13). Glycogen Analysis will be located in Blender's properties region (number 5 in Fig. 14).

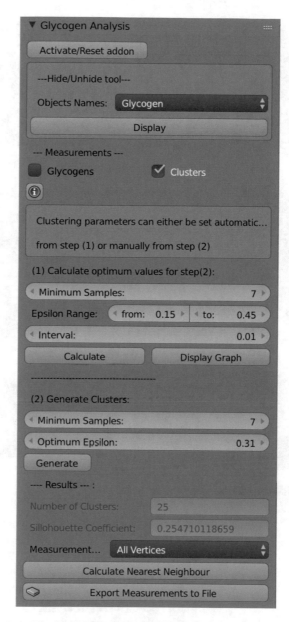

Fig. 15 Glycogen Analysis add-on user interface

While NeuroMorph will be located at the Misc tab in Blender's tool shelf (number 1 in Fig. 14). It is worth mentioning that NeuroMorph is required for the operation of Glycogen Analysis.

Figures 15 and 16 will demonstrate a step-by-step process on using Glycogen Analysis to perform clustering computations on our glycogen data followed by nearest neighbor lookup between

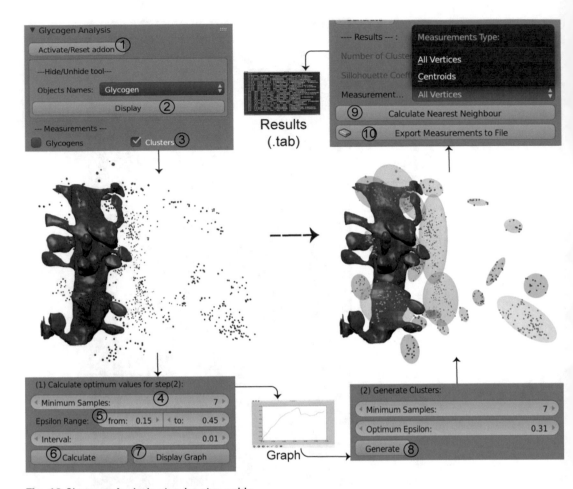

Fig. 16 Glycogen Analysis step-by-step guide

each cluster and the nearest synaptic element or cellular process of interest. These objects include pericytes, endothelial cells, axonal boutons, and dendritic spines. The latter two are features belonging to axons and dendrites, that are annotated and measured using NeuroMorph, whose functionality is well documented in a previous work [36]. In this analysis, all measurements are based on an "all-vertices" pattern, and depend on annotations, and surface area and volume functions from NeuroMorph.

Glycogen Analysis graphical user interface (GUI) is captured in Fig. 15. However, for clarity, we split up the GUI into subpanels that correspond to each analysis step (Fig. 16). The Glycogen Analysis use case scenario is laid out in a list of steps where analysis functions are executed accordingly and in that particular order. The numbers assigned to each panel on Fig. 16 correspond to that list, and they are as follows:

1. Upon startup, click on the Activate/Reset button. This will activate the add-on where an initialization function will be

called. This means everything will be deselected in the 3D-view window and scene variables will be initialized to zero/null. It is considered a useful function whenever a new analysis operation is initiated.

2. The Hide/Unhide tool will be initialized with all categories that group all objects in the file by their name. Select "Glycogen" (the naming convention for granule objects) from the dropdown list and click "Display." Glycogen granules should be displayed and selected in the 3D view window.

3. Check the box next to the option "clusters" to reveal all operators and parameter fields for the DBSCAN function. This panel allows for an automatic calculation for the optimum DBSCAN parameter value (epsilon) by using the silhouettes coefficient. The numerical values shown in 4, 5 in Fig. 16 are relative to the size of the stack and the glycogen granule count.

4. Input an integer value for Minimum Samples (e.g., 7).

5. Input range values for computing the epsilon (e.g., 0.15–0.45; the interval between loop cycles should be 0.01 in that case).

6. Once all input fields in the clusters panel are filled, click on "Calculate." This should take a few seconds depending on the size of the data. Progress of the computations during runtime should be visible in the terminal window. Note that during runtime, Blender will be locked from any user interaction.

7. Once the calculations finish, Blender will be released again. You can click on the "Display graph" to display a chart of all silhouettes values against epsilon range values. The maximum silhouette is our optimum epsilon. Those will be put automatically into the main fields in the "Generate Clusters" subpanel.

8. Click on "Generate" to begin modifying granules objects in the 3D view window, by assigning different mesh colors to each granule where it is correlated with the generated clusters. In addition to color codes, the minimum enclosed ellipsoid will be drawn to visualize our clusters. You will notice new objects added to the outliner list region with the naming convention "ellipsoid[1:n]". Progress of the computations will be displayed in the terminal window.

9. After the clusters are generated successfully, choose the option "All Vertices" from the dropdown menu to select a measurements type, and then click on "Calculate Nearest Neighbor" to execute the distance calculations.

10. Once the calculations are done, you will see the results displayed in the terminal window. Following a successful calculation, the "Export Measurements to File" button will be activated. Click "Export" to save the analysis results onto disk in tab delimited format.

4   Use Case Scenario II: ER and Mitochondria Analysis

Another analysis use case involves measuring distances between the ER in astrocytes and synapses. The analysis extends into finding the nearest ER mesh to each synapse and then calculating the cross-sectional area of that particular ER. The purpose of this analysis is to obtain the intracellular spatial information, allowing for modeling the mechanisms of Ca^{2+} release stored into the ER in astrocytes. In addition, mitochondria's spatial localization has been used as a marker for energy consumption site. In this use case scenario, we will target mitochondria and synapses as our two main objects of interest.

Installation and environment setup will be similar to Glycogen Analysis in Subheading 3. When activating ER-Mito Analysis add-on, you will find the panel located in an independent tab labeled "ER Analysis" in Blender's tools shelf (number 1 in Fig. 14). Table 1 references all links related to ER-Mito Analysis add-on including its dependencies.

Like in Subheading 3, ER-Mito Analysis user interface is displayed in Fig. 17a, while Figs. 17b and 18 will include subpanels with numbers to refer to each analysis step in the use case scenario. Correspondingly, the steps for using the ER-Mito Analysis add-on to perform nearest neighbor lookup between mitochondria and its closest synapse are as follows:

1. Click on the "activate/refresh" button to initialize the environment. This gives the same effect as the activate function in Glycogen Analysis.

2. Select the synapses objects from the dropdown list; the layers list will be updated automatically with the current layer(s) that contain all synapse objects. To control displaying and selecting of synapses click on the radio button (layers button) located on the left most bit of the three aligned buttons. The mouse arrow button (selection button) will select all the synapses and the "Move" button will relocate them to another layer. "Move" will work if you specify a different number than the current one in the layers dropdown list. This is the manual sort of objects and layers.

3. Another option for sorting objects into layers is the "Auto assignment" function (number 1 in Fig. 18). This will cause the add-on to automatically handle all objects by selecting them and moving them to one unique layer (if available) or a shared layer if all 12 layers were occupied. It sorts objects according to their categories.

   If the total number of categories extracted from objects exceeds 12 (the total number of layers in Blender), then category 1 and 13 will share layer 1, and so on.

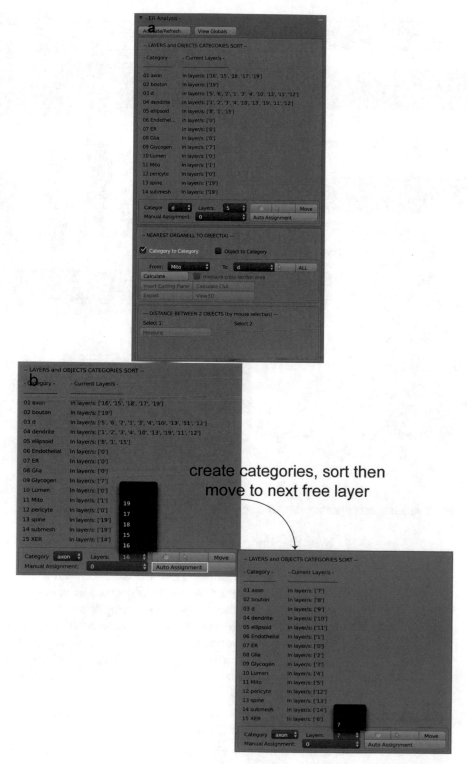

Fig. 17 ER-Mito Analysis. (**a**) Add-on user interface. (**b**) Layerwise moving tool

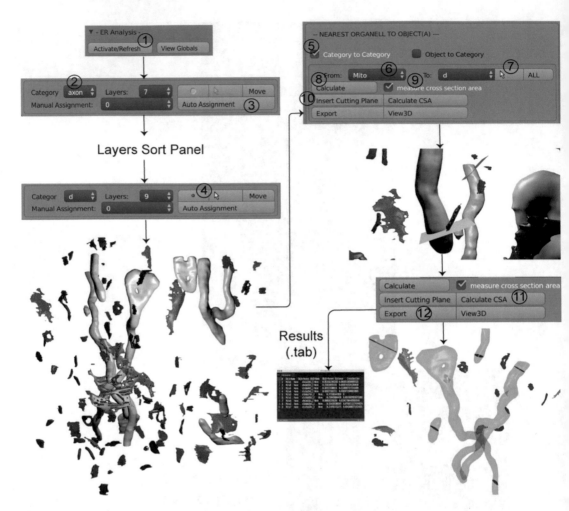

Fig. 18 ER-Mito Analysis user interface step-by-step guide

4. Now that our objects are sorted (Fig. 17b), click on the layers button then the selection button again to display the synapses in the 3D view window. This subpanel allows for the selection and display of more than one object by selecting a different category from the "Category" dropdown list. **Steps 4–7** insures that all objects of interest are selected and displayed in the 3D view window, as this is an essential step to do before executing the nearest neighbor look up functions. You must make sure that no other layers/objects are selected except for the targeted ones.

5. Select "Category to Category" option to execute a nearest neighbor look up for mitochondria and synapses.

6. The "From" list will contain all categories obtained from our dataset. Select "Mito" to indicate the origin points of measurements.

7. The "To" list is a replica of the layers sort panel dropdown list. If you selected synapses previously then this should be set automatically. The "To" value indicates the ending points of measurements. Click on the selection arrow button if it is not active.

8. Click on "Calculate" to launch nearest neighbor look up functions. Progress will be displayed in the terminal window.

9. When the calculations finish, the "measure cross section area" option will be active. This gives us the option to further calculate the cross-sectional area for each mitochondrion. If checked, it will activate two buttons, "insert cutting plane" and "calculate CSA."

10. Measuring cross-sectional areas is done in two steps. First step, you click on the "insert cutting plane" button. This will cause plane objects to be created (will show in the outliner region as plane[1:n]) and intersecting (cutting) each mitochondrion. The location and slope of the plane is calculated according to the nearest neighbor vertex belonging to each mitochondrion mesh as resulted from **step 8**. This step should be revised visually to ensure all cutting planes are at a correct angle relative to its closest synapse. Modifications can easily be made manually within Blender's 3D view by right clicking with mouse on the desired plane to select it. Then use keyboard shortcuts: r + [$x|y|z$] to indicate the axis of rotation, then move mouse to initiate rotation.

11. Once all planes are positioned correctly, the second step is clicking "Calculate CSA." This will involve using the planes to create an actual sliced section object. A set of new objects representing the cross-sectional area from each mitochondrion will be added to the outliner as CSA[1:n]. Figure 18 displays transparent mitochondria to show the cross-sectional objects within. Analysis results will be shown as output in the terminal window.

12. To save your analysis results, click on "Export" for Blender's files window to open. Navigate to a directory, and then type in the file name. Click on "Write data to file" and you will have a tab-delimited file saved to the disk.

5 Hardware

This section will provide overview information on hardware specifications and software setup used in our lab. Our computer systems include Mac OS, Linux, and Windows.

For Linux operating system, we use Supermicro workstations running scientific Linux 6.8, GNOME 2.28.2, with the following hardware specifications:

1. Fast access RAM with 16×8 MB DDR3 1600 MHz Samsung (total 128 GB).

2. NVIDIA K2000 GPU.

3. $2 \times$ Intel(R) Xeon(R) CPU E5–2680 v2 at 2.80 GHz (each processer has ten cores with a total of 20 threads).

4. 2 TB Seagate Constellation ES.3 ST2000NM0033 2 TB 7200 RPM 128 MB Cache SATA 6.0Gb/s 3.5″ Enterprise Internal Hard Drive Bare Drive.

5. Supermicro X9DA7/E motherboard.

6. High-speed gigabit network, all data are stored on NetApp FAS8080 filer using NFS.

For Mac operating system, we have two machines, a MacBook Pro 2015 and a Mac Pro 2015 workstation, with the following specifications:MacBook Pro:

(a) Intel Core i7, 2.5 GHz CPU with four cores.

(b) 16 GB DDR3 RAM.

(c) AMD Radeon R9 M370X GPU.

(d) 512 GB Apple Solid State Drive (SSD) SM0512G.

Mac Pro workstation:

(a) 8 Core-Intel Xeon E5 3 GHz CPU.

(b) 48 GB DDR3 RAM.

(c) AMD FirePro D700 6 GB GPU.

(d) 1 TB Apple Solid State Drive (SSD).

For Windows operating system, we have a Supermicro workstation with the following hardware specifications:

1. $2 \times$ Intel(R) Xeon(R) CPU E5-2667 v4 at 3.20 GHz (each processor has eight cores)

2. NVIDIA Quadro M6000.

3. 240 GB INTEL SSDSC2BB240G6.

4. Fast access RAM with total 1 TB memory.

It is important to know that hardware requirements are variable, as data varies in size and resolution. In this context, data specifications are fixed at: 220 μm^3 volume (7.07 μm × 6.75 μm × 4.75 μm) at a voxel resolution of 6 nm. The whole stack size on disk is 997.3 MB. Consequently, hardware specifications are customized accordingly.

6   Software

Our computer lab comprises Mac OS, Linux, and Windows machines. For Mac operating system, we have a MacBook Pro 2015 machine running OS × Yosemite (10.10.5) and a Mac Pro 2015 workstation running OS × El Capitan (10.11.6). For Linux operating system, we use Supermicro workstations running scientific Linux 6.8, GNOME 2.28.2. For Windows, we use a Supermicro workstation running Windows 10 Professional.

In addition, our analysis environment is equipped with the following software tools/dependencies (*see* Table 1 for download links):

1. Python 3.4.5 can be obtained easily using Anaconda distribution. Python is considered the backbone of most of our analysis tools and it is available across all three operating systems. With Anaconda, python packages can be easily obtained from command line. There is a list of python packages that are necessary to be installed as a prerequisite for the analysis tools to work properly, and they are as follows:

   (a) Numpy.

   (b) Scipy.

   (c) Scikit-learn.

   (d) Matplotlib.

   (e) Mathutils.

   (f) Math.

   One example to install scikit-learn library using Anaconda: from command line, you run the following:

       $ conda install scikit-learn.

2. Java SE 7 or JRE is an essential requirement for running Fiji (ImageJ) (Linux/Mac/Windows).

3. ImageMagick, is useful to configure/view the stack 2D images (Linux).

4. Blender 2.76 (Linux/Mac).

5. Fiji (ImageJ) (Linux).

6. Unity 5.6.3 pro (Windows OS).

7. Microsoft Visual Studio Pro (Windows OS).

8. Steam/VR client application (Windows OS).

References

1. Parekh R, Ascoli GA (2013) Neuronal morphology goes digital: a research hub for cellular and system neuroscience. Neuron 77:1017–1038. https://doi.org/10.1016/j.neuron.2013.03.008

2. Coggan JS, Keller D, Calì C et al (2018) Norepinephrine stimulates glycogenolysis in astrocytes to fuel neurons with lactate. PLoS Comput Biol 14(8): e1006392

3. Borrett S, Hughes L (2016) Reporting methods for processing and analysis of data from serial block face scanning electron microscopy. J Microsc 263:3–9. https://doi.org/10.1111/jmi.12377

4. Lichtman JW, Pfister H, Shavit N (2014) The big data challenges of connectomics. Nat Neurosci 17:1448–1454. https://doi.org/10.1038/nn.3837

5. Neuro Cloud Consortium. Electronic address: jovo@jhu.edu, Neuro Cloud Consortium (2016) To the cloud! A grassroots proposal to accelerate brain science discovery. Neuron 92:622–627. doi: https://doi.org/10.1016/j.neuron.2016.10.033

6. Vogelstein JT, Amunts K, Andreou A et al (2016) Grand challenges for global brain sciences. arXiv 2016:q-bio.NC

7. Calì C, Baghabra J, Boges DJ et al (2016) Three-dimensional immersive virtual reality for studying cellular compartments in 3D models from EM preparations of neural tissues. J Comp Neurol 524:23–38. https://doi.org/10.1002/cne.23852

8. Agus M, Boges D, Gagnon N, Magistretti PJ, Hadwiger M, Calì C (2018) GLAM: Glycogen-derived Lactate Absorption Map for visual analysis of dense and sparse surface reconstructions of rodent brain structures on desktop systems and virtual environments. Comput Graph 74:85–98

9. Coggan JS, Calí C, Keller D et al (2018) A process for digitizing and simulating biologically realistic oligocellular networks demonstrated for the neuro-glio-vascular ensemble. Front Neurosci 12:664

10. Boges D, Calì C, Magistretti PJ, Hadwiger M, Sicat R, Agus M (2019) Virtual environment for processing medial axis representations of 3D nanoscale reconstructions of brain cellular structures. 25th ACM Symposium on Virtual Reality Software and Technology, 1–2

11. Rich L, Brown AM (2016) Glycogen: multiple roles in the CNS. Neuroscientist. https://doi.org/10.1177/1073858416672622

12. Vezzoli E, Calì C, De Roo M et al (2019) Ultrastructural evidence for a role of astrocytes and glycogen-derived lactate in learning-dependent synaptic stabilization. Cereb Cortex 30(4):2114–2127. https://doi.org/10.1093/cercor/bhz226

13. Beyer J, Al-Awami A, Kasthuri N et al (2013) ConnectomeExplorer: query-guided visual analysis of large volumetric neuroscience data. IEEE Trans Vis Comput Graph 19:2868–2877. https://doi.org/10.1109/TVCG.2013.142

14. Denk W, Horstmann H (2004) Serial block-face scanning electron microscopy to reconstruct three-dimensional tissue nanostructure. PLoS Biol 2:e329. https://doi.org/10.1371/journal.pbio.0020329

15. Knott G, Marchman H, Wall D, Lich B (2008) Serial section scanning electron microscopy of adult brain tissue using focused ion beam milling. J Neurosci 28:2959–2964. https://doi.org/10.1523/JNEUROSCI.3189-07.2008

16. Hayworth KJ, Kasthuri N, Schalek R, Lichtman JW (2006) Automating the collection of ultrathin serial sections for large volume TEM reconstructions. Microsc Microanal 12:86–87. https://doi.org/10.1017/s1431927606066268

17. Titze B, Genoud C (2016) Volume scanning electron microscopy for imaging biological ultrastructure. Biol Cell 108:307–323. https://doi.org/10.1111/boc.201600024

18. Seymour K-B, Mike R, Narayanan K et al (2013) Mojo 2.0: Connectome Annotation Tool. Front Neuroinform. https://doi.org/10.3389/conf.fninf.2013.09.00060

19. Peng H, Hawrylycz M, Roskams J et al (2015) BigNeuron: large-scale 3D neuron reconstruction from optical microscopy images. Neuron 87:252–256. https://doi.org/10.1016/j.neuron.2015.06.036

20. Calì C, Kare K, Agus M et al (2019) A method for 3D reconstruction and virtual reality analysis of glial and neuronal cells. J Vis Exp

21. Calì C, Agus M, Kare K, Boges DJ, Lehväslaiho H, Hadwiger M, Magistretti PJ (2019) 3D cellular reconstruction of cortical glia and parenchymal morphometric analysis from Serial Block-Face Electron Microscopy of juvenile rat. Prog Neurobiol 183:101696. https://doi.org/10.1016/j.pneurobio.2019.101696. Epub 2019 Sep 21

22. Liu T, Jones C, Seyedhosseini M, Tasdizen T (2014) A modular hierarchical approach to 3D electron microscopy image segmentation. J

Neurosci Methods 226:88–102. https://doi.org/10.1016/j.jneumeth.2014.01.022

23. Kaynig V, Vazquez-Reina A, Knowles-Barley S et al (2015) Large-scale automatic reconstruction of neuronal processes from electron microscopy images. Med Image Anal 22:77–88. https://doi.org/10.1016/j.media.2015.02.001

24. Berning M, Boergens KM, Helmstaedter M (2015) SegEM: efficient image analysis for high-resolution connectomics. Neuron 87:1193–1206. https://doi.org/10.1016/j.neuron.2015.09.003

25. Kasthuri N, Hayworth KJ, Berger DR et al (2015) Saturated reconstruction of a volume of neocortex. Cell 162:648–661. https://doi.org/10.1016/j.cell.2015.06.054

26. Templier T, Bektas K, Hahnloser RHR (2016) Eye-trace. Proceedings of the 2016 CHI conference on human factors in computing systems – CHI '16. https://doi.org/10.1145/2858036.2858578

27. Scorcioni R, Polavaram S, Ascoli GA (2008) L-Measure: a web-accessible tool for the analysis, comparison and search of digital reconstructions of neuronal morphologies. Nat Protoc 3:866–876. https://doi.org/10.1038/nprot.2008.51

28. Billeci L, Magliaro C, Pioggia G, Ahluwalia A (2013) NEuronMOrphological analysis tool: open-source software for quantitative morphometrics. Front Neuroinform 7:2. https://doi.org/10.3389/fninf.2013.00002

29. Helmstaedter M, Briggman KL, Denk W (2011) High-accuracy neurite reconstruction for high-throughput neuroanatomy. Nat Neurosci 14:1081–1088. https://doi.org/10.1038/nn.2868

30. Cardona A, Saalfeld S, Schindelin J et al (2012) TrakEM2 software for neural circuit reconstruction. PLoS One 7:e38011. https://doi.org/10.1371/journal.pone.0038011

31. Schneider-Mizell CM, Gerhard S, Longair M et al (2016) Quantitative neuroanatomy for connectomics in Drosophila. elife. https://doi.org/10.7554/eLife.12059

32. Sommer C, Straehle C, Kothe U, Hamprecht FA (2011) Ilastik: Interactive learning and segmentation toolkit. 2011 IEEE international symposium on biomedical imaging: from nano to macro. https://doi.org/10.1109/isbi.2011.5872394

33. Oe Y, Baba O, Ashida H et al (2016) Glycogen distribution in the microwave-fixed mouse brain reveals heterogeneous astrocytic patterns. Glia 64:1532–1545. https://doi.org/10.1002/glia.23020

34. Asadulina A, Conzelmann M, Williams EA et al (2015) Object-based representation and analysis of light and electron microscopic volume data using Blender. BMC Bioinformatics 16:229. https://doi.org/10.1186/s12859-015-0652-7

35. Aguiar P, Sousa M, Szucs P (2013) Versatile morphometric analysis and visualization of the three-dimensional structure of neurons. Neuroinformatics 11:393–403. https://doi.org/10.1007/s12021-013-9188-z

36. Jorstad A, Nigro B, Cali C et al (2015) NeuroMorph: a toolset for the morphometric analysis and visualization of 3D models derived from electron microscopy image stacks. Neuroinformatics 13:83–92. https://doi.org/10.1007/s12021-014-9242-5

37. Dercksen VJ, Hege H-C, Oberlaender M (2014) The filament editor: an interactive software environment for visualization, proof-editing and analysis of 3D neuron morphology. Neuroinformatics 12:325–339. https://doi.org/10.1007/s12021-013-9213-2

38. Vandenberghe ME, Hérard A-S, Souedet N et al (2016) High-throughput 3D whole-brain quantitative histopathology in rodents. Sci Rep 6:20958. https://doi.org/10.1038/srep20958

39. Holst G, Berg S, Kare K et al (2016) Adding large EM stack support. In: *4th Saudi International Conference on Information Technology (Big Data Analysis) (KACSTIT)*, Riyadh, Saudi Arabia, 2016, pp 1–7. https://doi.org/10.1109/KACSTIT.2016.7756066

40. Barnes SJ, Cheetham CE, Liu Y et al (2015) Delayed and temporally imprecise neurotransmission in reorganizing cortical microcircuits. J Neurosci 35:9024–9037. https://doi.org/10.1523/JNEUROSCI.4583-14.2015

41. Foley JD et al (1997) Computer graphics: principles and practice, 2nd edn. Addison-Wesley Publishing, Reading, MA, pp 472–473

42. Totten C (2012) Game character creation with blender and unity. John Wiley & Sons, Hoboken, NJ, pp 10–13

43. Lorensen WE, Cline HE (1987) Marching cubes: a high resolution 3D surface construction algorithm. ACM SIGGRAPH Comput Graph 21:163–169. https://doi.org/10.1145/37402.37422

44. Bentley JL (1975) Multidimensional binary search trees used for associative searching. Commun ACM 18:509–517. https://doi.org/10.1145/361002.361007

45. Maneewongvatana S, Mount DM (1999) It's okay to be skinny, if your friends are fat. In:

Center for geometric computing 4th annual workshop on computational geometry, pp 1–8

46. Jones E, Oliphant T, Peterson P, et al. (2015) SciPy: open source scientific tools for python, 2001. 73:86. http://www.scipy.org

47. Pedregosa F, Varoquaux G, Gramfort A et al (2011) Scikit-learn: machine learning in python. J Mach Learn Res 12:2825–2830

48. DeFreitas T, Saddiki H, Flaherty P (2016) GEMINI: a computationally-efficient search engine for large gene expression datasets. BMC Bioinformatics. https://doi.org/10.1186/s12859-016-0934-8

49. Ester M, Kriegel HP, Sander J, Xu X (1996) A density-based algorithm for discovering clusters in large spatial databases with noise. KDD

50. Gan J, Tao Y (2015) DBSCAN revisited: mis-claim, un-fixability, and approximation. In: Proceedings of the 2015 ACM SIGMOD international conference on management of data. ACM, New York, NY, pp 519–530

51. Rousseeuw PJ (1987) Silhouettes: a graphical aid to the interpretation and validation of cluster analysis. J Comput Appl Math 20:53–65. https://doi.org/10.1016/0377-0427(87)90125-7

52. Lewis D (2014) The CAVE artists. Nat Med 20:228–230. https://doi.org/10.1038/nm0314-228

53. Magistretti PJ, Allaman I (2015) A cellular perspective on brain energy metabolism and functional imaging. Neuron 86:883–901

54. Balogh T, Zanetti G, Bouvier E et al (2006) An interactive multi-user holographic environment. ACM SIGGRAPH 2006 Emerging technologies on – SIGGRAPH '06. https://doi.org/10.1145/1179133.1179152

55. Agus M, Gobbetti E, Guitiàn JAI et al (2008) GPU accelerated direct volume rendering on an interactive light field display. Comput Graph Forum 27:231–240. https://doi.org/10.1111/j.1467-8659.2008.01120.x

56. Marton F, Agus M, Gobbetti E et al (2012) Natural exploration of 3D massive models on large-scale light field displays using the FOX proximal navigation technique. Comput Graph 36:893–903. https://doi.org/10.1016/j.cag.2012.06.005

57. Calì C, Agus M, Gagnon N, Hadwiger M, Magistretti PJ (2017) Visual analysis of glycogen derived lactate absorption in dense and sparse surface reconstructions of rodent brain structures. Eurograph Assoc. https://doi.org/10.2312/stag.20171224

58. Brooke J et al (1996) Sus—a quick and dirty usability scale. Usability evaluation. Industry 189(194):4–7

59. Salvatore L, Christina K (2008) Simple guidelines for testing vr applications. In: Advances in human computer interaction. InTech, London

INDEX

Irene Wacker et al. (eds.), *Volume Microscopy: Multiscale Imaging with Photons, Electrons, and Ions*, Neuromethods, vol. 155, https://doi.org/10.1007/978-1-0716-0691-9, © Springer Science+Business Media, LLC, part of Springer Nature 2020

Printed in the United States
by Baker & Taylor Publisher Services